# THRUST AND DRAG: ITS PREDICTION AND VERIFICATION

Edited by
**Eugene E. Covert**
Massachusetts Institute of Technology
Cambridge, Massachusetts

**Associate Editors:**
**C. R. James**
Vought Corporation
Dallas, Texas

**William F. Kimzey**
Sverdrup Technology AEDC Group
Tullahoma, Tennessee

**George K. Richey**
U.S. Air Force
Wright-Patterson AFB, Ohio

**Eugene C. Rooney**
U.S. Navy Department of Defense
Washington, D.C.

Volume 98
PROGRESS IN
ASTRONAUTICS AND AERONAUTICS

**Martin Summerfield, Series Editor-in-Chief**
Princeton Combustion Research Laboratories, Inc.
Monmouth Junction, New Jersey

Published by the American Institute of Aeronautics and Astronautics, Inc.
1633 Broadway, New York, NY 10019

American Institute of Aeronautics and Astronautics, Inc.
New York, New York

**Library of Congress Cataloging in Publication Data**
Main entry under title:

Thrust and drag: its prediction and verification.

  (Progress in astronautics and aeronautics; v. 98)
  Includes index.
  1. Drag (Aerodynamics) 2. Airplanes—Motors—Thrust.
I. Covert, Eugene E. II. Series.
TL507.P75     vol. 98     629.1s [629.132'34]     85-18681
[TL574.D7]
ISBN 0-930403-00-2

# Table of Contents

## Chapter III. Gas Turbine Engine Performance Determination.............47

W.F. Kimzey and S. Wehofer, *Sverdrup/ARO, Inc.,*
*Arnold Air Force Station, Tennessee,*
and E.E. Covert, *Massachusetts Institute of Technology,*
*Cambridge, Massachusetts*

# Preface

THIS book is the outgrowth of a project started by the Flight Mechanics Technical Committee under the chairmanship of C. R. James. The original intent was to sponsor a series of special sessions at specified meetings and to edit the material prepared for these sessions. As this plan evolved, it appeared timely to conduct a more detailed review of the state of the art of prediction and verification of the thrust and drag of aircraft in flight. The editorial committee* agreed that a more practical approach was to commission a set of essays dealing with specific topics. This proposal was submitted to both the Flight Mechanics Panel, then under the chairmanship of Dr. Donald Daniel, and to Dr. Martin Summerfield, editor-in-chief of the AIAA Progress in Astronautics and Aeronautics series. Both groups accepted the revised proposal, and understood the implication of increased time to publication.

It is a pleasure to report the outstanding support from industry, universities, and government laboratories. In particular, the moral support provided by Dr. Robert Whitehead of ONR was of great importance in the formative stages, as was the full cooperation of Dr. Robert Abernethy and his SAE Committee E-33. Indeed, the material summarized in Chap. VI was provided to us by Dr. Abernethy, with the approval of that committee and the Society of Automotive Engineers. We are also in debt to Mr. Brian Furness and the (British) Ministry Industry Drag Analysis Panel (MIDAP) for information that was exchanged. A complete listing of individuals who willingly helped us would be out of order here. We wish to thank authors of individual essays, Mr. Sam Wehofer, Chap. III; Dr. Charles Jobe, Mr. Wayne Baldwin, and Mr. Donald Kinsey for Chap. IV; and Mr. Douglas Bowers and Mr. Gordon Tamplin for Chap. V.

Of course I am indebted to the final editorial board: Mr. C. R. James (retired from Vought); Dr. William Kimzey, Director of Operations, Sverdrup Technology AEDC Group; Dr. Keith Richey, Chief Scientist, Flight Dynamics Laboratory, Air Force Wright Aeronautical Laboratories; and Mr. Eugene Rooney, Naval Air Systems Command who wrote Chap. II. These associate editors gave unstintingly of their time. Without their effort, the project would have never succeeded. Some of my travel and artwork was supported by the

---

*At that time the editorial committee consisted of Dr. Covert, Mr. James, Dr. Richey, Mr. Rooney, Mr. William Usab of Pratt & Whitney, and Mr. Harold P. Washington of NASA Dryden Research Center.

General Electric Company through the MIT Gas Turbine and Plasma Dynamics Research Fund, for which I am thankful. I would also like to acknowledge the assistance provided by Ms. Diana Park, who typed and retyped, carried out many important editorial tasks, and was otherwise of great value in preparing the typescript.

Finally, I would like to make it clear that while every effort was made to avoid errors of omission, co-mission, or by rewriting for clarification, ultimately the decisions were mine and I apologize in advance for those errors and oversights that may exist.

Eugene E. Covert
March 1984

# Introduction

Eugene E. Covert

THE purpose of this book is to provide a survey and critical review of the state of the art in prediction and verification of thrust and drag of jet-propelled airplanes. If successful, this critical review will help practitioners of the art to do their job better and to understand the evolution of their art. There will be aspects of this art for which some agreement exists among the practitioners. These particular aspects will not be discussed in great depth. Rather, it is the intent to treat those aspects that are controversial. A second purpose of the discussion is to delineate as far as possible those aspects that are deterministic in character from those that are purely random. In other words, any discrepancy between expectation and reality may be written

$$e = b + r \qquad (I.1)$$

Here $e$ is the discrepancy, $b$ the bias, and $r$ the random causes of difference that are nondeterministic. For example, in the case of airplane drag, random causes $r$ could be the drag increment due to manufacturing tolerance, meteorological effects, surface deterioration from service usage, or some combination of these causes. Initially, assume $b$, the bias, to be a deterministic factor that is well defined but perhaps not properly accounted for. For example, $b$ could be the drag increment due to error in determining the center-of-gravity location. Since there is a unique relation between drag increment due to trim and center-of-gravity location, that increment of drag will be considered to be a bias, not a random element. Clearly a similar discussion could be prepared about the engine. The important point is simple. The deterministic bias elements must be separated as clearly as possible from the nondeterministic elements.

Another example may help clarify this point. On a given day over a given route segment, the number of passengers in a particular aircraft can only be predicted statistically. The same statement holds for the center-of-gravity location. The same statement holds for the turbulence encountered in the second-segment climb, or the degree of icing. Nevertheless, just prior to takeoff, the aircraft load and center-of-gravity location could be known to a certain precision. Thus certain predictive statements can be made about the drag in the second-segment climb within some error. At the present time the effect of turbulence and icing on the performance during second-segment climb can only be estimated ex post facto, if adequate instrumentation is available during the interval in question.

This study of thrust and drag assumes a progression of knowledge from engineering research to understand physical phenomena. As understanding increases, one will be able to make deterministic statements which are not possible at this time. If this survey acts to stimulate deeper study, the survey will be out of date earlier.

## An Historic Perspective

### Performance Calculations

Long before the airplane was invented, the basic ideas behind equilibrium, or even accelerated flight, were conceptually understood. They followed from the general laws of motion, which can be written for the $i$th component of a vector in a space fixed Cartesian coordinate system as

$$F_i = m\frac{dV_i}{dt} \tag{I.2}$$

This is Newton's law applied along the $i$ direction. $F_i$ is the sum of all external forces and may be a function of position, velocity, and time. $m$ is the mass of the body at time $t$. The fact that an airplane is an extended body requires that the angular motion be described

$$M_i = I_{i_1}\frac{d\omega_1}{dt} + I_{i_2}\frac{d\omega_2}{dt} + I_{i_3}\frac{d\omega_3}{dt} \tag{I.3}$$

$M_i$ is the applied torque about each of the principal coordinate axes; $\omega_1, \omega_2, \omega_3$ are angular velocities about the three axes; and $I_{i_j}$ are the moments or products of inertia at time $t$. Additional equations may be written that describe the deflection of the airplane due to aerodynamic and inertial loads.

However, these equations are still too general. Consider an airplane whose thrust axis is inclined upward on angle $\varepsilon$ with respect to the fuselage reference axis. Resolving forces along and normal to the level equilibrium flight path gives, at a fixed flight condition,

$$L(\alpha, \delta_e, T) + T\sin(\varepsilon + \alpha) = W \tag{I.4}$$

$$T\cos(\varepsilon + \alpha) = D(\alpha, \delta_e, T) \tag{I.5}$$

$$M(\alpha, \delta_e, T) = 0 \tag{I.6}$$

Note that the lift $L$, the drag $D$, and the pitching moment about the center of gravity (or center of mass) $M$ are all dependent upon the angle of attack $\alpha$, the elevator required for trim $\delta_e$, and the thrust $T$. Note that $T$ is inclined at an angle with respect to the fuselage reference line. Equations (I.4)–(I.6) constitute a set of three simultaneous equations for $\alpha, \delta_e, T$ for each altitude and speed $V$ or Mach number. The primary problem is to determine the aerodynamic forces and moments, the propulsive forces, and the coupling between them.

The earliest consideration of power required seems to be Cayley's. According to Pritchard (Ref. 1, p. 87) Cayley made a number of estimates, which ranged from very pessimistic to very optimistic. Cayley seems to have regarded 20 lb/hp as the absolute upper bound and 4 lb/hp as an unattainable goal. Although estimates would be based upon knowledge of the lift-drag ratio deduced from the equilibrium glide angle of his gliders, he in fact used drag data from a whirling-arm experiment.[1] It is of interest that the general formulas for equilibrium level flight that Cayley used (in modern notation)

$$L = W \tag{I.7}$$

$$D = T \tag{I.8a}$$

were in general use up to about 1940, although in some cases, Eq. (I.8a) was written

$$D \cdot V = \mathrm{hp}_a \times 550 \tag{I.8b}$$

to reflect the fact that engine power was written in terms of horsepower.

In 1908 Lanchester knew that the drag must include two parts: parasite drag and drag due to lift.[2]

By 1910 Eiffel[3] used wind-tunnel data explicitly and the measured drag polar to determine the power required, i.e.,

$$\mathrm{hp}_r = (D \cdot V)/550 \tag{I.9}$$

Eiffel's collaborator, Rith, developed the logarithmic polar diagram. It was based on the formula

$$W = L'V^2 \tag{I.10a}$$

$$P = D'V^3 \tag{I.10b}$$

In modern notation

$$L' = \tfrac{1}{2}\rho(0.682)^2 SC_L \tag{I.10c}$$

$$D' = \tfrac{1}{2}\rho(0.682)^2 SC_D \tag{I.10d}$$

Here $V$ is in miles per hour, and $L'$ and $D'$ are lift and drag in pounds at a speed of 1 mph.

The great advantage of the Eiffel-Rith procedure was that it allowed wind-tunnel-defined drag polar (the curve of $L'$ vs $D'$ is, after all, proportional to $C_L$ vs $C_D$) to serve as the basis for power required. Note two implicit assumptions:

1) Trimming the airplane did not change drag.
2) Wind-tunnel drag equals flight article drag.

The next fundamental improvement was Bairstow's.[4] In modern notation, Bairstow assumed

$$C_L = ( 2W/S)/\rho_0\sigma(V)^2 \tag{I.11}$$

where $\sigma = \rho(h)/\rho_0$ density ratio at altitude $h$, and $\rho_0$ = sea-level density of air. The key here is the use of $\sqrt{\sigma V^2}$, the so-called "indicated speed," as the independent parameter. This implies for the same lifting force, the velocity at altitude is greater than at sea level in the ratio $1/\sqrt{\sigma}$. Similarly the drag coefficient is defined

$$C_D = 2 \text{ drag}/\rho_0\sigma(V)^2 \tag{I.12}$$

Thus the drag polar allows one to plot, once and for all, "indicated power" required. The performance for each altitude follows from knowledge of the engine performance.

Over the next few years the advantages of the use of "indicated power" required and the idea of graphical analysis became widespread. Typical of such analyses was that of Diehl.[5] The primary problem faced by Diehl, among others, was representation of propeller performance and engine performance in a relatively universal fashion.

The next step in making simplified and, at the same time, more sophisticated calculation was due to Oswald.[6] He separated out the parameters in a slightly different way and was able to develop charts that were essentially of universal use. This step required recognition of the approximation to the drag polar, namely,

$$C_D = C_{D_P} + \frac{C_L^2}{\pi A\!R_e} \tag{I.13}$$

But his charts were expressed in terms of

$l_s$ = effective span loading = $W/be^2$
$b_e^2$ = effective span squared = $A\!R_e \cdot S$, includes airplane efficiency factor which accounts for variation of parasite drag with angle of attack
$l_p$ = parasite loading = $W/f$
$f$ = sum of equivalent flat-plate area (Table I.4) = $C_{D_p} \cdot S$

In addition there is the parameter

$$l_t = W/(\eta \text{BHP})_{\text{max}}, \text{ the power loading}$$

where BHP is the engine brake horsepower and $\eta$ is the propulsive efficiency.

Oswald's charts provide a means of calculating performance in terms of groups of parameters, i.e.,

$$V_{max} = V_m\left(\frac{l_p}{l_t}, l_s l_t\right) \tag{I.14}$$

Further, Oswald assumed that the c.g. location of the minimum speed corresponds to trim. From this assumption one finds a correction for "unstable airfoil," i.e., the increment of change due to trim or elevator deflection. Oswald's procedure assumes that the effects of compressibility are negligible.

All these calculation procedures have in common the problem of estimating the drag. Hence it seems prudent to review historical developments of estimating drag.

## Estimation and Prediction of Drag

*From Antiquity Through the Seventeenth Century*

The ancients did not consider aerodynamic drag to be a matter of practical consequence.* Apparently it was not given much consideration until early in the sixteenth century, when Leonardo da Vinci suggested that atmospheric air was a source of resistance to motion. He based his argument upon an expected increase in density in front of the body. Over a hundred years later Galileo conducted an experiment using two pendulums that were started at different angles. He concluded that drag was proportional to velocity but doubted that this rule would hold at high speeds. There the matter rested until 1690, when Christian Huygens, who is better known for his studies in optics, published experimental results showing that drag (resistance) is proportional to the square of the velocity. Not long afterward, Newton addressed the problem of calculating resistance of moving bodies in fluids. He assumed drag to be primarily a function of density (inertia of fluid), velocity, and shape. He felt that other fluid properties, like viscosity (tensity and friction) and elasticity, would have a small effect. Newton calculated fluid forces using a discrete (free-molecular) model and asserted that in a continuum the drag would be half that found using a discrete fluid model. In 1713 Roger Cotes, an assistant of Newton's, argued that over the front half of a body the drag is due to momentum transferred to the fluid. Following this argument logically, he found that fluid transfers this momentum to the back half of the body. Hence the drag is zero. Cotes argued 1) that some momentum transferred to the fluid must be lost, hence there is drag, and 2) that the local velocity near the midsection of the body must be twice the forward speed if all the momentum is to be recovered. He maintained that, since this is clearly impossible, drag exists and should show up in the velocity in the wake following the moving body. Cotes clearly foresaw d'Alembert's paradox.

---

*The material in this section is based upon that given in Ref. 7, unless otherwise indicated.

*Rise of Experimentalism — The Eighteenth and Nineteenth Centuries*

In a remarkable series of studies, d'Alembert developed the continuity equation (1744) and found the drag was zero for a body moving in a perfect fluid. He was astonished by his results. He had expected to show

$$D \sim \rho V^2 S \qquad (1.15)$$

d'Alembert knew he was neglecting viscosity. He argued forcefully that viscosity was small so its effect would be small if the velocity was large. Nevertheless, in his subsequent publications (1752 and 1768) d'Alembert failed to resolve the paradox. We now know that the paradox is due to viscosity-induced separation, which changes the flow patterns markedly. This separation exists no matter how small the viscosity is. To be precise, we know that the flow pattern around a bluff body in a fluid having zero viscosity differs from the flow pattern about the same body in the limit as viscosity vanishes.

This paradox must have been a source of great frustration to d'Alembert because at the same time Robbins measured drag using a whirling arm. Robbins verified and extended Huygen's results, publishing his results in a book, *New Principles of Gunnery*, in 1743. Subsequently Hutton extended Robbins' work and reported this in *Philosophical Transactions* in 1798. George Cayley, who also conceived of an airframe made up of three fundamental parts —a fuselage to carry passengers, freight, and the propulsion system; a wing to generate lift; and a tail for stability—resolved the aerodynamic forces into lift normal to the flight path, and drag acting along the flight path. Cayley, using Robbins' data along with data from his own experiments, calculated his power requirement.[1] All this was published in *Nicholson's Journal* in 1808 and 1809. Incidentally Cayley reduced his data in the form (in modern notation)

$$\rho C_D = 2D/V^2 S \qquad (1.16)$$

For all practical matters, this situation existed for about the next century—but not for lack of attention to the problem.

From the mid-nineteenth century on, an ever-increasing number of people were studying problems of flight. Horatio Phillips invented the wind tunnel in the early 1880s. Lilienthal perfected hang gliders. Perraud built small powered models. All these processes refined knowledge of generation of lift and drag. These studies were all experimental in nature. Nevertheless, theoreticians were still concerned with d'Alembert's paradox. It was Helmholtz who, in 1868, first offered an alternative, the paper, "On Discontinuous Motion of a Fluid." Here he theorized that in a perfect fluid the tangential component of the velocity need not be continuous across a streamline. Kirchoff, in 1869, used Helmholtz' ideas to compute the pressure on a unit of surface of a lamina exposed to a stream; and he continued to apply these ideas to a variety of problems until his death in 1877. Rayleigh also was impressed by Helmholtz' approach and used it to compute drag also. Rayleigh made use of the idea that the fluid must leave the edge of the laminar tangentially to avoid the "infinite velocity...that would require a negative pressure of infinite magnitude." While Rayleigh's formula was at considerable variance with data, it did predict a finite drag. This line was followed at considerable length, particularly by Levi-Civita. However, for reasons that will be clear subsequently, this type of theory led to naught.

*Aerodynamic Reality — The Early Twentieth Century Through World War I*

Aerodynamics in the period from 1890 to 1920 was dominated by the ideas of four men: Lanchester, Prandtl, Eiffel, and von Kármán. F. W. Lanchester conceived of the relation between lift on finite wings and trailing vorticity, from which he concluded that three-dimensional wings experienced vortex drag, while this kind of drag did not exist for two-dimensional wings. In *Aerodynamics* (London, 1907), high aspect ratio elliptic wings are used for illustrations.[2] The implication surely must be that he thought, or knew, that this was the most favorable shape. In 1902 Kutta studied the problem of flow along a thin laminar that was a segment of a circular arc. He postulated incompressible, inviscid flow in two dimensions (infinite span) and smooth flow at the edges of the laminar. The result: lift and no drag. Further, the lift agreed within 22–27% of that measured by Lilienthal. In 1906 Joukowski calculated the lift per unit length of a cylinder having circulation. These two results tied in well with Lanchester's concept of circulation, lift, and trailing vorticity from a three-dimensional wing.

In 1909 at Göttingen, Prandtl conducted an experiment in which he introduced ammonia for flow visualization and was able to see clearly the trailing vortex pair that Lanchester had deduced. At this point the way was clear for Prandtl to relate theory and practice. In 1913 Prandtl published his airfoil theory in a chapter called, "Fluid Motion" for *Handwörterbuch der Naturwissenschaften* (Jena, 1913). Here, for the first time, was a demonstration of the hydrodynamic background for lift and induced drag. He also discussed form drag, in which he pointed out that a tapered fairing behind a hemispherical nose reduced the drag markedly. In the latter connection he referred to his 1910 review in *Zeitschrift für Flugtechnik and Motorluftschifahrt* (Nos. 1–7, 1910), in which he divided drag on a body into two parts, one due to friction and the second due to separation. Prandtl added that by careful design the drag due to separation can be reduced. In essence then, Prandtl implied that drag of a well-designed body is essentially due to friction. In modern notation, and for laminar flow, the friction drag per unit width may be written

$$D_F \sim \sqrt{\rho\mu}\,(lV)^{1.5} \tag{I.17a}$$

which is very small when compared to drag per unit width on a bluff body

$$D_s \sim \rho V^2 h_s \tag{I.17b}$$

where $h_s$ is the separation height.

Prandtl was now able to deal with d'Alembert's paradox. Even if a body is shaped so that the flow is unseparated ($h_s = 0$), there will be friction drag. To follow Cote's argument, it is friction that prevents all the momentum put into the fluid at the nose from returning it to the body at the base.

To follow both lines of thought further, as the viscosity is smaller, so is friction. So $D_F$ becomes small, vanishing as viscosity becomes zero—if the flow pattern is unchanged. But, as noted earlier, that condition is not usually satisfied.

It was largely in the decade between 1910 and 1920 that the modern view of drag prediction at low speeds was developed. Prandtl had demonstrated the modern formula for induced drag in terms of the aspect ratio. This was refined by Betz, who calculated the effect of nonideal loading, and Munk, who demonstrated experimentally that an elliptically loaded wing had a minimum induced drag. Further work by von Kármán, Fuhrmann, and Weiselsberger defined drag of bodies related to aeronautical needs. Only von Kármán's ideas will be outlined here. In 1908 the Frenchman, Bernard, published a note describing the motion of vorticity in the wake behind a bluff body. Independently in 1910 von Kármán studied the problem of a doubly infinite row of vortices and showed the staggered pattern was stable. Then, by means of the momentum theorem he showed

$$C_D = 1.42(h/b)(u/V) \qquad (I.17c)$$

where $h$ = plate width, $b$ = width of vortex wake, $V$ = flight speed, and $u$ = fluid speed in the wake centerline. In this publication von Kármán said the Helmholtz-Kirchoff-Rayleigh approach was not practical because free streamlines having a jump in tangential velocity across them are unstable to small disturbances. Hence the vortex wake is the more natural model. With the exception of von Kármán's vortex wake, the results of the activity at Göttingen remained unknown outside Germany until 1921, partly, of course, because of World War I.

Meanwhile similar studies were being carried out in England. In 1915 Lanchester arrived at the same formula for induced drag as Prandtl. Studies, including bluff body drag,[4] were similar to those carried out by Prandtl's group.

It may be interesting to note the raging argument over drag of the air vehicles that took place in roughly the decade from 1907 to 1917. The protagonists in this argument were Prandtl, Eiffel,[8] Riabouchinsky, Stanton, Lanchester,[2] and Villamil.[9] With the exception of Lanchester, all these men were experimentalists, and even Lanchester was an experimentalist on occasion. Lanchester, who tended to rely on hydrodynamic theory, and, not altogether appropriately, eschewed the use of mathematics. Prandtl was a great theoretician, but in this controversy he, along with Stanton and Riabouchinsky, were essentially wind-tunnel oriented. Eiffel did both free-fall and wind-tunnel experiments.

Eiffel maintained that one should define (in modern notation)

$$C_D = \text{drag}/\tfrac{1}{2}\rho V^2 S \qquad (I.18)$$

where $C_D$ is a function of geometry. (Although Eiffel argued forcefully for the case of $C_D$, he actually used $\rho C_D$ to conform to convention. He did, however, correct $\rho$ to standard conditions.) Further, Eiffel argued that $C_D$ was the same whether the air was in steady motion past a fixed object, or the object moved at constant speed through stationary air. He had data on disks, and flat plates of various aspect ratio, which were held so that their normal pointed into the wind. For the most part, all the data Eiffel studied to reach these conclusions were at relative speeds below 50 m/s. For high speed, up to 400 m/s, he quotes Colonel Duchemin (1842), who proposed the formula

$$D = av^2 + bv^3$$

Villamil held differing views. He had studied the data closely, he had studied existing theories closely, and he had studied experimental devices closely. Villamil concluded:

1) Most data from whirling-arm experiments was of doubtful value because the size of the body was such that the velocity variation across the body due to rotation could not be ignored.

2) Most of the wind-tunnel data did not support the supposition that relative velocity was the key parameter. Horizontal buoyancy was a very important consideration that had been overlooked (here he was correct for some cases and incorrect for others).

3) Villamil believed that those who felt $C_D$ was constant had not looked at the data closely enough; even $D \sim V^n$ was oversimplified.

Both Villamil and Lanchester argued that for a fixed body, the drag was proportional to velocity (Stokes flow), then the power increased to 1.5, then 2, finally at high speeds (transonic) to 8. Villamil showed this power reduces back to 2 at supersonic speeds. In fact, Villamil showed that this change could be partially correlated with a variable that is roughly the speed a gas expanded to a vacuum will have if it starts from static conditions.

Except for the effects of transition on sphere drag, a matter that Prandtl had settled in the years before World War I, all contenders were more or less correct. Villamil knew that "fairing" would reduce drag (Cayley chose a top view of a trout as the best shape for struts and bodies).[1]

In 1920 Bothegat, in NACA Report 28, referred to von Kármán and Lanchester but not Prandtl. In 1921 Pistolesi summarized the German work, which led in late 1921 to Prandtl's NACA Report 116, "Application of Modern Hydrodynamics to Aeronautics."[10]

However, in spite of earlier work by Rayleigh, who argued that if $Re$ = Reynolds number and $C$ = the maximum speed obtained from expansion,

$$D = \rho V^2 Sf(Re, V/C)$$

this idea would not be accepted until 1924 or 1925.[11] (Note that $V/C$ is a sort of Mach number.)

Indeed, the use of Mach number as the index of compressible flow was not common until after World War II, although the use of the local speed of sound as a reference was used by the late 1920s for propeller testing.

Thus we conclude that one of the tasks to be resolved in the 1920s was to define the drag coefficient properly.

*Period Between the Wars to the End of World War II*

One can assume that the influence of the aspect ratio on the induced drag coefficient was well known, that is,

$$C_{D_i} = \frac{C_L^2}{\pi \mathcal{R}}(1 + \delta) \qquad (I.19a)$$

where $\mathcal{R} = [(\text{span})^2/\text{area}]$ for monoplanes and equals equivalent monoplane aspect ratio for biplanes.

The correction for nonelliptic loading ($\delta$) was not yet in a simple form to use, although it would be in a matter of a few years. Alternatively the induced drag could be written

$$D_i = \frac{W^2}{\pi q b^2}(1 + \delta) \qquad (I.19b)$$

where $q = \frac{1}{2}\rho V^2$, and $b^2 = $ wingspan for a monoplane.

The induced drag coefficient $C_{D_i}$ was frequently found from wind-tunnel tests and could be corrected for change in aspect ratio following Prandtl.[10] Two serious problems remained to be attacked. The first, that of determining the power available and propeller efficiency, is not germane to our study. The second problem is finding the residual or the sum of wing profile and parasite drag coefficients. Early on, the former seemed to be more important than the latter. Nevertheless, detailed studies were made and the procedures were defined with increasing precision.

The first approach was to measure the drag of certain relatively standard components like fuselages, wheels, struts, undercarriage assemblies, and radiators. An example of drag vs airspeed is given in Ref. 12, and one of the fuselage tests is contained in Refs. 12 and 13. Typical results are shown in Figs. I.1–I.3.

As for wheels and undercarriage, an important additive drag ranging from 20 to 40% of the total must be applied. In Germany the whole assembly was tested, and $D/q$ was plotted vs $q$. This term, $C_D A$, tended to decrease slightly with increasing $q$. At $V = 31$ m/s the results are 0.175 m$^2$. Bairstow[4] gives the

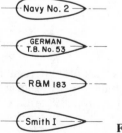

Fig. I.1  Standard strut profile.

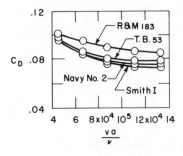

· Fig. I.2 Reynolds number effects on standard strut drag coefficient.

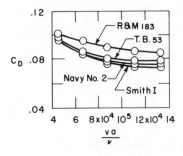

Fig. I.3 Estimated drag contribution of components to the drag of a sperry messenger fuselage at 100 mph.

following fractional breakdown for additive drag.

| | |
|---|---|
| Wheels | 0.39 |
| Axle | 0.10 |
| Struts | 0.10 |
| Wires | 0.11 |
| Shock absorbers and joints | 0.30 |
| | 1.00 |

By the mid-twenties and early thirties more data were available, but the difficulties were the same. The major contribution came from a better understanding of scaling. Diehl[14] has a graph for fuselages of

$$C_D = f\left(\frac{V \cdot l}{V_{\text{mod}} \, l_{\text{mod}}}\right) \tag{I.20}$$

which is roughly Reynolds number scaling, if the kinematic viscosity is constant. By 1930 three major results were available. First, there were good

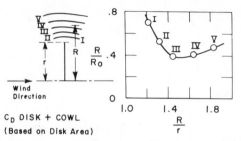

**Fig. I.4  Townend ring and drag coefficient reduction (ratio of resistance with ring $R$ to resistance of disk along $R_0$).**

data in various airplanes and assemblies; second, accurate experiments in aerodynamic interference were being conducted; and, third, the Towend Ring (cowling) was being applied to the problem of engine installation. Typical results of the first type were reported by Weik.[15]

The Aeronautical Research Committee tested a disk whose area was 1/30 the cross-section of a fuselage but whose drag was the same as the fuselage. The combined drag was generally greater than the sum of individual drags. When the disk was near the maximum thickness point on the fuselage, its drag was 1.8 times the free air drag. (ARC Technical Report, 1928-1929, p. 44.)

The Towend Ring, when applied to cowlings, can reduce fuselage drag by over 50%,[16] (Fig. I.4). As for the practical aerodynamics, Kerber[3] lists 12 NACA technical notes and technical reports that contain information on drag buildup (Fig. I.4). Tables I.1–I.3 are reproduced from Ref. 3.

NOTES:

1) The effect of roughness on struts at high Reynolds number is indicated by the relatively high drag for struts wound with tape and then varnished and by the much lower drag on struts with smooth polished metal surfaces. It is reasonable to assume that similar effects must take place on other parts of the airplane, and the condition of the finish of the exposed parts should be taken into account in determining the resistance.

2) The effect of protuberances on a streamline fuselage is very great, especially if they are located near the front. As an example, it can be cited that the drag of a good streamline body of fuselage form may be increased fivefold by mounting an uncowled radial engine. These effects must all be considered in estimating the drag of component parts.

3) The drag of a fitting including its interference drag is given approximately by considering the fitting of a flat plate of double its projected area. Miscellaneous projecting parts not streamlined should be treated as fittings.

4) The total parasite drag must be divided into two parts, the first variable with the lift coefficient and designated $P_1$, the second independent of the lift and designated $P_2$. If there is any doubt as to the classification of the fuselage or nacelle drag, it should be put in the $P_1$ class. The items of variable and constant drag can then be tabulated in the form shown in Table I.3. The parasite drag coefficient has the form

$$C_{D_p} = \frac{2D_p}{\rho V^2 S}$$

The variation of the $P_l$ part of the parasite drag coefficient, designated by $C_{D_{p_1}}$,

## Table I.1  Drag of various items

| Item | Drag in pounds at 100 mph |
|------|---------------------------|
| Flat plate (1 ft$^2$) | 32.8 |
| Control horn well streamlined | 1.0 |
| Tail skid | 2–5 |
| Compass | 4 |
| Searchlight | 13 |
| Radio generator | 20–30 |
| Machine gun | 20–30 |
| Torpedo and gear | 20 |
| Bomb (500 lb) and gear | 20–40 |
| Mud-guard | 5–10 |
| Standard wheel with usual fairing from hub to rim (Bendix): | |
| 26×3 | 7.5 |
| 28×4 | 10.0 |
| 30×5 | 12.5 |
| 32×6 | 16.0 |
| 36×8 | 24.0 |
| 44×10 | 36.6 |
| 54×12 | 54.0 |
| Low-pressure tire wheel 22×10 | 9.25 |
| Tail surfaces (per ft$^2$ —thin) | 0.35 |
| Tail surfaces (per ft$^2$ —average thickness) | 0.40 |
| Tail surfaces (per ft$^2$ —flat plate) | 0.30 |
| Fuselage, square (ft$^2$ of projected area) | 6–8[a] |
| Fuselage, round or elliptical (ft$^2$ of projected area) | 4–5 |
| Fuselage, well streamlined, with no irregularities (ft$^2$ of projected area) | 3–4 |
| Nacelles (ft$^2$ of projected area) | 4–8 |
| Radiator, free air (ft$^2$ of projected area) | 15–20 |
| Hull (ft$^2$ of projected area) | 5.0 |
| Hull, exceptionally clean (ft$^2$ of projected area) | 4.0 |
| Seaplane float, average shape (ft$^2$ of projected area) | 5.50 |
| Wing, tip float square (ft$^2$ of projected area) | 8.0 |
| Wing, tip float average shape (ft$^2$ of projected area) | 6.0 |
| Wing, tip float well shaped (ft$^2$ of projected area) | 5.0 |
| Air-cooled engines per cylinder | 8–10 |
| Landing gears[b] | |

[a] The area upon which the drag of fuselages, hulls, floats, etc., should be based is the maximum cross-section area in the front view and should include all projecting sections, windshields, etc., except the wing stubs which are included in the wing drag.

[b] See U.S. NACA Technical Reports Nos. 485 (1934); 518, 522 (1935).

Table I.2  Summation of component parasite drag at 100 mph (structure)

| Designation or description | Number $N$ | Length $L$, ft | Size $t$, in. | Total length $N \times L$, ft | Drag/ft at 100 mph, lb | Drag at 100 mph, lb |
|---|---|---|---|---|---|---|
| Interplane struts | 6 | 5.35 | 1.000 | 32.0 | 0.195 | 6.2 |
| Aileron connecting struts | 2 | 5.35 | 0.714 | 10.7 | 0.160 | 1.7 |
| Cabane struts | 4 | 3.83 | 1.000 | 15.3 | 0.195 | 3.0 |
| Flying wires | 6 equiv. | 12.83 | 0.375 | 77.0 | 0.150 | 11.5 |
| Landing wires | 4 | 9.17 | 0.313 | 36.7 | 0.140 | 5.1 |
| Cabane wires (lateral) | 4 | 3.83 | 0.313 | 15.3 | 0.250 | 3.8 |
| Cabane wires (fore and aft) | 4 | 3.83 | 0.313 | 15.3 | 0.250 | 3.8 |
| Front outer chassis struts | 2 | 3.92 | 1.500 | 7.8 | 0.280 | 2.2 |
| Rear chassis struts | 2 | 3.92 | 1.500 | 7.8 | 0.280 | 2.2 |
| Front inner chassis struts | 2 | 5.59 | 2.000 | 11.2 | 0.370 | 4.1 |
| Horizontal tail surface struts | 2 | 5.00 | 0.857 | 10.0 | 0.180 | 1.8 |
| Fin brace wires (streamline) | 4 | 5.20 | 0.250 | 20.8 | 0.160 | 3.3 |
| Tail skid | 1 | — | — | — | 5.000 | 5.0 |
| Control horns | 8 | — | — | — | 1.000 | 8.0 |
| Total structural drag at 100 mph | 61.7 | | | | | |

Table I.3  Summation of component parasite drag at 100 mph

| Designation or description of item | Number | Area, ft$^2$ | Drag per ft$^2$ at 100 mph, lb | Drag at 100 mph, lb |
|---|---|---|---|---|
| Parasite drag varying with angle of attack $P_1$ | | | | |
| Fuselage | 1 | 9.6 | 5.0 | 48.0 |
| Tail surfaces | — | 65.6 | 0.4 | 26.3 |
| Total $P_1$ at 100 mph | | | | 74.3 |
| Total $P_1$ at 100 mph | | | | |
| Parasite drag constant with angle of attack $P_2$ | | | | |
| Structure (Table 2) | 1 | — | — | 61.7 |
| Radiator (underslung) | 1 | 2.5 | 18 | 45.0 |
| Wheels | 2 | 32″×6″ | 16 each | 32.0 |
| Observer's gun and mount | 1 | — | — | 25.0 |
| Forward fixed gun, exhaust pipes, expansion tank | | 0.42×2 = 0.84 | 32.8 | 27.6 |
| Total $P_2$ at 100 mph | | | | 191.3 |

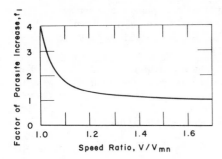

Fig. I.5 Scale effect on parasite drag.

is assumed to be dependent on the factor $f_1$, plotted against speed ratio in Fig. I.5 as shown by the equation

$$C_{D_{P_1}} \text{ at } V = C_{D_{P_1}} \text{ at } V_{\max} \times f_1$$

These values do not take care of interference of the fuselage with wings, landing gear, etc. In order to take care of such interference, both $P_1$ and $P_2$ should be increased 10–15%.

One of the important results of tests like those just referred to was the increasing emphasis upon the need for drag reduction. Two additional technical improvements made a major improvement. First, the cantilever wing construction eliminated struts and bracing wires. Second, retractable landing gear was introduced. The improved aerodynamic performance resulted in spite of increased weight, partly because improvements in the propulsion system became available.

In the last decade before World War II, the biplane gave way to the monoplane, at first externally braced, and then it was learned how to make efficient thick wings, which could be internally braced. These wings have very low parasite drag. Figure I.6 shows a low-drag airfoil of the late 1940s, which has the same drag as the cylinder also shown in the figure. The comparison is impressive.

The procedure that evolved to estimate drag was based upon drag buildup and use of a proper drag coefficient that related to frontal area. Millikan[17] shows typical results. The parasite drag was thus the sum of the profile drag

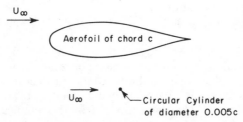

Fig. I.6 Effectiveness of a field of flow which is largely irrotational ( at high Reynolds numbers the drag of these two bodies is roughly equal ).

Table I.4    Proper drag coefficient of major airplane components
(Galcit statistical data, $R \sim 10^6$ to $2 \times 10^6$ based on wing chord)

| Element | Description | $C_{D_\pi}$ |
|---|---|---|
| Fuselage | Dirigible hull, circular section, tested alone | 0.072 |
| Fuselage | Large transports or bombers, no nose engine or turrets | 0.070 to 0.105 |
| Fuselage | Small plane, including nose engine, closed cockpit | 0.090 to 0.130 |
| Nacelle | Nacelle mounted above a wing | 0.250 |
| Nacelle | Leading-edge nacelles, small airplane, relatively large nacelles | 0.120 |
| Nacelle | Leading-edge nacelles, large airplane, relatively small nacelles | 0.080 |
| Tail surfaces | Single-engined, low-wing monoplane | 0.0085 to 0.0120 |
| Tail surfaces | Multiengined, low-wing monoplane | 0.0060 to 0.0110 |
| Tail surfaces | Single-engined, high-wing monoplane or biplane | 0.0120 to 0.0180 |

and the proper drag.

$$C_{D_p} = C_{D_0} + C_{D_\pi} \tag{I.21}$$

where $C_{D_\pi}$ was found by comparison with existing designs, as shown in Table I.4.

The induced drag contains a correction for nonideal load distribution

$$C_{D_c} = \frac{C_L^2}{\pi\!R} ( 1 + \delta ) \tag{I.22}$$

where $\delta$ is the correction for nonelliptical spanwise loading[18] (Fig. I.7).

Scaling of wind-tunnel tests was based upon an effective Reynolds number, which in turn depended upon the wind-tunnel Reynolds number at which a sphere had a drag coefficient of 0.3.

Two comments seem appropriate. To quote from Millikan,[17]

> For very high speed airplanes (of the order of 400 mph), another factor, in addition to those mentioned above, must be considered. This is the effect of compressibility, which may cause serious drag increases at such speeds, particularly at corners or places where the curvature is small. The data required in this connection is too meager as yet to permit any general discussion of the problems. Additional data should become available in the near future.

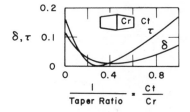

Fig. I.7 Effect of taper ratio on induced drag parameter $\delta$ and induced angle of attack parameter $\tau$.

The other is the use of fairings or "fillets" to reduce separation at wing-body junctions. This very useful geometric modification was, to be sure, found experimentally by trial and error. Perhaps Millikan, who was said to be an excellent practitioner of the art of shaping fillets, felt such empiricism was inappropriate in a textbook.

*The Jet Era — World War II to 1960*

The introduction of the jet engine made a very important change in the design of aircraft and the calculation of profile drag. Further, the jet engine led to high speeds, which led to additional complications. Consider first the change in aircraft configuration. The reasons for this are perfectly clear. In a propeller-powered airplane, thrust is produced by adding a little momentum to a large amount of air. Additional air is required for combustion and for cooling. This air essentially followed three parallel paths: flow outside the cowling and nacelle that passed through the propeller to produce thrust, flow inside the cowling but outside the cylinders that cooled the cylinders, and flow inside the cowling and inside the cylinder to oxidize the fuel. The jet engine passes, more or less, all the air through the same path. Hence, even if total quantity of air to be moved was about the same, great changes in ways of handling this air would be necessary. In fuselage-mounted engines, long ducts were required. In nacelle-mounted engines, the midwing mounting was no longer practical, so pylon-mounted nacelles appeared.

However important these changes, there was a further complication. This came about because of coupling that now existed between internal aerodynamic flow—the inlet, diffuser, engine, and exhaust nozzle—and the external aerodynamic flow, which includes all those items discussed above. To illustrate this issue, consider the following control volumes:

1) A plane across the inlet, the inside of the engine, and a plane across the exit.

2) The stream tube starting upstream at a great distance from the airplane, through which all the air that will pass through the jet is contained. This stream tube passes through the engine, emerges, and ultimately ends up a great distance downstream of the airplane. The dividing line on a stagnation line separates the outside from the inside flow.

By convention the following notation is assumed. The momentum flux from the engine at the exit plane is called the gross thrust, $F_g$, where $\dot{M}_e$ is the exit mass flux, $V_j$ the jet velocity (assumed uniform), $p_e$ the exit static pressure, and

## a) Internal Control Volume

## b) Complete Mass Flow Control Volume

**Fig. I.8    Alternate possibilities for defining engine-airframe interface.**

$A_e$ the exit area. Thus

$$F_g = \dot{M}_e V_j + (p_e - p_\infty) A_e$$

The inlet momentum flux is based upon the plane so far in front of the inlet that $p_i = p_\infty$, this term is just $\dot{M}_i V_\infty$. By definition the net thrust is

$$F_N = \dot{M}_e V_j - \dot{M}_i V_\infty + (p_e - p_\infty) A_e$$

The implications of this convention are discussed in detail in Chap. II.

It is helpful to look briefly at the early, classical approach to performance, using the idea of control volumes. These two control volumes are shown in Fig. I.8.

Assume the thrust and the flow far from the body are aligned with the $x$ axis. (See Ref. 18 for an excellent discussion of the implications from use of control volumes.) Applying the momentum theorem to the first control volume (Fig. I.8a)

$$T = \iint (\rho u_x u_x + p) \, dA_{\text{exit}} - \iint (\rho u_x u_x + p) \, dA_{\text{inlet}}$$

$$+ \iint (p + \tau_{xn}) \sin(n, x) \, dS_L \tag{I.23}$$

$\sin(n, x)$ is the sine of the angle between the normal to the lateral surface and the $x$ axis.

The second control volume leads to a slightly more complex form of the momentum integral

$$T = \iint \left( \rho_D U_D^2 - \rho_u V_\infty^2 \right) dA_u + \iint \left( p + \tau_{xn} \right) dS_{L_{\text{inside}}}$$

$$+ p_\infty \left( A_D - A_\infty \right) + \iint p \, dS_{L_{\text{outside}}} \tag{I.24}$$

where the subscript $D$ implies downstream conditions and the subscript $u$ implies upstream conditions.

In either case the drag is simply the integral of pressure and shear over the outside area of the airplane

$$D = \iint \left( p + \tau_{xn} \right) \sin(n, x) \, dS_A \tag{I.25}$$

Note that in Eq. (I.23) the outside of the airplane is defined by geometric conditions and in Eq. (I.24) it is defined by the location of stagnation lines in the inlet and a Kutta condition at the exhaust. This all seems straightforward enough. However, the pressure and shear around the inlet, on both the outside and inside, depend upon throttle setting; somehow that must be accounted for.

The most serious problem comes from the difference that is clear from the control surface used in Eq. (I.24)

$$\rho_\infty U_\infty^2 u_v + p_\infty A_u + \iint p \sin(n, x) \, dS \big|_{\text{outside}} - \iint \left( \rho u_x u_x + p \right) dA \big|_{\text{inlet}} \tag{I.26}$$

This represents a net drag on the body due to momentum acting between a far distance upstream and the inlet for the calculation of Eq. (I.23). This drag is normally accounted for in wind-tunnel tests, if there is flow through the inlet and if this flow is properly scaled. If, however, the wind-tunnel model has no flow ducts, then the measured drag will not include this throttle-dependent or additive drag, as it is sometimes called.

At first this additive component of drag was omitted entirely. By the late 1940s or early 1950s, there was a common agreement on the existence of this force. Then the argument was whether the drag component should be charged to the airframe—hence the term "additive drag"—or whether the thrust given by Eq. (I.23) should be reduced by this amount. We espouse no preference one way or the other; we merely note that the existence of such a term is important and must be accounted for. Note that in supersonic flight, with an attached shock fully internal diffusion inlet, the integrals defining additive drag sum to zero.

Millikan's prediction that data on effect of compressibility on drag coefficient was needed to resolve this problem is still appropriate. The present state of the art is discussed in Chap. V. The appearance of Perkins and Hage's work (Ref. 19, 1949, pp. 90–113) presented a wealth of data on compressibility effects on drag. The drag coefficient is defined as

$$C_D = C_{D_{P_{\text{min}}}} + k'' C_L^2 + \frac{C_L^2}{\pi \mathcal{R}} (1 + \delta) \tag{I.27}$$

where $C_{D_{p_{\min}}}$ is the minimum parasite drag, $k''C_L^2$ represents unavoidable parasite drag change with lift, and the last term is the same as defined 10 to 15 years earlier. The advice is given to increase $C_{D_{p_{\min}}}$ for each component by 5% to allow for interference, and then to increase $C_{D_{p_{\min}}}$ by 5–10% to allow for antennas, radomes, and the like. Perkins includes an increment to the drag due to flap settings. But he fails to discuss drag due to trim or additive drag. That was resolved subsequently. Nevertheless, Perkins does allow for drag due to cooling of reciprocating engines.

One useful chart uses the log of equivalent parasite area $f = C_F \cdot A_W = (C_{D_{p_{\min}}} \cdot S)$ as the ordinate and log of wetted area as the abscissa. Lines of constant $C_F$ can be drawn to readily compare relative skill of designers in reducing parasite drag.

The calculation of aerodynamic drag and the development of a theoretical foundation for some of the more esoteric aspects of drag were served more in the breach than directly in this period, with a single exception. The exception is Hawthorne's paper on drag due to secondary flow in 1954.[20] The importance of this paper really lies in the clues it gives to indicating the flowfield in the boundary layer near a wing-body junction, and hence the noting of geometry that will avoid premature separations in the wing-body junction. Toward the end of this period, D. Küchemann started an ambitious basic program at the Royal Aeronautical Establishment (RAE) of Great Britain, Farnborough, to study the details of flow about swept wings, using body junctions and the intersection of horizontal and vertical tails. Results from this program allowed us to appreciate subtle interferences in aerodynamic drag. Early results are given by Thwaites,[21] and a complete summary is given by Küchemann himself[22] about 20 years after the program was first in full swing. One of the keys to this program was the interaction of theory, experiment, and computations. In the United States the computational effort was less concentrated, though a leading proponent was A. M. O. Smith at Douglas.[23]

In the United States much similar information was developed by the NACA and by several companies by means of careful parametric studies. Much of this information is still held as proprietary. However, it appears that to a certain extent the tendency existed to substitute larger engines for reduced drag, a step the purist may find objectionable. Nonetheless the jet transport was part of a system and, if it was demonstrable that close attention to details would increase the manufacturing and overhead costs to the extent that the total cost of ownership was increased, any operator would opt for the less finely tuned design that was cheaper to own and operate.

By the close of this era, standard procedures were developed[24–26] and are under continual revision as technical skills advance.

*Modern Times — 1963 to the Present*

The modern era can be said to start, at least from one viewpoint, with the appearance of two documents. The first is Horner's *Fluid Dynamic Drag*,[26] and the second is Haines' "An Appreciation of Aerodynamic Drag."[27] If the first is a tour de force, and assembling an encyclopedia of drag should be so regarded, the second is also a tour de force—an elegant, detailed study of how to approach the calculation of drag. The discussion includes all the elements that determine drag, including drag creep (or what to do if you have a small but consistent transonic drag rise that begins at too low a Mach number), favorable and unfavorable aerodynamic interference, and effects of antennas. Naturally Haines includes trim drag. If there is a weakness in both these references, it is in the lack of discussion of throttle-dependent drag and drag due to aeroelastic deformation.[28,29] These matters are acknowledged in more recent books,[30,31] but not treated in depth. We will be more explicit on these topics in Chaps. IV and V.

## A Short History of the Turbojet Engine

The literature of jet engines and of fanjet engines is less rich. Some details are provided by Perkins,[19] Kücheman,[22] and Kerrebrock,[32] as well as in the newer design books.[30,31] These following remarks are based largely upon Boyne and Lopez[33] and upon Constant.[34]

The jet engine as discussed here is an assembly of five parts. The first and last are installation-related, and the middle three constitute an internal-combustion gas engine. These parts are the inlet, the compressor, the combustor, the turbine, and the exhaust. There are differences in detail depending upon whether all the working fluid passes through the combustor or whether some bypasses it. In the latter case, the engine is called a fanjet, or bypass jet, not a pure turbojet. Further, the combustion process does not usually deplete the working fluid of oxygen. Hence fuel can be added downstream of the turbine and upstream of the exhaust, with additional combustion and hence thrust. This process is called "afterburning" in the United States and "reheat" in Great Britain. In operation, the compressor draws air in from the outside and pressurizes it. Fuel is added to the high-pressure air and the mixture burned. This very high-pressure air passes through a turbine, which extracts the power and drives the compressor. The working fluid, while cooled somewhat by this process, is still at a relatively high pressure and temperature. Hence the turbine is followed by an expansion process, which converts the thermal energy into kinetic energy. The exhaust gases have a pressure that is on the average less than the pressure acting on the inlet and compressors. Further, the exhaust gases have a momentum flux that exceeds the inlet momentum flux. Both these factors contribute to the production of thrust, although which factor is dominant depends upon the flight regime and detailed design.

In writing a short history of the turbojet, it is easiest to look separately at the components that make up the internal-combustion gas turbine engine. Then the matter of jet propulsion will be discussed separately.

### The Turbine

The "aeolipile" of Heron, or Hero, of Alexandria represents the first use of a heated working fluid under pressure to generate rotary motion. The birth and death dates of this Greek mathematician and mechanician are unknown, but he seems to have lived sometime between the second century B.C. and the third century A.D. Heron's idea of using a jet to generate rotary motion is used today to drive lawn sprinklers. Perhaps a truer antecedent of the gas turbine is the "smokejack" which was invented in medieval times (Ref. 34, p. 63). This was essentially a single-stage turbine or fan that took energy from the smokestack of a kitchen and converted the energy into rotary motion to drive the spit.

In a technical sense, the development of waterwheels preceded the development of water turbines, which preceded the development of gas turbines. Simple impulse turbines were built by Branea (1629). In 1784 James Watt considered the turbine as a prime mover for steam power. According to his calculations, tip speed perhaps as great as 1000 ft/s would be necessary for a steam turbine to compete with the water turbine. Watt felt this was beyond the state of technology at that time. In 1853 Tournai read a paper to the French Academy of Science that contained an outline of the principles for designing both steam and internal-combustion turbines (Ref. 34, p. 66). He recognized the need for staging but encountered practical difficulties and never built an effective machine. W. J. M. Rankine, in his books *The Manual of the Steam Engine and Other Prime Movers* (1859) and a *Manual of Applied Mechanics* (1864), indicated that the thermodynamic processes were known and understood to the extent that turbine development depended upon practical considerations. Indeed about 20 years later de Laval invented the high-speed single-stage impulse steam turbine (1883), followed closely by Parson's multiple-stage (long-lived) reaction turbine (1884), Rateau's compound multistage impulse turbine (1894), and Curtis' two-stage impulse turbine (1897) whose development was completed at the General Electric Company (two years before Sanford Moss was hired—but more on that later). Parson's turbines were used for marine propulsion, with more than 300 in service by 1889. Constant felt that the impetus to provide shaft power from the turbine delayed the invention of jet propulsion (Ref. 34 p. 82).[†]

## Compressor

The history of compressors is much shorter. Rateau built the first axial flow compressor as a ventilation system. Parson built axial flow compressors in 1884 and centrifugal compressors in 1887. Many unsuccessful attempts were made to build unsuccessful internal-combustion turbines before 1900.

It is said that the modern turbojet engine really started in 1903 (a striking date!), when Sanford Moss succeeded in operating a turbine by burning fuel in a chamber under pressure. At this time Moss was a graduate engineering student at Cornell, and that quiet campus was said to be shaken by the noise of his experiments. His professor wrote about the experiment, "Whatever he thinks he is doing, it is not likely worth the noise, smokes, and smells. In fact, it is very likely to be worth nothing at all." This professor was W. F. Durand,

---

[†] Constant also says that A. A. Griffith, who studied axial flow compressors and gas turbines at RAE in the late 1920s for use as a turboprop, not a turbojet, felt Whittle's jet propulsion was at best suited for a short-range high-speed aircraft, if it was useful at all.

who in World War I would call upon Moss[‡] and the General Electric Company to develop the turbo supercharger. In 1909 Brown Bovari Company in Switzerland built a Hans Holzmark-designed constant-volume internal-combustion turbine. This 4000-hp design only produced 200 hp and was dropped.

The idea of jet propulsion in the form in which we know it today emerged through inventions of Sir Frank Whittle in England and Dr. Hans von Ohain in Germany in the late 1930s. Earlier the possibilities of jet propulsion were dismissed as impractical. Buckingham[35] felt the idea may be practical at higher speed, if they ever come about, but noted the idea was impractical "At such flying speeds as are now in prospect" (250 mph). Nevertheless both Whittle and von Ohain saw a future for jet propulsion. These men used a centrifugal compressor to ensure an adequate pressure ratio. von Ohain used a radial turbine to simplify matching problems. By the early 1940s these engines had flown. The first gas turbine that flew with an axial flow compressor followed principles laid down by Herbert Wagner in 1937. Dr. Anselm Franz designed this machine, which first flew in 1943 in the Me-262. [E. W. Constant credits Helmut Schelp, of the research division of the Reichsluftfahrt Ministerium (RLM), for seeing the value of jet propulsion and keeping the project alive.] Subsequent development was so rapid that by 1950 all first-line fighters were propelled by turbojet engines and, within the next half decade, the transition to jet-propelled transports and bombers was well under way.

Gas turbine engines in aircraft use in the 1980s have achieved a level of power density, fuel economy, and reliability unthought of 30 years earlier, for either gas turbine or reciprocating engines.

### Flight Validation

The need for verification of the airframe performance was real and continuing. It is clear by 1928 that serious test facilities were developed to this end.[15] The 40×80 tunnel at NASA Ames Research Center was built to allow engineers to carry out drag cleanup on real airplanes and at flight (relatively low speed, to be sure) Reynolds numbers. This clearly indicates that detailed prediction of real details on practical configuration was not considered reliable.

The earliest discussions of these problems in connection with jet-propelled aircraft were given by C. L. (Kelly) Johnson in 1947.[36] He discusses many of the problems of performance that are still unresolved, including inlet diffusor matching and stability and engine-to-engine variation in thrust. Incidentally the procedure used at the time to determine thrust was a correlation of thrust with burner pressure. Naturally corrections needed to be made to correct the airflow to standard conditions. Drag was determined from tail-pipe thrust and by dive-angle measurements. The effect of rain on changing surface smoothness was clearly noted, discussed, and overlooked for 40 years.

The problem of correlation remains a difficult one that is not fully solved. The Flight Mechanics Panel of AGARD has sponsored a series of procedural

---

[‡]Moss successfully applied the high-temperature technology developed for turbo superchargers to Whittle's configuration. It is a little-known fact that Moss used a gas turbine as a test stand to develop turbochargers in the mid-1930s.

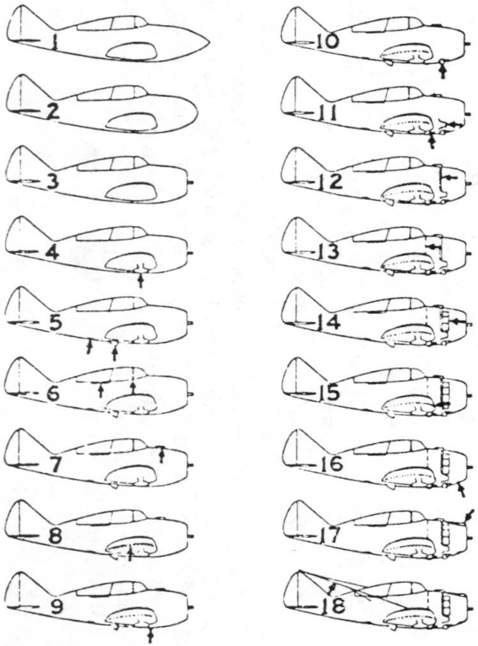

| Condition number | Description | $C_D$ ($C_L = 0.15$) | $\Delta C_D$ | $\Delta C_D$, percent[a] |
|---|---|---|---|---|
| 1 | Completely faired condition, long nose fairing | 0.0166 | | |
| 2 | Completely faired condition, blunt nose fairing | .0169 | | |
| 3 | Original cowling added, no airflow through cowling | .0186 | 0.0020 | 12.0 |
| 4 | Landing-gear seals and fairing removed | .0188 | .0002 | 1.2 |
| 5 | Oil cooler installed | .0205 | .0017 | 10.2 |
| 6 | Canopy fairing removed | .0203 | -.0002 | -1.2 |
| 7 | Carburetor air scoop added | .0209 | .0006 | 3.6 |
| 8 | Sanded walkway added | .0216 | .0007 | 4.2 |
| 9 | Ejector chute added | .0219 | .0003 | 1.8 |
| 10 | Exhaust stacks added | .0225 | .0006 | 3.6 |
| 11 | Intercooler added | .0236 | .0011 | 6.6 |
| 12 | Cowling exit opened | .0247 | .0011 | 6.6 |
| 13 | Accessory exit opened | .0252 | .0005 | 3.0 |
| 14 | Cowling fairing and seals removed | .0261 | .0009 | 5.4 |
| 15 | Cockpit ventilator opened | .0262 | .0001 | .6 |
| 16 | Cowling venturi installed | .0264 | .0002 | 1.2 |
| 17 | Blast tubes added | .0267 | .0003 | 1.8 |
| 18 | Antenna installed | .0275 | .0008 | 4.8 |
| | Total | | 0.0109 | |

a Percentages based on completely faired condition with long nose fairing.

Airplane Condition

Fig. 1.9 Experimental study of drag sources on Seversky XP-41.

manuals. Poisson-Quinton[37] in a masterful survey shows that handling qualities are much easier to determine than performance.

Poisson-Quinton has a good breakout of drag as: 1) elliptic vortex drag, 2) nonelliptic vortex drag, 3) trim drag, and 4) profile drag: friction drag, pressure drag, and equivalent roughness (steps, gaps, and protuberances). He shows that the profile drag in the Falcon was reduced 6% by attention to detail. He does not discuss aeroelastic drag or throttle-dependent drag. Treon and Steinle[38] show that there is wind-tunnel–to–wind-tunnel variation that needs accounting for. The state of the art is advancing slowly. If one compares the data given in Ref. 39 with current parameters, it is clear that the problem comparing measured with estimated drag needs further attention. This will be discussed in detail in Chap. IV. Conclusions from Ref. 39 are listed below:

1) Reynolds number effects are unresolved except for special situations.
2) Modeling techniques are inadequate in many areas.
3) Tunnels may have idiosyncrasies that can be allowed for, if enough experience and data comparison have been made.
4) Wall effects are not fully understood.
5) More complete comparison with wind tunnels and flight is needed.

The last recommendation is partly supported by Loftin,[40] who has collected a great deal of practical data in this regard, particularly in Secs. 5.4, 5.6.1, and 6.5.4. An interesting illustration of drag sources is given in his Table 5.111 (p. 318), reproduced here as Fig. I.9.

### Closing Comment

In the chapters that follow, it is intended to a present a critical review of thrust and drag bookkeeping, prediction of installed thrust, prediction of drag, origin and propagation of error, and consequences of powered-flight, throttle-dependent effects.

Finally, this book will be concluded with some examples of confirmation or lack thereof, together with recommendations for future work.

### References for Chapter I

[1] Pritchard, J. L., *Sir George Cayley*, Horizon Press, New York, 1962.
[2] Lanchester, F. S., *Aerodynamics*, 4th Ed., Constable and Company, Ltd., London, 1918. I quote from the Preface, "Except for correction of errata, revisions of a minor character, and some few references to later work, no change has been made in the work since the first edition (1907)."
[3] Kerber, L. V., "Airplane Performance," Vol. V, Division O, Durand's *Aerodynamic Theory* (1934) contains a detailed discussion of the Eiffel-Rith procedure.
[4] Bairstow, L., *Applied Aerodynamics*, Longmans, Green, and Co., London, New York, and Toronto, 1920.
[5] Diehl, W. S., "Charts for Graphical Estimation of Airplane Performance," NACA Report 192, 1924.
[6] Diehl, W. S., "Three Methods of Calculating Range and Endurance of Airplanes," NACA Report 234, 1926.

[7] Giacomelli, R. and Pistolesi, E., "Historical Sketch," Vol. I and Division D, Durand's *Aerodynamic Theory*, 1954.

[8] Eiffel, G., *La Resistance de l'air*, edited by H. Donod and E. Pinat, Paris, 1910.

[9] Villamil, R. de, *Resistance of Air*, E. and F. N. Spon, Ltd., London, 1917.

[10] Prandtl, L., "Application of Modern Hydrodynamics to Aeronautics," NACA Report 116, 1921.

[11] Taylor, D. W., "Some Aspects of the Comparison of Model and Full Scale Tests," NACA Report 219, 1925.

[12] Pannell, J. R. and Lavender, T., "Experiments on Forces Acting on Airplane Struts," ARC R&M 183, 1915.

[13] Jones, R. and Pell, G. N., "Air Resistance of Flying Boat Hulls," ARC R&M 461, 1918–1919.

[14] Diehl, W. S., *Engineering Aerodynamics*, Ronald & Co., New York, 1928.

[15] Weik, F. E., "Full Scale Drag Tests on Various Parts of a Sperry Messenger Airplane," NACA Report 271, 1928.

[16] Townend, H. C. H., "The Townend Ring," *Journal of the Royal Aeronautical Society*, Vol. XXXIV, Oct. 1930, pp. 813–848.

[17] Millikan, C. B., *Aerodynamics of the Airplane*, John Wiley and Sons, New York, 1941.

[18] MIDAP Study Group, "Guide to In-Flight Thrust Measurements," AGARDO-GRAPH AG-237, Jan. 1979.

[19] Perkins, E. D. and Hage, R. F., *Airplane Performance Stability and Control*, John Wiley and Sons, New York, 1949.

[20] Hawthorne, W. R., "The Secondary Flow About Struts and Airfoils," *Journal of the Aeronautical Sciences*, Vol. 21, No. 9, 1954, pp. 588–608.

[21] Thwaites, B., ed., *Incompressible Aerodynamics*, Oxford University Press, 1963.

[22] Kücheman, D., *The Aerodynamic Design of Aircraft*, Pergamon Press, London, 1978.

[23] Hess, J. L. and Smith, A. M. O., "Calculation of Potential Flow About Arbitrary Bodies," *Progress in Aeronautical Sciences*, Vol. 8, D. Kücheman, (ed.), Pergamon Press, New York, 1967.

[24] Malthan, L. V., Hoak, D. E., and Carlson, J. W., *The USAF Stability and Control Handbook (Dat Com)*, Air Force Flight Dynamics Laboratory, Wright-Patterson Air Force Base, Ohio, 1960.

[25] Royal Aeronautical Society Data Sheets, *Wings and Airfoil*, Vols. 2c, 3b, 3c, 4a, 4c, 4d. There are also data given in *Aircraft* Magazine.

[26] Horner, S. F., *Fluid Dynamic Drag*, published by the author, 1964. (The first edition, *Aerodynamic Drag*, was published in 1951.)

[27] Haines, B. A., "Subsonic Drag: An Appreciation of Present Standards," *Journal of the Royal Aeronautical Society*, Vol. 12, 1968, p. 253.

[28] Ferri, A., "Airframe-Engine Integration," *AGARD Lecture Series*, LS-53, 1972.

[29] Antonatos, P. P., "Aerodynamic Drag," *AGARD Conference Proceedings*, CP-124, 1973.

[30] Torrenbeck, E., *Synthesis of Subsonic Airplane Design*, Delft University Press, The Netherlands, 1976.

[31] McCormick, B. W., *Aerodynamics, Aeronautics and Flight Mechanics*, John Wiley and Sons, New York, 1980.

[32] Kerrebrock, J. L., *Aircraft Engines and Gas Turbines*, MIT Press, Cambridge, Mass., 1977.

[33] Boyne, W. J. and Lopez, D. S., *The Jet Age*, National Air and Space Museum, Washington, D. C., 1979.

[34] Constant, E. W. II, *The Origin of the Turbojet Revolution*, Johns Hopkins Univer-

sity Press, Baltimore, Md., 1980.

[35] Buckingham, E., "Jet Propulsion for Airplanes," NACA Report 159, 1923.

[36] Johnson, C. L., "Development of the Lockheed P-80A Jet Fighter Airplane," *Journal of the Aeronautical Sciences*, Vol. 14, No. 12, pp. 659–679.

[37] Poisson-Quinton, P., "From Wind-Tunnel to Flight, the Role of the Laboratory in Airplane Design," *Journal of Aircraft*, Vol. 5, May–June 1968, p. 193.

[38] Treon, S. L. and Steinle, F. W., "Data Correlation of a High Subsonic Space Transport Aircraft in Three Major Transonic Wind Tunnels," AIAA Paper 69-794.

[39] Treon, S. L. and Steinle, F. W., "Further Correlation of Data from Investigation of a High Speed Subsonic Transport Aircraft Model in Three Major Transonic Wind Tunnels," AIAA Paper 71-291, *Journal of Aircraft*, Vol. 12, No. 4, June 1975.

[40] Loftin, L. K., "Subsonic Aircraft, Evolution of the Matching of Size to Performance," NASA Reference Publication 1060, Aug. 1980.

# Thrust-Drag Accounting Methodology

Eugene C. Rooney*
*Naval Air Systems Command, U.S. Navy,
Department of Defense, Washington, D.C.*

## Introduction

$T$HE accurate prediction of air vehicle performance and correlation between the prediction and flight documentation phases of air vehicle development programs require definition of a thrust-drag accounting or "bookkeeping" system which clearly defines the treatment of all the aerodynamic and propulsion-related forces acting on the system. The need for a bookkeeping system in the performance prediction phase of an aircraft development program arises largely from the inability to determine the performance of the complete airplane system, with simultaneous real inlet and an exhaust system operation, in a single test or calculation. Furthermore, it is often desirable to optimize the inlet and exhaust system in separate tests which are independent of the general aerodynamic testing of the basic airplane configuration. The implicit assumption exists that the effects of the inlet and exhaust nozzle can be measured separately and combined linearly. This is valid for buried configurations near the ideal angle of attack. The assumption may or may not be valid for some configurations and certain flight conditions. For pod-nacelle installations where the inlet and nozzle are aerodynamically close-coupled, the assumption is rarely valid. For purposes of illustration, our discussion will be based upon a buried installation, so the assumptions of independence and linearity will be used. Even so, a well-defined performance integration system is required to ensure that the various elements (i.e., inlet, exhaust, airframe, turbomachinery) of the airplane system are combined properly to yield an accurate prediction of overall system performance. Also, comparison of aircraft performance predictions with results of the flight-test documentation process requires an understanding of the elemental thrust and drag forces to validate the aircraft

---

EDITOR'S NOTE: The material in this chapter is taken primarily from Refs. 1–3.
*Head, Aircraft Performance.

performance model for conditions not actually tested or to isolate the source of any aerodynamic or propulsion differences that may exist.

The variety of actual and possible aerodynamic and propulsion system configurations makes it impractical to specify a single rigorous accounting system. Therefore, to obtain the most accurate and efficient bookkeeping system, the accounting methodology is tailored to the requirements and resources of each individual aircraft development program. To accomplish these objectives, the accounting system should address the following characteristics:

1) Requirement for consistency. All forces must be accounted for once and only once. Consistent and precise definitions and terminology must be associated with all force terms used.

2) The bookkeeping procedures should afford as much visibility as feasible to the performance of the elements of the airplane system.

3) Selection of reference conditions, although somewhat arbitrary, should provide for a way to correct the airplane drag polar to realistic inlet and exhaust system operating conditions. (The selection of the reference conditions is no less arbitrary than selection of a bookkeeping system. Nevertheless, certain guidelines exist. For example, the reference conditions should correspond as closely as possible to flight conditions. Ideally, the inlet mass flow ratio should be nearly one. The flight conditions should correspond to reference Reynolds number and Mach number, when possible.)

4) The thrust-drag accounting methodology must be suitable and consistent for tracking of integrated propulsion/airframe performance throughout the aircraft development program. Thus, for example, the drag polars generated theoretically in the earliest phases of a program, those generated experimentally in wind-tunnel programs, and those generated in flight-test programs should all be based on the same accounting rules. This is the only way in which drag polars can be compared with each other in a meaningful way.

The remainder of this chapter addresses these concepts in the development of thrust-drag accounting procedures and methodology for a particular aerodynamic/propulsion configuration—the fully integrated (burried) propulsion system shown in Fig. II.1. The assumptions and changes required to adapt this accounting system to the podded engine nacelle propulsion configuration, shown in Fig. II.2, are also discussed.

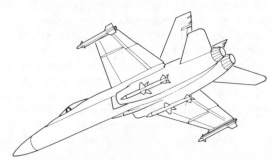

Fig. II.1   Integrated (burried) propulsion configuration.

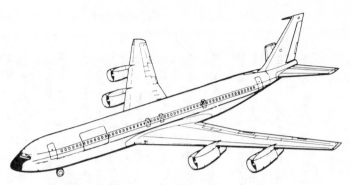

**Fig. II.2   Podded nacelle propulsion configuration.**

## Thrust-Drag Accounting Approach

Considering an aircraft in level flight, the simplified force equation applied in the flight direction takes the form

$$F_{EX} = F_{IPF} - D_{AFS} \qquad (II.1)$$

where

$F_{EX}$ = imbalance (total force) between the engine thrust and aircraft drag forces in the flight direction at given attitude, altitude, and Mach number conditions

$F_{IPF}$ = installed net propulsive force that is obtained from net thrust, with adjustments for deviation from full scale operating reference conditions (a specification engine thrust in the prediction phase but is the actual engine thrust in flight)

$D_{AFS}$ = airframe system drag force reflected in the vehicle drag polar at full scale operating reference conditions

The approach presented here establishes three distinct aircraft and engine performance elements: aircraft drag (Chap. IV), engine thrust (Chap. III), and throttle-dependent drag (Chap. V). Although summarized separately in this chapter and in detail in subsequent chapters, the elements of thrust and throttle-dependent drag, because they are both throttle-dependent, are combined together as a single force, $F_{IPF}$.

Additional forces included in the airframe system drag and the net propulsive force terms account for defined or chosen reference full scale operating conditions and excursions from the full scale operating conditions. The breakout of the additional forces, definition of terms and reference conditions for a thrust-drag accounting system applicable to the fully integrated propulsion configuration like that in Fig. II.1 are contained in Eq. (II.2) and in the following list:

$$F_{EX} = \underbrace{F_N + \Delta F_{INL} + \Delta F_{EXH} + \Delta F_{TRIM}}_{\text{installed net propulsive force }(F_{IPF})} \quad \underbrace{-D_{REF} - \Delta D_{INL} - \Delta D_{EXH} - \Delta D_{TRIM} - \Delta D_{RN}}_{\text{airframe system drag }(D_{AFS})} \qquad (II.2)$$

where:

$F_N$ = installed engine net thrust: net force generated by the engine, accounting for the effects of inlet internal performance, nozzle internal performance, engine compressor bleed, and power extraction

$\Delta F_{INL}$ = throttle-dependent external force increment due to the change in inlet operation from the operating reference condition to the actual operating condition

$\Delta F_{EXH}$ = throttle-dependent external force increment due to the change in exhaust system operation from the operating reference condition to the actual operating condition

$\Delta F_{TRIM}$ = throttle-dependent external force increment due to the change in control surface position from that required for trim at the operating reference inlet, exhaust, and center-of-mass location to the control surface position required for trim at the actual inlet and exhaust operating conditions

$D_{REF}$ = external force (associated with aerodynamic force and moment model) at aerodynamic reference conditions and with corrections for full scale effects

$\Delta D_{INL}$ = external force increment due to the change in inlet operation from the aerodynamic reference condition to the operating reference condition

$\Delta D_{EXH}$ = external force increment due to the change in exhaust system operation from the aerodynamic reference condition to the operating reference condition

$\Delta D_{TRIM}$ = external force increment due to the change in control surface position from that required for trim at the aerodynamic reference conditions to the control surface position required for trim at the propulsion operating reference conditions and actual center-of-mass location. This increment may be further divided into the trim increment for the change from the aerodynamic reference conditions to the operating reference conditions $(\Delta D_{TRIM}^{(1)})$ and the trim increment for the change in center-of-mass variations at the propulsion operating reference conditions $(\Delta D_{TRIM}^{(2)})$

$\Delta D_{TRIM}^{(1)}$ = external force increment due to the change in control surface position from that required for trim at the aerodynamic reference inlet, exhaust, and center-of-mass location to the control surface position required for trim at the operating reference and inlet, exhaust, and center-of-mass location

$\Delta D_{TRIM}^{(2)}$ = external force increment due to the change in control surface position from that required for trim at the operating reference inlet, exhaust, and center-of-mass location to the control surface position required for trim at the actual center-of-mass location

$\Delta D_{RN}$ = external force increment due to the change from the full scale reference Reynolds number to the actual flight operating Reynolds number (function of altitude, Mach number, temperature, surface roughness, and boundary-layer conditions)

Figures II.3a and II.3b illustrate the integration of this force accounting system for use in the wind-tunnel prediction and flight documentation phases on an air vehicle development program. Figure II.3a illustrates the integration of the various wind-tunnel model tests (aerodynamic force and moment, inlet and powered nacelle/afterbody models) required to implement this bookkeeping system for accurate prediction of aircraft performance. Figure II.3b shows the interrelationship of this accounting system with the flight documentation process to provide a consistent tracking system throughout the aircraft development program. The incremental forces indicated by the dashed lines in Figs. II.3a and II.3b are elements determined for the prediction process, based on model data, and assumed applicable to the flight documentation process.

The preceding list is not all inclusive but is characteristic of the kinds of items that must be included. Additional detailed items such as drag increment due to inlet and exhaust of secondary airflows will be included in Chap. IV and increments like throttle-dependent trim drag will be included in Chap. V.

The various elements of this accounting system are categorized relative to strictly defined reference and operating conditions including aerodynamic (aero) reference, full scale geometry reference, and full scale operating conditions. The establishment of full scale reference conditions requires the selection of several variables, including inlet mass flow ratio, inlet geometry, nozzle geometry, secondary airflows, and aircraft trim setting. Since these variables influence the installation drag, a fixed set of reference conditions must be identified.

Additional wind-tunnel tests are conducted to determine the incremental drag forces attributable to the inlet and afterbody geometry, inlet mass flow ratio, and nozzle pressure ratio. A geometric issue of great importance here is the location of the part line delineating the different models. These models allow for estimation of the three primary components that enter into performance analysis. The three distinct aircraft/engine performance elements—aircraft drag, engine thrust, and throttle-dependent drag—are discussed in detail in the following paragraphs.

## Airframe System Drag

During the performance prediction phase of an aircraft development program, the aircraft drag polar is generally obtained through wind-tunnel testing of a subscale aerodynamic force and moment (or aero-reference) model, an example of which is shown in Fig. II.4. This model is designed to be as representative as possible of the full scale aircraft but usually differs in several respects. The aero-reference model has no engines but instead may have flow-through ducts or nacelles installed. While flow-through ducts or nacelles provide the simplest and most accurate means of simulating an engine installation, they do have limitations. Since the flow-through nacelle does not simulate the work added by the engine to the captured stream tube, it cannot simultaneously simulate the full scale engine inlet mass flow ratio, inlet geometry, nozzle pressure ratio, and nozzle geometry. Although the flow-through duct/nacelle does have associated with it a force, the internal drag, this force

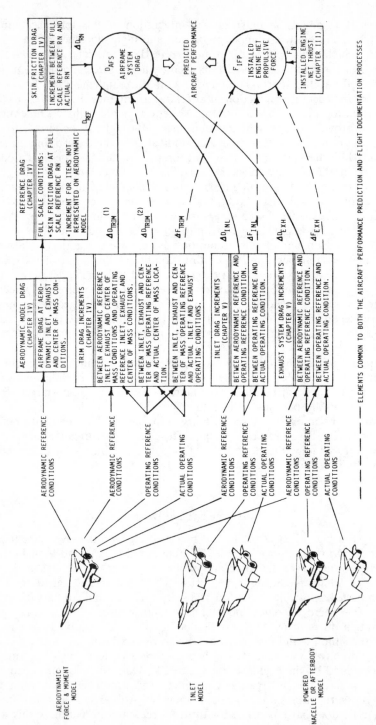

**Fig. II.3a  Aircraft performance prediction process.**

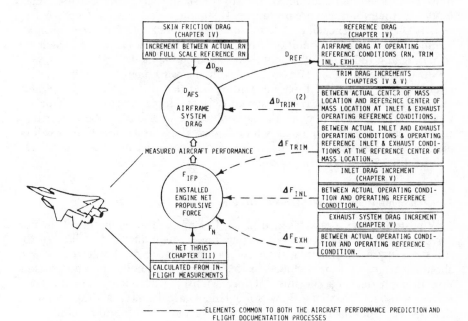

SKIN FRICTION DRAG
(CHAPTER IV)
INCREMENT BETWEEN ACTUAL RN AND FULL SCALE REFERENCE RN
ΔD RN
REFERENCE DRAG
(CHAPTER IV)
AIRFRAME DRAG AT OPERATING REFERENCE CONDITIONS (RN, TRIM INL, EXH)
D REF
D AFS
AIRFRAME SYSTEM DRAG
TRIM DRAG INCREMENTS
(CHAPTERS IV & V)
ΔD TRIM (2)
BETWEEN ACTUAL CENTER OF MASS LOCATION AND REFERENCE CENTER OF MASS LOCATION AT INLET & EXHAUST OPERATING REFERENCE CONDITIONS.
BETWEEN ACTUAL INLET AND EXHAUST OPERATING CONDITIONS & OPERATING REFERENCE INLET & EXHAUST CONDITIONS AT THE REFERENCE CENTER OF MASS LOCATION.
ΔF TRIM
MEASURED AIRCRAFT PERFORMANCE
F IFP
INSTALLED ENGINE NET PROPULSIVE FORCE
INLET DRAG INCREMENT
(CHAPTER V)
ΔF INL
BETWEEN ACTUAL OPERATING CONDITION AND OPERATING REFERENCE CONDITION.
EXHAUST SYSTEM DRAG INCREMENT
(CHAPTER V)
ΔF EXH
BETWEEN ACTUAL OPERATING CONDITION AND OPERATING REFERENCE CONDITION.
F N
NET THRUST
(CHAPTER III)
CALCULATED FROM IN-FLIGHT MEASUREMENTS
ELEMENTS COMMON TO BOTH THE AIRCRAFT PERFORMANCE PREDICTION AND FLIGHT DOCUMENTATION PROCESSES

**Fig. II.3b Aircraft performance flight documentation process.**

AERO FORCE AND MOMENT
CHARACTERISTICS
• AIR FLOW NORMALLY LIMITED
• AIR FLOW CONTROL USUALLY LIMITED
• POSSIBLE SUPPORT INTERFERENCE AND GEOMETRIC DISTORTION
• COMPLETE MODEL
INLET DRAG
CHARACTERISTICS
• PARTIAL MODEL
• WIDE RANGE AIR FLOW CONTROL
• STING SUPPORT
JET EFFECTS
CHARACTERISTICS
• PARTIAL MODEL
• HIGH PRESSURE AIR SUPPLIED FROM EXTERNAL SOURCE
• HOT OR COLD AIR MAY BE USED
• FAIRED-OVER INLETS

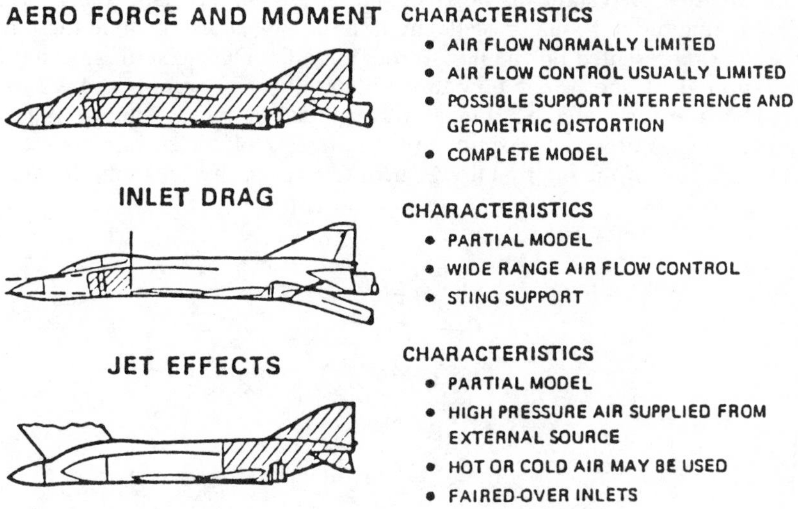

**Fig. II.4 Example wind-tunnel models.**

must be isolated and subtracted. Due to the small scale of the aero-reference model, auxiliary airflows are not normally simulated in the aero-reference drag test. The aero-reference drag data must also be corrected to full scale Reynolds number and reference conditions (of Figs. II.3a and II.3b) and adjusted for trim drag and full scale details not represented on the model.

Most of the deficiencies associated with the aero-reference model can be circumvented by proper force accounting and additional test programs and models (as shown in Fig. II.4), designed to isolate force increments as much as possible. Those corrections to the aero-reference drag data to be reflected in the full scale aircraft drag polar will be discussed in this section. Others will be accounted for as throttle-related drag which impacts net propulsive force and will be discussed in a later section.

### Flow-Through Model Internal Drag Correction

Wind-tunnel drag data taken using a flow-through airplane model include forces acting on the internal walls of the flow-through duct or nacelle. For the full scale engine installation, these forces are replaced by the net thrust and should therefore not be included in the airplane drag polar. For this reason, flow-through duct/nacelle internal forces must be deducted from the aero-reference model force data. Flow-through duct/nacelle internal drag should be determined experimentally on a static thrust stand or calibration facility.

For the case of a short-duct turbofan installation, shown in Fig. II.5, the engine gross thrust includes fan air scrubbing drag on the pylon and external core cowl as well as primary air scrubbing drag on the external plug. These effects are measured statically on an engine thrust stand (Chap. III). Therefore, the short-duct flow-through nacelle internal drag must also include these terms if they are represented on the flow-through model. During static internal drag calibration of a short-duct flow-through nacelle, any scrubbed areas represented on the aero-reference model are included as part of the flow-through internal drag. Further, for simplicity, the effects of jet entrainment on the nacelle are part of the internal flow. The interference effects among the nacelle,

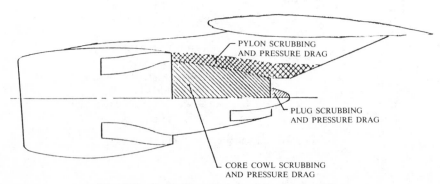

**Fig. II.5   Podded nacelle propulsion configuration.**

the pylon, and the wing are termed throttle-dependent and discussed in Chap. V.

### Scale-Model/Full Scale Geometry Correction

The aero-reference model geometry usually differs from the full scale airplane. The aero-reference configuration is selected to minimize flow separation that may be experienced during off-design inlet mass flow ratio operation and to minimize the effects of exhaust nozzle pressure ratio associated with a flow-through duct/nacelle installation. Therefore, the aircraft drag polar must be adjusted to reflect any geometry-related drag increment. This incremental drag may be evaluated with scale-model inlet and/or afterbody drag tests of the two different designs. Normally this can be included in the inlet and nozzle drag testing to be discussed later. In the small-scale aero-reference model, laminar flow separation should be ruled out by some diagnostic tool, unless the model and flight Reynolds number are similar. If unsought laminar separation is detected, then additional tests at a higher Reynolds number are required. Typically the inlet and nozzle drag testing is conducted with a larger-scale model than the aero-reference model. In this way the Reynolds number and hence the flow more closely match the full scale flight conditions. These steps reduce the problem of laminar separation.

### Trim Drag

In the wind tunnel, aircraft drag polars can be generated without regard for aircraft center of gravity or pitching moments. In steady flight, however, the total pitching moment must be zero to maintain equilibrium. The incremental pitching moment needed to balance that acting on the airplane in flight is generated by deflecting aerodynamic control surfaces. The resulting incremental change in aircraft drag, called trim drag, is a separate drag correction term which varies with aircraft center of gravity and may vary with thrust. Trim drag is usually evaluated in the wind tunnel utilizing the aero-reference model.

### Reynolds Number Correction

Due to the small scale of high-speed wind-tunnel airplane models, full scale Reynolds numbers cannot usually be achieved. This leads to differences between scale model and full scale in both skin-friction coefficients and the location of boundary-layer transition (Chap. IV).

Transition from a laminar to a turbulent boundary layer on the scale model at low Reynolds number is achieved through the use of grit strips placed on the fuselage, wings, tail, and nacelles. Location of the grit strips is designed to match the expected full scale transition point. While the use of transition strips is usually adequate for achieving the desired turbulent skin-friction coefficients, the location of flow separation on the inlet lip (at low mass flow ratio) or on the wing leading edge (at high angle of attack) cannot be confidently duplicated at low Reynolds number. Separation characteristics may differ between the full scale and the low Reynolds number model with transition strips. Therefore, the aero-reference model may incorporate some design modifications or test limitations to minimize flow separation.

The problem of adjusting aircraft drag to account for the lower Reynolds number of the scale model can be approached in two ways. One is to test the scale model at varying tunnel Reynolds number and extrapolate the results to the full scale Reynolds number. An alternate approach is to apply the incremental difference between the ratio of scale-model and full scale predicted flat-plate skin-friction coefficients to the wind-tunnel drag data. However, this Reynolds number correction procedure applies only to the portion of drag caused by skin friction and not to vehicle pressure drags or lift-induced drags (Chap. IV).

**Full Scale Drag Polar**

The basis for the full scale aircraft drag polar is the flow-through aero-reference model data. These scale-model lift and drag characteristics are corrected for internal flow, inlet, or afterbody geometry differences between the aero-reference model and the full scale aircraft, the Reynolds number difference, and aeroelastic effects to obtain the full scale drag polar. Additional adjustments are required to account for items not represented or impractical for representation on the model, such as surface roughness, protuberances and excrescences or secondary air inlet and exhausts. These adjustments are usually developed analytically. This full scale drag polar applies to a fixed altitude (i.e., specific Reynolds number range) and has associated with it the full scale reference inlet mass flow ratio and flow-through nozzle pressure ratio.

To evaluate aircraft performance, the full scale drag polar need only be adjusted to compensate for altitude (Reynolds number) effects and trim drag. Any other changes in drag associated with variable inlet or nozzle geometry, inlet mass flow ratio, or nozzle pressure ratio do not influence the drag polar. As shown in Figs. II.3a and II.3b, these forces are accounted for as propulsion-related drags and are part of the net propulsive force.

**Engine Thrust**

In-flight thrust is based on net thrust generated by the engine, accounting for the effects of inlet internal performance, nozzle internal performance, engine compressor bleed airflow, and power extraction.

For single stream, gross thrust is defined as the gage stream force at the engine nozzle exit station [see Eq. (III.4) of Chap. III]. Net thrust is defined as the difference between gross thrust and the entering freestream momentum.

For dual-flow engine, using the engine station numbering system outlined in Fig. II.6[4]:

$$F_N = F_{G_9} + F_{G_{19}} - F_{G_0} \tag{II.3}$$

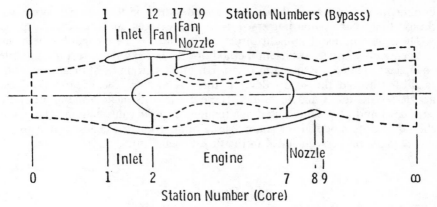

**Fig. II.6 Engine station numbering system.**

where

$$F_{G_9} = W_9 V_9 + A_9 \left( P_{S_9} - P_{S_0} \right) \tag{II.4}$$

$$F_{G_{19}} = W_{19} V_{19} + A_{19} \left( P_{S_{19}} - P_{S_0} \right) \tag{II.5}$$

$$F_{G_0} = W_1 V_0 \tag{II.6}$$

$W_1$ = captured engine airflow

$19$ = fan nozzle exit station conditions

$9$ = core nozzle exit station conditions

$0$ = upstream infinity conditions

For single-flow engine the station $19$ terms are deleted. In the above definition of net thrust, the gross thrust and ram drag terms are assumed to act collinearly. For cases where the gross thrust vector is not in the flight-path direction due to engine or nacelle incidence/toe in or aircraft angle of attack, the gross thrust vector must be factored into axial and normal components. For practical applications in calculating airplane performance, ram drag and gross thrust must be considered individually and separated into flight-path and normal to flight-path components (i.e., the lift-drag coordinate wind axes system). For simplicity the remainder of this discussion will assume collinear ram drag and gross thrust vectors.

For most installations, in-flight thrust is consistent with the above definition of net thrust. One common installation, the short-duct turbofan shown in Fig. II.5, represents a notable exception. It is generally accepted that the drag on external surfaces immersed in the fan exhaust or core exhaust jet will be bookkept as engine gross thrust. This includes drag on the external core cowl, the portion of the pylon immersed in the fan exhaust, and the external plug.

These definitions of in-flight thrust are intended to be compatible with the gas generator/nozzle coefficient method of calculating gross thrust. The nozzle coefficient method, discussed in Chap. III, utilizes ground-test or altitude-test-facility-measured thrusts to develop thrust and flow coefficients based on nozzle temperatures and pressures. These coefficients include the appropriate core cowl and plug scrubbing drags for the short-duct installation. The pylon should also be modeled for static testing of short-duct nacelles so that pylon scrubbing is properly included in the thrust coefficient.

## Throttle-Dependent Drags

In previous sections the terms "in-flight thrust" and "full scale operating aircraft drag" have been developed. If the engine were operating at full scale reference conditions (i.e., the reference nozzle pressure ratio and reference inlet mass flow ratio), all of the forces acting on the full scale aircraft would be known (neglecting any auxiliary flows which are not simulated in the aero-reference model but which produce thrust or drag in the full scale aircraft). Actual engine operating conditions are variable and often significantly different from the operating reference conditions (in fact the operating reference condition may not be an obtainable engine condition). To account for the differences between real engine operating conditions and full scale operating reference conditions, the concept of throttle-dependent drag has been developed. Since throttle-dependent drags vary with engine throttle setting, it is inconvenient to include them in the airplane drag polar. Therefore, these terms are normally charged against in-flight thrust. In-flight thrust, adjusted for throttle-dependent drag, has been termed net propulsive force. Net propulsive force and the full scale aircraft drag (adjusted for trim drag, Reynolds number, and other real airplane effects) should account for all forces acting on the aircraft/engine system.

Throttle-dependent drags fall into one of four general categories: inlet spillage drag, exhaust system or jet-effects drag, auxiliary airflow drag, and throttle-dependent trim drag.

### Inlet Spillage Drag

Throttle-dependent inlet spillage drag is defined as the change in aircraft drag resulting from the difference between the operating conditions and the operating reference condition inlet mass flow ratios. Spillage drag, which results from unrecovered inlet additive drag, varies with inlet mass flow ratio as illustrated in Fig. II.7.

Inlet tests may be conducted to evaluate the throttle-dependent spillage drag characteristics. Drag of the aero-reference model inlet (Fig. II.7) (1) is included in the aero-reference drag. The incremental drag between (1) and the full scale operating reference conditions (2) represents the scale-model to full scale drag correction $\Delta D_{INL}$. Drag differences between engine operating condition

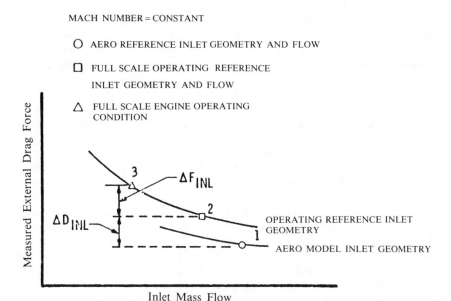

Fig. II.7  Drag variation with inlet mass flow.

(3) and full scale engine reference conditions (2) are accounted for as throttle-dependent inlet drag $\Delta F_{INL}$ to be included in the net propulsive force.

For subsonic, podded installations, the inlet forebody is generally designed to minimize spillage drag over the operating range of the real engine; however, if desired, spillage drag for this type of installation can be investigated by testing isolated flow-through nacelle models with internal geometry variations to alter inlet mass flow ratio. Nacelle fan cowl geometry should be correctly simulated due to possible inlet/afterbody coupling effects. As discussed earlier, each flow-through model must be calibrated individually to determine the internal drag correction to the force data to obtain external drag. The isolated nacelle drag test is used primarily to determine difference in drag due to mass flow ratio rather than absolute nacelle drag levels.

For integrated supersonic aircraft installations, the inlet is often susceptible to high spillage drag levels at low inlet mass flow ratios. An inlet spillage drag test is essential for this application, but due to the interaction between inlet and fuselage, an isolated test is not practical. The inlet and forward fuselage should be tested as a unit in order to accurately assess mass flow ratio effects on inlet spillage drag. Again, a flow-through model with internal duct alterations or an internal flow-metering device could be used to conduct this test. Note that if some air is bled off and injected into the external flow, an additional drag could result. Since inlet and afterbody drags are not closely coupled, as they are in podded installations, there is usually no requirement to exactly duplicate the afterbody geometry during integrated nacelle inlet spillage drag tests.

It is not unusual for inlets on high-speed aircraft ($M > 2.0$) to have variable inlet geometry. This can be in the form of variable ramps, a translating spike, a deflecting cowl lip, or a rotating cowl. Inlet tests must be conducted to determine drag sensitivity to geometry variation as well as inlet mass flow ratio.

To summarize, inlet testing is conducted to determine the variation in drag resulting from inlet mass flow ratio and geometry. Measured drag at the full scale operating reference geometry and mass flow ratio is considered a baseline level which is included in the full scale drag polar. Incremental drags from this baseline level define the throttle-dependent inlet drag characteristics. Since the throttle-dependent inlet drag is predominantly a pressure drag rather than a friction drag, no Reynolds number correction is made to relate scale-model and full scale characteristics. However, if inlet separation is a dominant factor in the throttle-dependent spillage drag, the inlet tests must be conducted at nearly full scale Reynolds number to ensure proper flow separation characteristics. Applicability of the subscale-model-derived throttle-dependent spillage drag to the full scale vehicle (for use in the flight documentation bookkeeping process, as indicated in Fig. II.3b), can be achieved through correlation of discrete static-pressure measurements on the inlet lip and cowl of both the model and flight vehicle.

## Exhaust System Drag

Throttle-dependent exhaust system or jet-effects drag is defined as the change in aircraft drag resulting from the difference between the actual operating condition and the operating reference condition in terms of nozzle pressure ratio and area. For integrated exhaust systems, nozzle pressure ratio can influence the afterbody pressure distributions and drag. Changing nozzle area alters the afterbody closure and pressure distribution, which in turn affects aircraft drag. For podded installations, nozzle pressure ratio may influence pressure distributions on the nacelle, pylon, wing, or fuselage. Figure II.8 illustrates typical variations in aircraft drag with nozzle area and pressure ratio.

Jet-effects or powered model tests are conducted to evaluate throttle-dependent exhaust system drag characteristics. Drag of the aero-reference model nozzle (Fig. II.8) (1) is included in the aero-reference drag. The incremental drag between (1) and the full scale operating reference conditions (2) represents the scale-model to full scale drag correction $\Delta D_{EXH}$ to be included in the full scale drag polar. Drag differences between engine operating condition (3) and full scale operating reference conditions (2) are accounted for as throttle-dependent exhaust system drag $\Delta F_{EXH}$ to be included in the net propulsive force.

If exhaust gas temperatures are not too high, throttle-dependent exhaust system drag can be investigated for an integrated propulsion system using a blown afterbody model or a blown aircraft model. The term "blown" refers to the use of a high-pressure external air source to vary nozzle pressure ratio. The airplane model inlets are usually faired over, since external rather than inlet

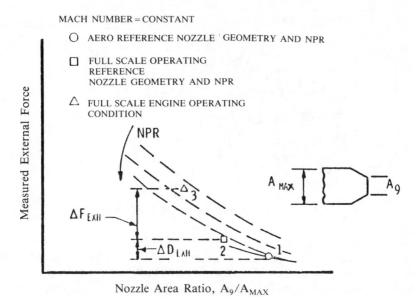

Fig. II.8   Drag variation with nozzle area and nozzle pressure ratio.

airflow is exhausted. The afterbody model simulates only the aft geometry (from maximum cross-sectional area aft) and would therefore have no air-intake system. The baseline afterbody drag level is measured at the full scale operating reference flow-through nozzle pressure ratio and geometry. The variation in drag from this level due to nozzle pressure ratio and nozzle area changes constitutes throttle-dependent afterbody drag. Naturally the exhaust system thrust would have to be evaluated to convert scale-model thrust minus drag force balance readings to drag values. This can be done by metering the flow into the jets and establishing nozzle thrust coefficients through static thrust stand testing of the scale-model nozzles. In the case of an afterburning engine, it may be necessary to simulate the pressure ratio, the temperature ratio, and the mass flux.

For podded installations, the aero-reference drag already includes a baseline level of interference drag associated with a given flow-through nacelle placement and flow-through nozzle pressure ratio. Interference drag is defined as the difference in drag between isolated nacelle drag plus clean airplane drag and total installed drag. In some installations the level of interference drag varies with nozzle pressure ratio. Again, this throttle-dependent drag can be evaluated using blown nacelles installed on the airplane model.

Application of the blown nacelle test method to short-duct installations where inlet/afterbody drags are interrelated may be somewhat misleading since the faired-over inlet is not representative of the real inlet flowfield. Conceptually the use of turbo-powered simulators, as illustrated in Fig. II.9, to simultaneously model inlet and nozzle flow conditions is much more representative of full scale engine operation. Turbo-powered simulators are

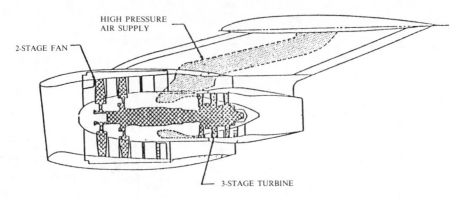

**Fig. II.9   Turbine powered engine simulator.**

extremely expensive to operate but usually provide accurate data. At supersonic conditions, temperature, swirl, and distortions can be important.

As with the inlet test procedure, afterbody or jet-effects drag tests are conducted to establish a baseline drag level at full scale operating reference nozzle area and pressure ratio which is assumed to be included in the full scale drag polar. Incremental drags from this baseline level define the throttle-dependent jet interference drag characteristics. Since the throttle-dependent drag is predominantly a pressure drag, no Reynolds number correction is usually required to relate scale-model to full scale characteristics. Validation of the applicability of the model-determined throttle-dependent nozzle/airframe drag to the full scale vehicle for use in the flight documentation bookkeeping process can be achieved through correlation of discrete static-pressure measurements on the nozzle and afterbody or nacelle forward of the nozzle.

### Auxiliary Airflow Thrust Drag

As stated previously, the aero-reference drag model usually does not simulate auxiliary airflows. Some of these effects are accounted for in the in-flight thrust. For example, engine bleed for environmental-control system (ECS) and engine cooling is incorporated in the in-flight thrust calculation. Other airflows such as bypass airflow, bay cooling airflow, ramp bleed airflow, and ECS airflow dump may or may not be accounted for. Ram drag (if the airflow is not part of the engine airflow) and thrust (if the airflow is exhausted through an exit other than the main engine nozzle) must be included in the thrust-drag accounting system. Also, any impact these flows may have on the operating inlet mass flow ratio and resulting spillage drag must be considered. Since auxiliary airflows are often throttle-related, the net thrust or drag of each represents an additional adjustment to in-flight thrust rather than to the aircraft drag polar. As such, they are usually accounted for in net propulsive force. These additional throttle-dependent thrust-drag terms may be evaluated experimentally with separate model tests, or by analytical calculations.

## Net Propulsive Force

Installed net propulsive force has been defined as net thrust adjusted for deviations from full scale operating reference conditions. While the definition of net thrust is relatively standard, the throttle-dependent drag adjustments rely completely on selected reference conditions. In this discussion, the aircraft/engine operating reference condition that separates throttle-dependent drag from aircraft drag can be described as follows:

1) Fixed full scale geometry.
2) Selected inlet mass flow ratio.
3) Selected exhaust nozzle pressure ratio.
4) No auxiliary airflows (bypass, boundary-layer bleed, ECS dump).
5) Selected full scale Reynolds number.

The full scale aircraft drag polar is also determined at the above operating reference conditions. The only adjustments necessary to apply this polar to actual in-flight drag are 1) the trim drag required to maintain desired aircraft attitude in flight and 2) the actual flight Reynolds number drag correction. Drag increments induced by other aircraft/engine operating variables are propulsion system-related drags which impact net propulsive force. Through proper accounting and a consistent reference system, aircraft drag, net propulsive force and accountability for aircraft weight, provide all the elements required for aircraft performance calculations.

## References for Chapter II

[1] Marshall, F. L., *Element Performance Integration Techniques for Predicting Airplane System Performance*, Boeing Co., Seattle, Wash., D162-10449-1, June 1972.

[2] Dunbar, D. and Ruehr, W., Unpublished material on Thrust/Drag Accounting. We thank the authors of the General Electric Co., Cincinnati, Ohio, for the permission to so use their material.

[3] Abernethy, R. B. et al., Society of Automotive Engineers, Inc., Aerospace Information Rept. AIR 1703, "In-Flight Thrust Determination," 1985.

[4] Society of Automotive Engineers, Inc., Aerospace Recommended Practice ARP 755A, "Gas Turbine Engine Performance Station Identification and Nomenclature," Nov. 30, 1973.

# Chapter III

# Gas Turbine Engine Performance Determination

William F. Kimzey* and Sam Wehofer[†]
*Sverdrup/ARO, Inc., Arnold Air Force Station, Tennessee*
and
Eugene E. Covert[‡]
*Massachusetts Institute of Technology, Cambridge, Massachusetts*

## Nomenclature

$a$ = speed of sound, ft/s
$A$ = area, ft$^2$
$\bar{A}$ = averaged area, ft$^2$
$B$ = bleed
$C_D$ = either discharge coefficient or drag coefficient
$C_F$ = wall friction factor
$C_G$ = thrust coefficient
$CV$ = control volume, ft$^3$
$d$ = derivative indicator
$D$ = diameter, ft
$f(\ )$ = function of argument within ( )
$FG$ = gross thrust, lbf
$FGC$ = engine corrected gross thrust, lbf
$FN$ = net thrust between stations $0$ and $9$ for single-stream engine, lbf or kN
$F'N$ = overall net thrust between stations $0$ and $00$, lbf
$FNC$ = corrected net thrust, lbf
$FR$ = ram drag, lbf
$FS$ = scale force, lbf
$g$ = gravitation constant, lbm-ft/lbf-s$^2$
$L$ = length, ft

*Director, Planning and Operations.
[†] Technical Specialist.
[‡] Professor, Dept. of Aeronautics and Astronautics.

$M$ = flight Mach number
$N1$ = low-pressure compressor rotor speed, rpm
$N1C$ = low-pressure compressor corrected rotor speed, rpm
$N2$ = high-pressure compressor rotor speed, rpm
$N2C$ = high-pressure compressor corrected rotor speed, rpm
$P$ = total pressure, lbm/ft$^2$
$\bar{P}$ = averaged total pressure, lb/ft$^2$
$P_S$ = static pressure, lb/ft$^2$
PLA = power lever angle, deg
$P_0$ = freestream static pressure, lb/ft$^2$
$q$ = dynamic pressure, lbf/ft$^2$
$r$ = radius, ft
$R$ = gas constant, ft-lbf/lbm°R
SFC = specific fuel consumption, lbm/lbf-h or kg-h/N
SFCC = corrected specific fuel consumption, lbm/lbf-h or kg-h/N
$t$ = time, s
$T$ = total temperature, °R
$T_s$ = static temperature, °R
$\bar{T}$ = averaged total temperature, °R
$T_0$ = freestream temperature, °R
$v$ = volume, ft$^3$
$V$ = velocity, ft/s
$W$ = mass flow rate, lbm/s
$Wa$ = airflow rate, lbm/s
$WC$ = corrected mass flow rate, lbm/s
$WF$ = fuel mass flow rate, lbm/s
$x$ = linear distance, ft
$X$ = engine variable
$\phi$ = axial gage force on a body or stream-tube surface, lbf
$\gamma$ = ratio of specific heats
$\delta$ = nondimensional pressure = $P/P_{S0}$
$\varepsilon$ = fan nozzle expansion pressure ratio
$\theta$ = exit flow angle, deg
$\theta$ = nondimensional temperature, $T/T_{S0}$, deg
$\rho$ = density, lbm/ft$^3$
$\tau$ = turbine discharge temperature time constant, s
$\tau$ = viscous drag force (Fig. III.10), lbf

*Subscripts*

amb = ambient
ex = exit
$i$ = initial
in = inlet
NAC = nacelle
SLS = sea-level-static
SS = steady-state value
th = throat

TRAN = transient value
        $w$ = wall
$1, 2, 3, \ldots$ = station designations (see Fig. III.1)
        $\infty$ = freestream infinity

## Introduction

IN the past several years, major advances have been made in the ability to determine the performance of aircraft during flight tests. These advances have been made possible largely by the increased reliability of calculated and experimental engine performance information and are primarily the result of progress in: 1) analytical approaches, especially computer codes and math models which characterize the basic internal performance of jet engines; 2) the basic instrumentation, data-acquisition, and data-processing systems; and 3) ground- and flight-test techniques.

While much progress has been made with respect to aircraft propulsion performance evaluation, there are still many problems confronting the engineer. For example, improvements are still needed to make computational methods more usable, reliable, and cost-effective in calculating overall engine and engine component performance. Continued development of analytical and experimental methods for the evaluation of complex engine systems having variable cycles and geometries, interactive digital electronic controls, and thrust vectoring exhaust nozzles is required. One of today's most difficult challenges is to develop methods for accomplishing more complicated testing at reduced overall test costs.

This chapter looks at these problems and presents a critical review of current methods presently in use for estimating the performance of gas turbine engines. For purposes of discussion, the subject is separated into three subtopics. The subtopics are performance estimation by *analysis and calculation*, by *ground testing*, and by *flight testing*. While this classification seems straightforward and complete at first thought, these topics are, in fact, strongly related and overlap in many areas. Nevertheless, this classification scheme will be used because it is not only convenient but also conforms to practice.

## III.1  Engine Performance Estimation by Analysis and Calculation

This section describes the general considerations and approaches used in current performance estimation by analytical methods. First, limitations preventing practical application of modern multidimensional computational fluid-dynamic approaches to the engine as a whole are discussed. Second, the general application of classical thermodynamic methods is treated. Third, and following from the general thermodynamic application, is the need to employ approximations and correlations. Fourth, because the most common current framework of overall engine cycle and performance analyses is based on quasi-one-dimensional and steady flow, thermal and mechanical equilibrium considerations and methods of averaging across nonuniform flowfields are described. Finally, the key present-day tool for performance estimation, the engine computer simulation or math model, is discussed.

### III.1.1  Practical Computational Limitations

While the principles of operation of a gas turbine engine can be outlined in a deceptively simple manner, the process of carrying out the details is not only complicated but also needs application of judgment that reflects considerable experience. The complexity of the machine itself forces the analyst to adopt a variety of approximations even before the study is started. At this point in time it is ridiculous to discuss the complete engine as a mere extension of the state of the art of modern computing. Indeed, it seems unlikely that even a reasonable estimate could be made of the number of grid points that would be needed to describe the geometry.[§] We feel that this matter lies sufficiently beyond the state of the art and that little is to be gained in discussing it in any detail. On a much more practical level, computing the aerodynamic character-istics of a fan rotor either is possible now or will be in the near future. This will not be treated here, since it too is a "detail" for our purposes.

For the purpose of estimating engine performance, the engine is regarded as a collection of components. These components conform to the definition of the stations used in the one-dimensional calculation of aerothermodynamic perfor-mance (Fig. III.1). The fact that the basic framework is one-dimensional will not limit our remarks to a one-dimensional flow model, or even to steady flow. Rather the implication is that this description fixes the definition of the components and provides a convenient basis for component performance description and integration for overall performance estimation.

### III.1.2  General Application of Thermodynamic Approach

The operation of the engine involves burning fuel to produce thrust. Because thermodynamic processes allow one to describe the interchange of mechanical and thermal energy, it is therefore necessary to invoke thermodynamic princi-ples. As before, the question is really how to carry out the calculation. Accepted experimental evidence allows one to draw the inference that flow processes in gas turbine engines are sufficiently close to local equilibrium on a streamline that thermodynamic arguments are valid for most purposes. De-tailed calculation of the processes in a combustor requires recognition of

---

[§]To avoid a rather pedantic argument on this matter, we grant that in principle one could seek out the smallest geometric scale in the gas path, for example. This might be the dimension of the knife edge on a seal, or a gap between a moving and stationary part, or such. After suitable subdivision of the length, a uniform grid scale would be defined. Potentially then, the entire gas path could be defined by the grid. Of course, this geometry needs modification to allow for the relative motion of the rotors and the stators. It is not clear that the results of such an exercise would be worth the effort. Many other more practical problems would seem to us to have a higher priority.

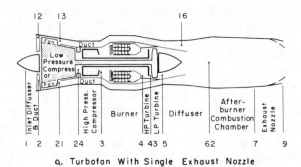

q. Turbofan With Single Exhaust Nozzle

This figure shows a low-bypass fan, afterburning engine. In the core flow the low-pressure compressor lies between stations 2 and 24, the high-pressure compressor between 24 and 3, the burner between 3 and 4. Note the aircraft inlet 1 and exit nozzle processes 9 are discussed subsequently. Station 0 is at "upstream infinity." The bypass duct stations are preceeded by 1, thus 12 is the fan inlet station, 13 is the fan discharge (tip section), and 16 is the bypass duct mixer inlet.

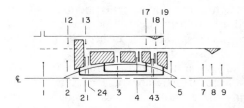

b. Turbofan With Dual Exhaust Nozzle

NUMBER/IDENTIFICATION: 1-Engine/inlet interface; 2-Fan inlet (hub section); 3-High-pressure compressor discharge; 4- Burner discharge; 5-Low-pressure turbine discharge; 7-Engine/exhaust nozzle interface; 9-Exhaust nozzle discharge; 12-Fan inlet (tip section); 13-Fan discharge (tip section); 16-Bypass duct mixer inlet; 17-Bypass duct/exhaust nozzle interface; 19-Bypass exhaust nozzle discharge; 21-Fan discharge (hub section); 24-High-pressure compressor inlet; 43-High-pressure turbine discharge; and 62-Augmenter inlet.

Fig. III.1   Engine station identification.

chemical rate processes and thus generalization beyond elementary thermodynamics. The thoughtful reader realizes that once again some sort of limitation must be made or the discussion will again treat the subject in too much detail. To resolve this issue we will first consider the performance of a gas turbine engine from the simplest point of view and then proceed with greater detail. This discussion will provide the background needed to discuss each of the individual subtopics.

Consider initially the performance of a gas turbine engine from the standpoint of cycle analysis. The thermodynamic cycle used to describe the performance of a gas turbine is the well-known Brayton Cycle. It consists of isentropic compression, isobaric heat addition, and a two-step isentropic expansion. The first step extracts the work necessary to drive the compression process, and the second does the useful work through shaft power and/or thrust. The analysis of this cycle is very helpful because, among its other virtues, it provides a relation between the several variables that define the cycle. Thus it allows one to estimate the best performance for a given compression ratio and either heat release or turbine inlet temperatures. The effects of inefficiencies at any point along the cycle can be illustrated. A detailed discussion on the application of classical thermodynamic analysis is presented in Ref. 1.

This kind of analysis has several weaknesses as well as strengths. For example, one can measure the power required to drive a compressor with some degree of accuracy. Further, one can vary the mass flow throttle setting at constant compressor angular velocities and measure the pressure rise across the compressor. (Note that care must be taken to ensure that pressure upstream and downstream has really reached equilibrium.) The means for accomplishing this equilibrating process is understood conceptually, but care is needed in practice. Now this data can be compared with the Brayton Cycle analysis at a fixed pressure ratio. The result of the comparison will show that more power is required to drive the compressor than is indicated by the thermodynamic analysis. In practice, this difference is used to define efficiency of the compressor. Indeed, this process is much like determining the orifice coefficient of the mass flow throttle. In either case, the flowfields depart from the ideal due to natural imperfections. In the latter case, depending upon the orifice design, the imperfections include viscous losses at the wall, the so-called boundary-layer effect that gives rise to a nonuniform velocity distribution, and a vena contracta if sharp edges exist in the metering valve, or orifice. Similarly the compressor is subjected to a variety of losses due to the boundary layer, to leakage across the gas path seals, to flow separation, and to windage in several regions—and, of course, shock wave losses if there are transonic flow regions in the compressor passages. Naturally the extent of all these sources of loss depends upon the operating state of the compressor. Similar statements can be made for the other components in a gas turbine engine, that is, turbines, nozzles, diffusers, and mixers.

To summarize, the thermodynamic approach to performance analysis is superb for defining relations between the several performance parameters in an engine and in that way providing guidance toward the "best" balance for a given purpose. The real weakness lies in its inability to predict the losses or the real operational characteristics of an engine or its components. It is clear that shifting from the components to the subcomponents, down to a stage consisting of a rotor and a stator, will not change this situation. Something else is needed to predict the losses. What is needed turns out to be nothing less than the details that we eschewed in the introduction to this chapter. That is, the flow into the inlet guide vanes, their ability to turn the flow appropriately, the effect of this flow upon the aerodynamics of the rotor, and the recovery of the momentum added in the rotor by the stator in the case of a compressor, and the turning and desired expansion in a turbine stator must be determined in sufficient detail that the desired predictions can be made. However, in most cases this level of detail is difficult to pursue accurately. If the circulation distribution in the radial direction along a blade decreases as the reciprocal of the radius, one can, in incompressible flow, determine a reasonable solution to the problem of finding the resulting potential flowfield. The next step is that of correcting for the effects of the viscosity of the fluid. The classical boundary-layer approach is difficult to apply here because of the coupling between the pressure gradient and the growth of the displacement thickness in the channel made of the blade surfaces and the hub and tip spinners, or casing. Unlike the case of external flow, in which the distribution of the displacement thickness has a very small effect on the external flow, the pressure gradient in the potential flow region of a compressor duct is strongly coupled to the growth of the displacement thickness, and conversely. Hence it is not surprising that few detailed calculations are available in the literature.

### III.1.3 The Need for Approximations and Correlations

The analyst is forced to approximations and correlations. Some of these correlations are very sophisticated indeed, requiring extensive use of large scale computing and deep understanding of the physical processes that occur in the flowfield. Details of many representative analyses leading to such approximations, correlations, and understanding are given in Ref. 1. These processes include radial flow and secondary flow in which radial vorticity is shed as axial vorticity due to nonuniform radial blade loading. Note that the radial circulation interacts with the azimuthal flow and is turned into axial vorticity. The radial and secondary flows also have a profound effect upon the viscous part of the flowfield. The boundary-layer flow may not be confined to a thin layer near the wall but may, in fact, be swept into the so-called potential core. This phenomenon is more common in turbines than in compressors. As a further complication, the wakes from the upstream rotating blades can be shown to lie across the passage of the stator as the wakes are convected downstream. Similarly the rotor wakes are not convected evenly through the stator. These unsteady processes are treated on the average; that is, in two steady frames of reference, one in the stator fixed to the stators, and one in the rotor fixed to the rotor. In this way the designers are able to trace average streamlines and account for the radial forces. It is then customary to iterate a boundary-layer-type calculation, frequently using an integral method, to correct approximately for the viscous effects. The point of this description is simple. At the present time there is not a general-purpose analysis procedure that one can recommend for our purpose. Thus the conclusion is reached that the best prediction of performance is based upon the experimentally determined compressor and turbine performance maps,[¶] where lines of constant speed are drawn on a plane whose ordinate is pressure rise, or pressure ratio, and whose abscissa is mass flow, or nondimensional axial velocity. Once one knows the characteristics of the limiting mass flow control in the system, one can determine the operating line of these components, and hence the system.

Thus, if the total temperature or enthalpy and the total pressure can be measured with sufficient accuracy at each of the defining planes (Fig. III.1),

---

[¶]In some cases, these performance maps are modeled by appropriate formulas that contain unknown constants. The constants are defined from experimental data. Thus, even though a model, or computer deck, exists from which one can calculate performance, the map is in this sense experimentally defined.

and if one knows the rotation speed of each of the components, the "map" allows one to deduce the performance of the gas turbine engine.

Strictly speaking, this statement is subject to the following restrictions:

1) The engine is in thermal equilibrium. That is, the temperature of each of the parts is not changing with time. Note this also implies no recent throttle transient. This will be discussed more fully in later sections of this chapter.

2) The measured velocities, pressures, and temperatures of the working fluid are uniform or may be appropriately averaged in the plane of the measurement. The consequences of this requirement will be discussed below.

3) The data have sufficient accuracy for the purpose of the calculation. This is a critical point because the calculation of the net thrust or shaft power depends upon the difference of large numbers. It is desirable that any performance calculation scheme minimize this difficulty to the extent possible.

The nature of the difficulty can be illustrated by a simple example. The gross thrust from a fan can be calculated from the formula

$$FG \simeq \frac{Wa}{g}\sqrt{\gamma R T_{19}}\sqrt{\frac{2}{\gamma-1}\left\{\left(\frac{P_{19}}{P_0}\right)^{(\gamma-1)/\gamma}-1\right\}} \qquad \text{(III.1)}$$

The first term represents the mass flow through the fan, the second the exit speed of sound, and the third the fan exit Mach number, if the fluid is expanded isentropically to ambient pressure. Suppose the expansion ratio is written in the form $1+\varepsilon$; then

$$M_{19} \simeq 1.23\sqrt{\varepsilon}\,(1-0.16\varepsilon) \qquad \text{(III.2)}$$

We will assume the engine is in an aircraft flying at 36,000 ft and that the corresponding ambient temperature is 390°R, so the exit speed of sound can be calculated from the formula

$$a_{19} \simeq 967\sqrt{1+0.286\varepsilon-0.102\varepsilon^2} \qquad \text{(III.3)}$$

It is clear from an examination of these two expressions that the speed of sound is relatively insensitive to fan pressure ratio, and hence in errors in determining the fan pressure ratio. The same cannot be said about the fan exit Mach number. First the expansion ratio is raised to the $(\gamma-1)/\gamma$ power (0.286), which drives the value toward one. (If the fan expansion ratio is 1.6, then after being raised to the 0.286 power, one obtains 1.13.) The next step is that of subtracting one. This process has the practical consequences of magnifying the effect of the error. Taking the final square root helps some. Nevertheless a 5% error on the fan expansion ratio results in an 8.5% error in the fan exit Mach number calculation for the case cited. Inspection shows that errors in the fan mass flow have a unit sensitivity weighting. If the airframe is flying in the isothermal layer, the net fan thrust may be calculated from the formula

$$FN_{19} \simeq \frac{Wa}{g}(967)\sqrt{\frac{T_{19}}{T_0}}(M_{19} - M_\infty)$$   (III.4)

Note that again a subtraction is involved, which implies another magnification of errors. Clearly a basic issue facing the analyst is accuracy, which implies the need for extreme care that the data have sufficiently low uncertainty to be of use in elementary calculations of gas turbine engine performance.

### III.1.4  Thermal and Mechanical Equilibrium Considerations

A gas turbine engine has a number of characteristic time scales corresponding to a change in fuel flow. The parts with relatively small "thermal inertia"** adjust quickly to the change in conditions. For example, the burner can is generally in thermal equilibrium in 10–15 s after a throttle change at sea level, though a longer time is needed to reach the new temperature at high altitudes, if the airspeed is held constant. On the other hand, the relatively massive parts like the disks and shafts heat up more slowly. These parts may require an elapsed time of 10 min or more to reach thermal equilibrium. This situation is illustrated in Fig. III.2 from Ref. 2. One important aspect of thermal equilibrium lies in the sensitivity of the leak rate to the relative position of parts in the seals and gaps. If the parts are not in their equilibrium position, then the leak rates do not correspond to those previously present when the performance was determined in the past, and errors can result. Another aspect is the transient storage and release of heat in the engine structure and gas stream during transient events. Without care, data recorded in engine test cells for the purpose of determining baseline performance, or that data recorded in flight either automatically or manually, may be incorrectly used for the purpose of performance analysis, diagnostic analysis, or trending. This is particularly true for rapid transients such as throttle burst and chops. If slow, smooth transients are considered, then a combination of experimental techniques and analytical extrapolation can shorten the wait for the long-time constant parts to come to equilibrium temperatures. This process can be used to generate quasi-steady-state performance information. Such techniques exist and have been verified in Ref. 2. Nonetheless, application of these techniques requires an accurate data base and understanding of the thermal time constants in question and how they change with a change in operating conditions.

Another nonequilibrium topic is that of the mechanical transient that is the change in rotor speed following throttle movement. This transient is due to the rotational inertia of the spool or spools, that is, the compressor and turbine

---

**Thermal inertia is defined here as the volume integral of the material density, specific heat product divided by the surface integral of the heat-transfer coefficient over the area bounding the volume in question. It is thus the ratio of the ability of the material in the volume to store heat to the ability of the volume to be heated. The thermal inertia is the thermal time constant for thin-shelled structures.

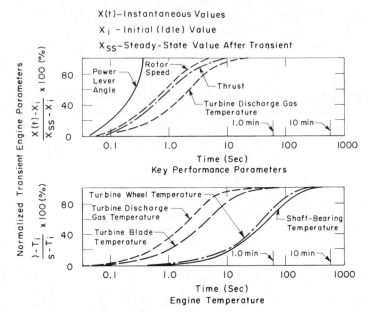

**Fig. III.2   Behavior of key performance parameters and temperatures during a rapid throttle transient for a typical turbine engine at $M = 1.0$, altitude $= 36,000$ ft.**

and other parts on a single shaft, including the shaft. For a simple single-spool system, several seconds are required for the rotational speed to change following a throttle motion. During this time interval the relation between the pressure rise and the mass flow is not the same as in equilibrium. The matter is more complex if there are two or more spools. Since there is no reason to believe the torque acting on each shaft scales with the inertia of each shaft, there will be a change in the relative speed between the shafts during the transient. Under these circumstances the steady-state operating line is not a reliable indication of either the mass flow or the pressure rise at a particular rotational speed. However, if the transient in question is suitably smooth and slow, the mechanical nonequilibrium effect can be minimized and in some cases nonanalytically compensated. Acceleration times between idle and maximum power on the order of 1–2 min or more have been demonstrated to be sufficiently slow to reduce this effect so that acceptable performance data may be obtained.

### III.1.5   Nonuniform Flowfields

*III.1.5.1   Averaging*

As mentioned above, the concept basing calculations on following streamlines is an extremely useful one in the design and analysis of performance of gas turbine engines. In the presence of spatial nonuniformities, the problem is made more difficult. One really cannot be sure how to define the pressure and temperature to be able to use a Mollier diagram. Consider the problem of defining the proper average value of the total pressure that can be used in a thermodynamic analysis. Note that a nonuniform distribution of total pressure really implies the flow is not in equilibrium, and so it is doubtful that thermodynamics can be used for an "in the large" analysis. Practical experience shows that suitable averages can generally be used, if care is taken in defining the average. Consider total pressure, averaging total pressure over several stream tubes gives

$$\bar{P} = \frac{1}{A} \int P_S \, \mathrm{d}A + \frac{1}{2A} \int \rho V^2 \, \mathrm{d}A = \frac{1}{A} \int P_S \, \mathrm{d}A + \frac{1}{2A} \int V \mathrm{d}W \qquad \text{(III.5)}$$

Here $P_S$ is the local static pressure and may not be constant, even approximately so, even if the curvature of the stream tubes is small. Note that the static pressure $P_S$ is averaged by an area integral while the dynamic pressure $\rho V^2$ is equivalent to a linear velocity averaged by a mass integral. If one attempts to make both terms area-weighted, then the velocity weighting is quadratic rather than linear. Evaluation of the surface integral at the inlet and the outlet planes of a component may be formally carried out. There is, however, no guarantee that the map of total pressure at the outlet plane looks at all like the one seen on the inlet plane. This difficulty is most severe when there is mixing between several streams in the component. If one intends to use thermodynamics and plans to use the Mollier diagram to analyze the cycle, then one must recognize that the analysis is based on a mass weighting. Several different procedures are used to overcome this problem and are discussed in Ref. 3.

The most common method of averaging is that of mass flux averaging, which is more commonly called mass weighting. In this case

$$\bar{P} = \int P \, \mathrm{d}W / \int \mathrm{d}W \qquad \text{(III.6)}$$

and

$$\bar{T} = \int T \mathrm{d}W / \int \mathrm{d}W \qquad \text{(III.7)}$$

This is more or less equivalent to matching a uniform mass flux with an integrated mass flux. The method of Dzung[3] involves simultaneous equations for properly weighted mass, momentum flux (sometimes called dynalpy), instead of momentum, and enthalpy. Livesey's method[3] involves entropy weighting. A new method due to Pianko[3] is based upon the idea of functional matching of certain quantities appropriate to each component. In applying Pianko's method, one starts at the exhaust nozzle and works forward, using an equivalent flow-averaging value that most represents the functional characteristics of the component being considered (i.e., the exhaust nozzle thrust, mass flow, and energy rate; the rotating machinery enthalpy change and mass flow rate; the combustion enthalpy entropy, mass flow, and exit kinetic energy).

Typical results of applying different pressure-averaging procedures are compared below. Percentage differences, taken from Ref. 3, referred to the Pianko

**Table III.1  Comparison of flow pressure-averaging methods**

| | Percent difference in average total pressure from Pianko method | | |
| | Method | | |
| Component | Mass weighting | Dzung | Area weighting |
|---|---|---|---|
| Compressor inlet | 0.1 | −1.3 | −1.8 |
| Outlet | −0.2 | 0.2 | −0.4 |
| Turbine inlet | 0 | 0 | 0 |
| Outlet | 0.1 | −0.1 | −0.1 |
| Compressor inlet | 0.1 | −0.1 | −0.2 |
| Outlet | −0.2 | −0.2 | −0.7 |
| Turbine inlet | 0.4 | 0 | −0.3 |
| Outlet | −0.5 | −1.9 | −3.1 |
| Reheat inlet | 0.3 | −0.1 | −0.5 |
| Outlet | 0.1 | −0.1 | −0.2 |

**Table III.2  Comparison of component efficiency using different flow pressure-averaging methods**

| | Percent difference in efficiency from Pianko method | | |
| | Method | | |
| Component | Mass weighting | Dzung | Area weighting |
|---|---|---|---|
| Compressor[a] | −0.2 | +1.0 | −1.3 |
| Turbine | −0.1 | −0.2 | +0.1 |
| Compressor | −0.1 | 0 | −0.2 |
| Turbine | −0.5 | −1.1 | −1.5 |

[a]With uniform inlet flow.

mean, are summarized in Table III.1. Note the table also includes area weighting, even though there is no theoretical basis for its use. The larger differences occurred when there were large Mach number differences across the section (as after the high-distortion intake and at the turbine outlet).

The Dzung method was usually but not always lower; the area method was always the lowest.

Although the differences were small, they were sufficient to lead to significant differences in component efficiency figures derived from them. Again using Pianko's efficiency as the data, the percentage differences shown in Table III.2 were found. In the case of multiple stream nozzle flow, we find differences in the results due to the originator's choice of the process for allowing for the

**Table III.3   Comparison of exhaust nozzle properties
using different averaging methods**

| Method | $\bar{P}$, kPa | $\bar{T}$, K | Thrust, kN |
|---|---|---|---|
| Mass flow weighted | 350 | 600 | 120.46 |
| Livesey | 430.2 | 600 | 128.22 |
| Kuchar and Tabakoff | 350 | 582.8 | 118.73 |
| Pianko | 335.3 | 600 | 118.73 |
| Dzung (mixing at $M = 0.36$) | 349.06 | 600 | 120.33 |
| Dzung (mixing at $M = 0.71$) | 346.51 | 600 | 120.04 |

entropy use due to mixing. To quote from Ref. 3, p. 8:

> To clarify this point, consider the simple example of a nozzle receiving two inlet streams both with $\gamma = 1.4$, both with stagnation pressure 350 kPa, expanding to 100 kPa and both having 100 kg/mass flow. One stream is at stagnation temperature 400°K and the other at 800°K. Table III.3 gives the various averages and the ideal thrust calculated by expanding the "average" flow isentropically using different flow averaging methods:

> Dunham explains that both Kuchar and Pianko derive their "ideal thrust" by assuming the two streams expand separately, which ignores the fact that a higher thrust can be obtained by mixing them before expansion. As a result, the Kuchar method gives a total temperature that is lower than other methods and the Pianko method gives a low value for the average total pressure. This example also shows that mass flow weighting, in effect, assumes that mixing takes place at such a low Mach number that the fundamental pressure loss attributable to mixing (which is accounted for in the Dzung method) is negligible.

Reference 3 found significant differences between the various averaging methods and explained why they occur. It has led to a single recommendation for temperature averaging (mass averaging) but not to a single recommendation for pressure averaging. For pressure averaging, the individual user must make the choice best suited to his particular application in the light of the findings of the study.

*III.1.5.2   Recommendations for Averaging*

The following methods have been found for averaging across nonuniform flow streams. Stream stagnation temperature should be averaged by mass-weighting stagnation enthalpy. In many practical cases, mass-weighting stagnation temperature will be adequate. In many test rigs, where stagnation temperature is uniform, area averaging of probe readings is acceptable.

More work on exhaust nozzles is needed to resolve some of the anomalies noted in Ref. 3, particularly to find a way of characterizing the entropy of a nonuniform stream, to examine by systematic examples the relationship between "true" efficiency and the efficiency derived by other methods in those special cases mentioned in the conclusions, and to find effective ways of using the extra information available when the total mass flow is separately measured.

Except in the case of propelling nozzles, stream stagnation pressure should be averaged by one of two methods, the Pianko or Dzung method. The choice depends on where the user wishes to debit the mixing loss implied by the nonuniformity. Pianko's method debits it to the components downstream of the traverse plane whereas Dzung's method debits it to the component upstream of the traverse plane. Mass weighting of the total pressure may be considered an acceptable approximation to Pianko's method and is easier to specify.

Area weighting should only be used when the static-pressure distribution cannot be estimated reliably and when the Mach number is small.

Since in most locations in a gas turbine the static pressure is much greater than the dynamic pressure (i.e., low Mach number), then if no static-pressure distribution data are available, area-weighting stagnation pressure can be used. Area weighted results are usually closer to the Dzung average than the Pianko average.

The entropy-weighting method should not be used unless the stagnation temperature of the flow is uniform.

In a gas turbine engine, in many cases, when heterogeneity is not too large (except for the case of the inlet of a mixer and a nozzle), all the studied methods may be used and will lead to comparable results.

In the case of an inlet plane of a mixer, the Dzung method should not be used because this method will credit this component with smaller losses than actually exist.

### III.1.6   Engine Computer Math Models

*III.1.6.1   Steady-State Math Model*

The key tool in analytical determination of engine performance is currently the computer simulation or performance math model. As discussed earlier in this chapter, the model is semianalytical in that much of the information stored in the model comes from experimental or empirical results and correlations. Analytical, component rig test, and full engine test results which are progressively obtained as the engine is designed, developed, tested, and put into service are used to continuously improve and update the model. The math model is therefore a dynamic, living analytical tool which represents the current best state of knowledge from all sources for a given engine model at a given point in the engine history. The model may, in principle, range from a

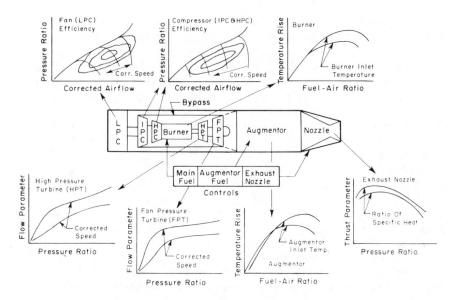

**Fig. III.3    Component stacking method.**

simple combination of component information to a sophisticated representation of engine cycle and performance data.

Analytical models or simulations have been in common use since the advent of the digital computer as a practical engineering tool in the late 1950s and early 1960s,[4] and primitive analog and even mechanical simulators were available as early as the late 1940s.

The function of the computer performance simulation is to satisfy (path) gas mass and mechanical energy and momentum-conservation equations throughout the engine. In so doing, a unique match of the engine components' operating characteristics (as shown in Fig. III.3) can be established. Appropriate control and variable geometry schedules can also be included as part of the model. This process is often referred to as the component-stacking method for performance estimation because the state of the gases exiting from the engine is the result of the stacked effects of the components operating on the gas. Early models used a nested-loop interactive approach to stack the components so as to satisfy mass, momentum, and energy. Most recent steady-state models use a matrix method to arrive more efficiently at component match points based on the work of Ref. 4 or similar efforts.

The engine computer model simulation method requires a relatively complex computer code. As already noted, the computer code must be amply supported by data from component rig tests and model component tests early in the development of the code. Later in the life cycle of the engine development and code development, substantial reductions in the uncertainty of the code values are possible if the results of full scale engine tests are used to tailor the characteristics of the code. Early in the development cycle of an engine when both the engine components and the engine computer code are relatively immature, it is customary to allow the computer to simply stack all components to arrive at the engine thrust values. Analytical or empirical adjustments to account for component interactions are applied or, in some cases, interactions are ignored at this point. As was noted in the introduction to this chapter, the division of performance determination by analytical, ground, and flight tests is not totally independent. The development cycle of the math model clearly presents the connection among the three divisions. Math-model development and usage transcend all three. The primary uses and development status of engine math models and the interaction with the engine development cycle are summarized in Table III.4.

As can be deduced from Table III.4, the engine math-model requirements are not constant for all uses. For example, in early evaluations of tradeoffs between the installed propulsion system and engine internal characteristics, the absolute values predicted by the model are of less importance than the predicted differences between configurations. For some phases (e.g., early development), a validated model is not required. Thrust-drag determination flight testing, however, does require the use of a validated engine computer model.

Math-model validation refers to the measure of the model's fidelity to represent the engine hardware performance over the engine operating envelope. Detailed methods for validation are discussed in Refs. 5 and 6. The level of fidelity is usually established by comparison with ground-test results. Validation is possible throughout the development cycle. However, because early development engines usually have relatively loose configuration control of both hardware and software, it is often not cost-effective to validate a model during these phases.

Engine configuration control is essential during the math-model validation process. Whenever validation data are being obtained, the engine configuration must be frozen. No arbitrary modifications can be permitted, even if these changes portend moderate improvements in engine characteristics. Variation of engine configuration during the validation test series introduces an unknown into the math-model data for which no satisfactory accounting technique exists. Of course, hardware and software failures during validation testing must be accommodated. At each occurrence of a failure, the need for repetition of previous validation testing must be addressed and the reasons for and against must be documented along with the decision. For flight testing, the engine math model should be validated with engine data that correspond to the "bill-of-material" flight engine hardware.

When conducting engine math-model validation ground tests, accurate engine test data must be obtained at sea-level-static and over a significant portion of the engine usage envelope. The test conditions should be selected to allow determination of uninstalled performance and to permit calculation of installation effects. Testing must address the effects of Reynolds number, Mach number, control schedule aspects, power level, horsepower extraction, and compressor bleed on engine performance. It is also necessary to examine the engine operational envelope in terms of temperature and pressure or combina-

**Table III.4  Summary of engine math-model status and usage**

| Engine status | Math-model status | Math-model usage |
|---|---|---|
| Preliminary design | Based on analytical results, empirical correlations, and scaled data from earlier engines | Engine sizing, cycle optimization, and early airframe matching |
| Component assembly and integration<br>Initial engine demonstrator<br>Development engines<br>Sea-level and altitude ground tests | Refined component data and engine data used to improve model. Model validated against sea-level and altitude test cell results | Integrate components<br>Evaluate tradeoffs<br>Monitor progress/status<br>Solve developmental problems |
| Flight test and problem resolution in ground tests | Validated but further refined through experience and further flight test and ground tests | Demonstrate limited envelope flight readiness<br>Problem resolution<br>Input to flight-test evaluation<br>In-flight thrust-drag determination<br>Aircraft flight handbook development |
| Production | Validated but refined for changes between experimental and production engines | Basis for contractual compliance<br>Provide full envelope performance information<br>Provide production engine characteristics for testing and maintenance requirements |
| Mature/refurbished | Operational service monitored and updated for engine deterioration and overhaul effects | Preserve engine performance baseline<br>Monitor engine performance for deterioration<br>Develop and test overhaul procedures<br>Operational problem resolution |

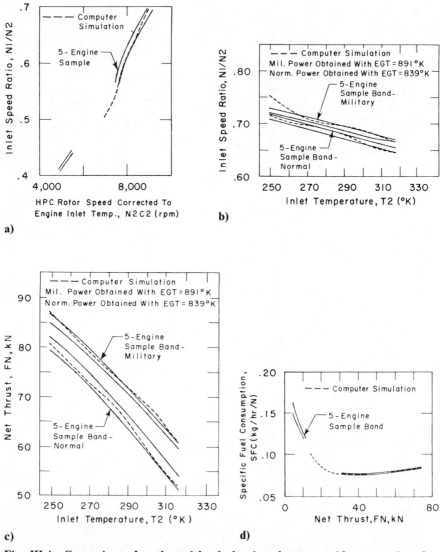

**Fig. III.4   Comparison of math model calculated performance with measured performance from a five engine sample (Ref. 7). a) Rotor speed ratio with corrected high-pressure compressor rotor speed. b) Rotor speed ratio at fixed power with inlet temperatures. c) Net thrust with inlet temperature. d) Net thrust specific fuel consumption.**

tional parameters such as Mach number and Reynolds number index (RNI), rather than the conventional altitude and Mach number envelope, in order to assure full envelope coverage for the parameters of prime influence on the engine. Since most engines will operate at both hot and cold day conditions (and all conditions in between), this will significantly alter the engine inlet pressure and temperature envelope resulting from standard day conditions.

As can be seen, the validation of turbine engine math models must be a detailed, systematic process requiring a great deal of controlled testing and documentation. Engine math models, however, are the only practical tool for "matching" the various engine components at specific engine operating conditions, and this is an important consideration if the installed engine rematches in flight or for engine operation with an unchoked exhaust nozzle.

Figure III.4 shows representative agreement between a validated math model for a J-75-P-17 engine from Ref. 7. Table III.5 gives representative levels of agreement for several engine types. The results of Table III.5 are for engine models that were validated with emphasis on intermediate power setting at sea-level-static. Agreement may not be as good at other points in the flight map. Unvalidated models may easily vary as much as $\pm 5$–15%. Also, engine-to-engine variations may cause as much as $\pm 4$–6% difference, even with validated models. As will be discussed later, the math model becomes a most powerful tool for in-flight thrust determination when it is specifically validated against the engine or engines to be used in flight.

*III.1.6.2   Transient Math Models*

The mainstay of aircraft thrust-drag determination is principally steady, 1-g flight which is well accommodated by the steady-state model. However, for purposes of data acquisition beyond easily reached steady aircraft operating conditions and for reasons of economy in both flight and ground tests, transient and dynamic tests are sometimes performed. In the section of this chapter on ground testing, short-duration turbine engine testing is discussed as a means for economic thrust and fuel consumption performance determination. Also, in flight tests, transient and dynamic aircraft tests, as described in Chap. VI, may be used. Therefore, the need exists for math models capable of providing thrust performance information on a transient basis.

Such models have been developed by inclusion of unsteady terms in the satisfaction of the mass-, energy-, and momentum-conservation laws. References 8 and 9 discuss various aspects of constriction validation and use of transient models. Transient influences including mechanical rotor inertia, transient heat storage and release in and between the engine structure and gas path, unsteady mass accumulation effects, control system dynamics, and variable geometry system dynamics are all included. Some models have been transiently calibrated in altitude test cells with the result that flight and propulsion system transients planned for flight tests could be reproduced to within approximately 1–2% of gross thrust during throttle transient.[10] Figure III.5 illustrates comparisons between a turbojet transient math model and experimental results prior to and after validation. The results indicate the importance of the validation process. During the process, thermal and mechan-

**Table III.5   Representative agreement between engine math models and experimental results**

| Engine type | Agreement at sea-level-static (SLS) intermediate power, % | | Thrust agreement over flight map, % | Comments |
|---|---|---|---|---|
| | FN | SFC | | |
| Low-bypass ratio afterburning turbofans (2 different engine models) | ±0.2 | ±0.3 | ±2.0<br>±0.5 over power range at SLS | Models validated through testing in sea-level and altitude test cells |
| Single-spool afterburning turbojets (2 different engine models) | ±0.3 | ±0.6 | ±2.0<br>±0.5 over power range at SLS | Models validated in sea-level and altitude test cells |
| Dual-spool afterburning turbojet (1 engine model) | ±3.0 | ±0.4 | ±3.0 | Validated through limited comparisons with sea-level and altitude data |
| Low-bypass ratio afterburning turbofan (1 engine model) | ±0.9 | ±0.3 | Unknown | Special model adjusted against other validated model |

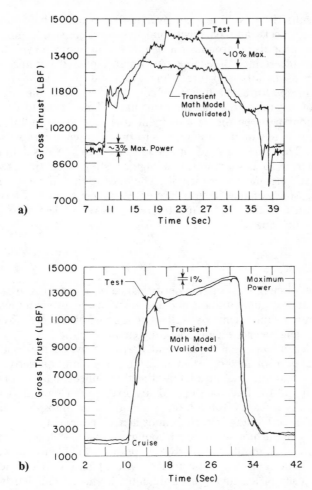

**Fig. III.5   Gross thrust from transient math model compared to experimental results (Ref. 10). a) Rapid power transient prior to validation. b) Rapid power transient after validation.**

ical time constants may be determined which allow significant improvements in transient model accuracy.

## III.2   Performance Estimation by Ground Testing

The direct measurement of engine thrust performance for aircraft in flight is generally not considered feasible. Performance is instead deduced from the measurement of related engine parameters which are evaluated using measurements from ground-test facilities. Therefore, an understanding of propulsion ground testing should be considered a prerequisite for evaluating aircraft flight-test results. In this section an overview of types of hardware systems, basic theories, checks and balances, and considerations of engine measurements used in propulsion ground-test facilities is presented. Typical values of measurement uncertainty of engine performance-related parameters and an identification of factors and procedures required to achieve reliable and accurate engine performance measurements are discussed. The strengths and weaknesses of several of the measurement methods are examined, and favorable interactions between the various methods reviewed, including considerations required to use model test results to predict engine airflow and thrust performance. The final topic of discussion is an engine developmental test technique referred to as short-duration engine testing, which integrates engine analytical modeling and experimental techniques to describe engine transient performance. Much of the material presented in this section is based on the work described in Refs. 11–13, and the reader is referred to these references for additional information.

### III.2.1  Test Configurations

Prior to the discussion of engine performance measurements, it is necessary to examine the test configuration and the key considerations imposed by the system program requirements as well as requirements imposed by the basic physics of the engine and engine operating environment.

The so-called direct-connect test configuration provides the best means of measuring the performance of jet engines. This configuration derives its name from the fact that the engine inlet is directly connected to a controlled-air supply system and the free-standing exhaust exits into a separately controlled environment. For a sea-level stand, the engine takes in atmospheric air directly and exhausts directly back into the atmosphere. The essential features of the different types of turbine engine test configurations are shown in Fig. III.6. Although there are a number of hardware options available to implement each of the key functions in a direct-connect test configuration, it is nevertheless essential that each of the functions represented by the specific hardware items identified in Fig. III.6 be successfully implemented.

First, the flow of air through the engine must be known very accurately. The venturi and inlet bellmouth represent two of the devices used to accomplish this measurement. After the flow of working fluid is carefully measured, it may be necessary to condition the flow profiles of the air entering the engine; this conditioning is often accomplished with flow-straightening screens (Fig. III.6d). It is important to note that frequently the test will require nonuniform aerodynamic profiles entering the engine. In a like manner, these properties can be obtained by the use of flow "unstraightening" devices such as nonuniform screens, air-injection systems, or valves placed near the engine inlet face.

If thrust is to be determined from the sum of all body forces acting through the engine mounting trunnions, then a key element in the direct-connect test installation is the interface between the test cell structural ground plane and the metered plane on the engine thrust stand. It is a challenging problem to provide an interface plane that is free of mechanical forces and, at the same

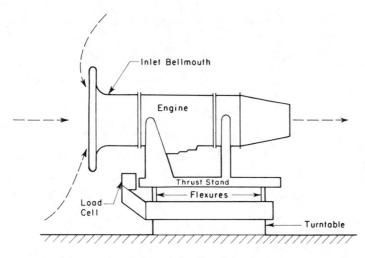

a. Sea-Level Open Test Stand

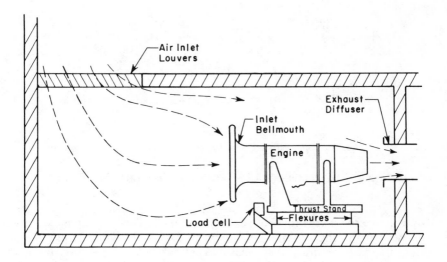

b. Ground Level Test Cell

Fig. III.6   Direct-connect engine installation (Ref. 12).

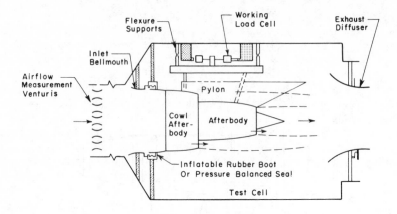

c. Altitude Test Cell With Overhead Thrust Measuring Systems

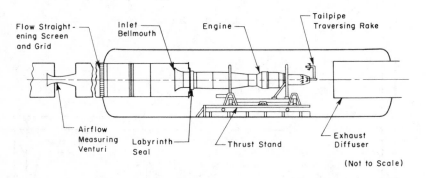

(Not to Scale)

d. Altitude Test Cell With Under-Carriage Thrust Measuring System

**Fig. III.6 (continued)  Direct-connect engine installation (Ref. 12).**

time, has zero leakage for those cases where the airflow measurement system is located off the engine thrust measurement system. Perhaps the most essential element in the entire thrust measurement system is the thrust stand which supports the test engine during operation. The force measurement subsystem within the thrust stand provides a direct measurement of the forces applied to the thrust stand.

If, however, thrust is to be inferred from the sum of all the aerodynamic forces acting on the engine tail-pipe inlet and exit interface planes, then the most critical measurement component is the traversing system used to measure the flow properties in the tail-pipe section (Fig. III.6d). The measuring system must accurately define the spatial flow properties (total pressure and temperature) without significantly altering the flowfield.

Finally, for control of the engine exit environment in enclosed test installations, it is necessary to utilize some type of exhaust diffuser to collect the engine exhaust gases and direct them away from the test cell. If adverse test cell flow recirculations are to be avoided in altitude test cells, it is necessary to match the exhaust diffuser to the test cell exhauster equipment and the test article flow rates. Test cell flow recirculations stem from the viscous mixing of the engine exhaust jet with the ambient air and the resulting flow impingement of the exhaust jet with the diffuser walls. For some afterburning engine operations, for highly canted engine exhaust nozzles, and for large, high-bypass airflow turbofan engines, test cell temperatures and flow pressure gradients resulting from the recirculated flows may require special cell cooling and data evaluation methods. In ground-level test cells, the diffuser size and diffuser placement have a marked effect on the test cell secondary flow rates and windage drag on the engine installation. Other diffuser design considerations are the diffuser aerothermal stress requirements, the required diffuser pressure rise, and the diffuser flow acoustical interactions with the test article.

### III.2.2  Factors Affecting Engine Performance Measurement

Of the wide range of the configuration parameters, six must be carefully evaluated to provide suitable tradeoffs for optimum engine performance measurements that meet all the program requirements, engine physical arrangements, and management constraints for tests in ground-test facilities. These six factors are: 1) engine cycle, 2) engine geometry, 3) engine suspension system, 4) operating environment, 5) measurement uncertainty, and 6) program limitations. The first three factors primarily affect the mechanical options available for the test. The next two, operating environment and measurement uncertainty, dictate the instrumentation options available but, at the same time, have a marked influence on the mechanical options. It goes without saying that the sixth factor, program limitations, affects everything involved in the selection of the test configuration, the test techniques, and the extent of the test program.

Examination of these six factors reveals the key dimensions that influence the selection of an appropriate measurement method.

1) *Engine Cycle.* The key dimensions are bypass ratio and afterburner augmentation ratio because of their significant influence on the air specific impulse (thrust produced per unit of inlet airflow). The achievable measurement uncertainty is a strong function of air specific impulse. Secondly, the engine cycle affects the local heat transfer through the engine skin and thus the external temperature of the engine skin, which translates directly into the important consideration of test cell heating and cooling, and dimensional stability of the measurement systems. For purposes of information presented in this chapter, a low-bypass ratio may be assumed to be 2 or less, and a high-bypass ratio may be assumed to be 4 or more. Also, an afterburning engine is assumed to be capable of reaching approximately stoichiometric fuel-air ratios.

2) *Engine Geometry.* The primary dimensions are single- or multiple-engine inlets and the Mach number level within the engine inlet. The engine inlet

configuration dictates, in some tests, the kinds of measurement methods that may be used. In all cases, the mass density (density × velocity) of the inlet air has a pronounced effect on the measurement uncertainty. Another key dimension relative to the inlet configuration is the degree of streamline curvature induced in the flow by the mechanical configuration of the engine intake. Similarly the exhaust nozzle configuration determines the thrust measurement methods that can be applied in some tests. The number, size, and shape of the exit nozzles and fixed geometry vs variable geometry are the key dimensions of the exhaust nozzle geometry. Ejector nozzles, particularly the type with blow-in doors, require special thrust measurement considerations because of the need to supply a separate airflow tailored to simulate the flow from an upstream bleed system or the flow coming off the airframe fuselage.

3) *Engine Suspension System.* The stiffness and deflection characteristics are the most important dimensions of the engine suspension system. These characteristics have a strong influence on the change of engine alignment between the nonoperating condition (as installed) and the operating condition (as tested). The other dimension for the engine suspension system is the symmetry or asymmetry of the engine thrust takeout points. These features of the suspension system determine, to a large degree, the engine-to-test cell misalignment that occurs as a function of engine operating conditions. This value of misalignment can dictate the range of options available for thrust measurement.

4) *Operating Environment.* The level and range of the simulated altitude and Mach number are key factors affecting the selection of a performance measurement method. The range of these variables, together with the range of engine power settings, dictates to a very large degree the levels of thrust and airflow measurement uncertainty that can be achieved. To a lesser degree, the range of environmental variables also affects the thermal loadings and hence the dimensional stability (drift) of the measurement systems. Another key dimension relative to the operating environment is whether aerodynamic distortion of the inlet airflow is required. The auxiliary test equipment required to generate certain nonuniform engine inlet flow patterns may well limit the options available for thrust measurement.

5) *Measurement Uncertainty.* The most important consideration is simply whether the program objectives can be satisfied with a given measurement method. In those cases where only relative engine performance is to be examined, measurement uncertainty requirements are easier to achieve.

6) *Program Limitations.* The key factors are the funds and test time available to achieve an understanding of the engine performance.

### III.2.3   Measurement Schemes

The primary methods used in test facilities to determine engine internal performance can basically be classed as:

1) Airflow-metering nozzle methods.
2) Tail-pipe continuity/momentum balance method.
3) Scale-force measurement method.
4) Fuel-metering method.

It is important to recognize that there are overriding factors in every test program, and the good tester is able to select and tailor the measurement system to meet specific test requirements. For example, where an engine performance verification test may require an extensive pretest measurement design and certification effort involving a complex force-balance arrangement, an engine operational test may only require minimal internal performance verification using aerodynamic probe measurements.

A brief summary of the measurement methods used in propulsion ground-test facilities follows.

### III.2.3.1  Airflow-Metering Nozzle Methods

Airflow rate measurements for gas turbine engines are generally made with a critical flow (sonic) venturi (Fig. III.7a) and/or a calibrated engine inlet bellmouth (Fig. III.7b). Critical flow venturis are designed to operate with sonic flow at the nozzle throat plane and, as such, are very accurate flow measurement devices because there is no requirement to calculate the flowfield velocity. On the other hand, a critical flow venturi is not well suited to measure subsonic flows because of the difficulty in defining the throat Mach number. Another disadvantage is that in the "just-choked" mode of operation, a venturi will have a 5–10% larger pressure loss than a direct-connect engine bellmouth flow-metering system. The disadvantage of the direct-connect engine bellmouth flow-metering system is that a measurement of the subsonic flowfield velocity is required, and this leads to an additional error source not present with the sonic venturi (refer to Sec. III.2.4.1 to quantify the general level of error). The use of the engine bellmouth to measure the engine flow rate does simplify the test installation by not requiring a separate flow-metering nozzle or a zero-leakage slip joint between the engine bellmouth and the engine metric free-body force measurement system. The variables involved in the calculation of engine airflow are presented in Fig. III.7.

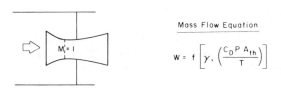

Mass Flow Equation

$$W = f\left[\gamma, \left(\frac{C_D P A_{th}}{T}\right)\right]$$

a. Isolated Critical Flow Venturi

Mass Flow Equation

$$W = f\left(\gamma, \frac{P_s}{P}, \frac{C_D P A_{th}}{\sqrt{T}}\right)$$

b. Direct-Connect Calibrated Engine Bellmouth

**Fig. III.7   Engine airflow measurement systems.**

*III.2.3.2   Continuity/Momentum Balance Method*

The balance method for airflow and thrust determination requires one small but complex set of hardware and two relatively complex computer codes. The essential components of the continuity/momentum balance method are listed in Fig. III.8, together with the equations applicable to this method for three different positions of the survey rake system. It is very important to note that the example shown in Fig. III.8 is for a single-nozzle exit only; if multiple nozzles are installed on the engine, the entire system must be repeated for each of the engine exhaust nozzles.

The complex piece of hardware provides for the survey of the total temperature, total pressure, and composition of the nozzle exhaust gases as a function of radius across the entire cross section of the flow at the chosen plane of interest. Three planes of interest are typically available, i.e., the nozzle entrance, the nozzle throat, or the nozzle exit. In the example chosen in Fig. III.8, the nozzle throat survey configuration is shown. The survey probes must have the capability to be positioned accurately within the nozzle flowfield despite the gross movements caused by thermal growth and deflections of the flow areas, as well as the planned variation of the nozzle throat and nozzle exit dimensions as a function of engine control operation. The probes also must be able to survive extended exposure to the aerodynamic loads and thermal loads imposed by the high-speed exhaust gases, and the probes must also be of sufficiently small cross section not to alter materially the nozzle flowfield or, even more importantly, the engine matchpoint relationships which are under the control of the engine control system.

One set of software required to support this method is a model of the flowfield which will define the additional properties that cannot be measured directly; the two most important such properties are the static-pressure profiles and the flowfield angularity. A second set of software is also required to define the nozzle geometry as a function of each specific engine operating condition.

The output from the survey probe system is then combined with the outputs from the two computer codes to provide inputs to the data reduction equations shown in Fig. III.8.

A number of tradeoffs must be made in the continuity/momentum balance method regarding the selection of the survey plane. As a general condition, the mechanical difficulties are maximum at the nozzle throat plane and minimum

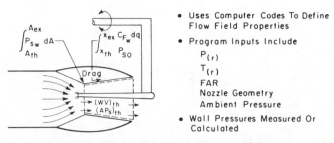

**Fig. III.8   Continuity/momentum balance method (rake located at nozzle throat).**

at the nozzle exit. A fixed rate can be used at the nozzle entrance plane. By contrast, the difficulties of interpreting the aerodynamic measurements of total pressure and total temperature are minimum at the nozzle entrance and maximum at the nozzle exit. At the present time, experience indicates that the nozzle throat plane is generally the preferred location, on the basis of overall flow measurement accuracy, for the survey.

Tail-pipe measurement methods have been primarily confined to single-exhaust nozzle configurations because of the difficulty of surveying a nozzle flow embedded within another flow and of accounting for the external flowfield effects on the inner nozzle flowfield. Tail-pipe measurements can, however, with proper attention, be successfully applied to dual-nozzle systems.[14] Ejector nozzles are another configuration not well suited to tail-pipe balance methods because the flowfield measurement and calculation process requires evaluation of a highly viscous flowfield with steep radial velocity gradients.

### III.2.3.3   Scale-Force Measurement Method

The essential characteristic of the scale-force measurement method is that the engine is installed so that it may be handled as a free body, and the net forces acting around the free body provide a measurement of the engine gross thrust. The scale-force measurement method requires the use of a thrust stand which attaches directly to the engine-mounting hardware. The essential functions of a thrust stand are shown schematically in Fig. III.9.

In the altitude test cell configuration, the engine thrust stand measures only a portion of the axial forces present on the metric engine free body. The total forces acting on the free body for a planar nozzle engine, the simplest of the engine arrangements, a nonplanar nozzle and ejector nozzle engine, are shown in Fig. III.10. Gross thrust, as defined for the examples of Fig. III.10, is always equal to the sum of the momentum term, the pressure area term referenced to

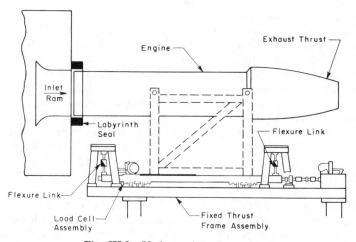

**Fig. III.9   Undercarriage thrust stand.**

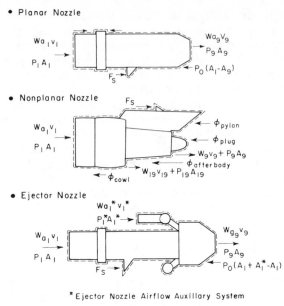

* Planar Nozzle

* Nonplanar Nozzle

* Ejector Nozzle

*Ejector Nozzle Airflow Auxillary System

**Fig. III.10   Scale-force thrust equations.**

cell pressure at the forward control volume boundary, and the scale force measured from the engine thrust stand. This result may be demonstrated by application of the momentum theory using the control volumes indicated in Fig. III.10 and setting the terms which comprise the gross thrust definition equal to $FG$.

In the nonplanar nozzle case of Fig. III.10, the gross thrust definition includes pressure-area and drag forces, $\phi_{\text{aftbody}}$ and $\phi_{\text{plug}}$, caused by the propulsive streams from the fan and gas generator. These forces are part of the gross thrust definition and, strictly speaking, do not have to be isolated for determination of gross thrust. However, it may be advantageous for diagnostic purposes to determine their contribution to gross thrust. Reference 14 illustrates the influence of such effects and is an example of the use of $\phi$ isolation for performance diagnostic purposes. Detailed evaluation of the $\phi$ integral requires a large number of surface pressure measurements and extensive analysis effort. In practice, the decision to expend the required resources must depend on the agreement of overall performance results with pretest expectations. In those cases where $\phi$ isolation is desired, static-pressure measurements are made over the plug and aftbody and integrated over the surfaces. The viscous part of $\phi$, which is the shearing term ($\tau$) of the surface area integral of Fig. III.10, may be determined from semianalytic estimation methods such as the half-concentric tube method used in Ref. 15 or by the patch-integration half-concentric tube method. The half-concentric tube method uses an assumed friction coefficient and models the flow so as to represent the drag as the friction along the inner wall of a diverging, circular concentric tube.

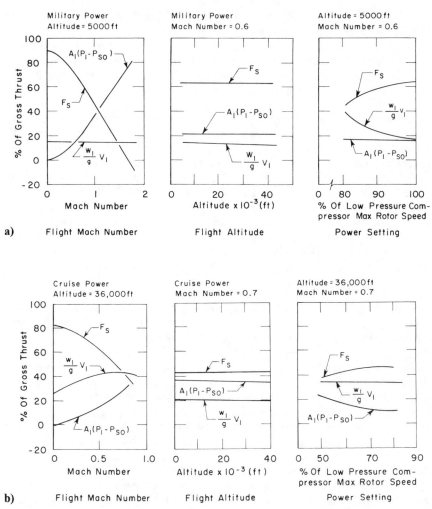

**Fig. III.11  Relative magnitude of the force components in the scale-force thrust equation; a) turbojet or low-bypass turbofan, b) high-bypass turbofan.**

The patch-integration method uses the fan or core nozzle discharge stagnation pressure, the measured surface pressures, and assumed friction coefficients to compute the friction over patches which surround each static-pressure tap. The total drag is then determined by summing the patches. Scale-model or purely analytical methods can also be used to estimate the surface area drag integral. All these methods, however, can suffer from unanticipated flow separations and uncertainty of the bulk flow angle out of the nozzles.

Another problem specific to the nonplanar nozzle case of Fig. III.10 is definition of the split line on the pylon which divides the propulsion system from the airframe for thrust-drag bookkeeping purposes. The division is somewhat arbitrary but, in general, is defined such that the propulsion side encloses the outer boundary of the fanjet over the entire flight map and in this way a fixed line along the pylon is defined. In ground testing, the $\phi$ integral above the split line but still on the metric pylon is estimated, and measured gross thrust is then adjusted accordingly. In pylon-mounted high-bypass installations, this type of split-line adjustment is less than 0.1% of the engine gross thrust.[16]

Finally, when using the scale-force measurement method, it is very important that the tester quantify the relative magnitude of the terms that make up the gross thrust equation so that the proper emphasis is applied to the dominant terms. The general trends of the gross thrust force terms as a function of flight altitude, flight Mach number, and engine power setting are shown in Fig. III.11. Figure III.11 shows the trends for both a low- and high-bypass airflow turbofan engine. In general, the scale-force term is the dominant force term, particularly in the subsonic flight regime. However, in the high transonic and supersonic flight regimes, the dominant factors are the pressure area and inlet flow momentum terms. At the higher supersonic flight Mach numbers, the scale force will actually read in the drag direction.

*III.2.3.4   Fuel-Metering Method*

The most common method of measuring engine fuel flows is with calibrated turbine-type flow meters which produce an output signal whose frequency is proportional to the volume flow rate through the meter.

Turbine meters are calibrated by three methods or a combination thereof. The first is volumetric calibration. It is accomplished by flowing a measured volume of fluid through the meter and recording the total number of turbine meter cycles (pulses) generated. The second method, gravimetric calibration, is accomplished by flowing a measured mass of fluid through the meter, determining fluid density through the measurement of fluid temperature and pressure, and converting the mass to volume to establish a pulse-per-gallon factor. The third method is comparative calibration in which the meter being calibrated is compared against a working standard turbine meter.

To minimize the uncertainty of the fuel-flow measurements, up to three ranges of flow meters are used to cover the engine's total fuel-flow operating range. This arrangement ensures that the turbine meters are only used over the linear range portion of their frequency calibration curve. A typical engine fuel-flow installation, along with the variables required to calculate fuel flow, is shown in Fig. III.12.

Another consideration with respect to fuel is the determination of its lower heating value. The standard method of testing the heat of combustion of aviation turbine fuels is with a bomb calorimeter. The heat of combustion is determined by burning a weighted sample in an oxygen bomb calorimeter under controlled conditions. The temperature is measured by means of a platinum resistance thermometer. The heat of combustion is calculated from temperature observations before, during, and after combustion, with proper allowance for thermochemical and heat-transfer corrections. Either isothermal or adiabatic calorimeters may be used.

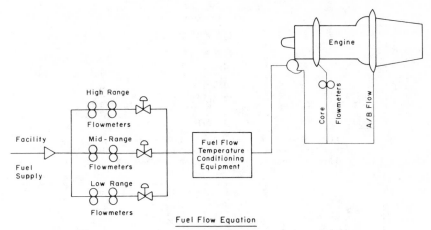

W = f (Meter Flow Factor, Fuel Temperature, Meter Output Frequency, Fuel Viscosity, Fuel Specific Gravity)

**Fig. III.12    Fuel-flow measurement system.**

### III.2.4    Measurement Error Sources

One of the challenges that befalls the engine ground tester is the task of translating engine pretest performance uncertainty requirements into measurement hardware, software, and resource requirements. In order to accomplish this task, the tester must be capable of: 1) making pretest measurement uncertainty estimates and relating the results to engine performance to ensure that an acceptable measurement system is selected, 2) identifying those measurement errors that must be worked in order to meet the program uncertainty requirements, and 3) estimating the man-hour resources required to meet both the test program uncertainty and resource objectives. To generate the required information, historical data relating engine measurement system uncertainties to engine test factors are compiled in the form of error charts.

In the following sections, ground propulsion test error charts for engine airflow, thrust, and fuel-flow measurement systems are presented and discussed.

*III.2.4.1    Airflow*

Error charts for three engine airflow measurement methods are presented in Fig. III.13. The chart abscissa lists the error sources that influence the measurement of the performance parameter and the chart ordinate specifies the level of this influence. An open or unhatched bar depicts the nominal range of the systematic errors, and the crosshatched bar depicts the error range that can generally be achieved with a normal effort to account for correctable errors. Thus, the magnitude of the crosshatched bar corresponds to the magnitude of combined error sources which are considered random in nature and not readily correctable in the measurement system.

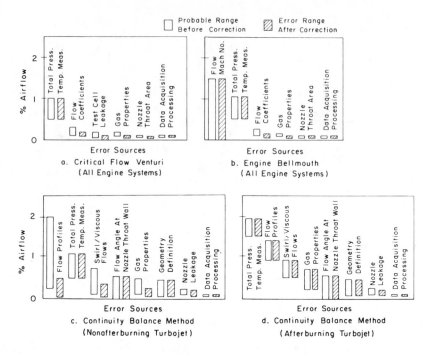

**Fig. III.13   Airflow measurement error charts.**

The critical flow and engine bellmouth methods are independent of the engine cycle, and therefore these charts are applicable to all engine systems. The continuity balance method is engine cycle-dependent. The results in Fig. III.13c and III.13d are for a turbojet engine operating at military (intermediate) power and full power, respectively. The largest error sources noted in Fig. III.13 are generally associated with the pressure and temperature measurement system and with flow nonuniformities. These are the areas that must be worked if a significant reduction in engine airflow accuracy is to be achieved.

As discussed in Sec. III.2.3.1, the critical flow venturi measurement system has the potential for the lowest value of airflow measurement uncertainty. For this method the largest error source is the ability to measure the freestream total pressure and temperature.

The largest error source for the engine bellmouth measurement system is the flow Mach number effect. Figure III.14 shows quantitatively how flow Mach number affects the measurement uncertainty for a flow-metering nozzle, assuming that the measurement system error for static and total pressures is equal.

The most significant error for the continuity balance (nonafterburning) is the effect of the nonuniform flow profiles of total pressure, total temperature, static pressure, and flow angularities. Fortunately these effects are amenable to theoretical and experimental treatment, which substantially reduces their impact. The nozzle throat flow angle error in the continuity balance is only present for convergent nozzles being operated above a nozzle total-to-ambient

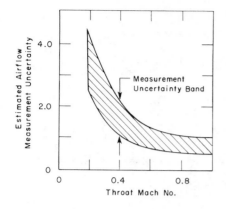

Fig. III.14  Effect of flow Mach number on airflow uncertainty (standard day, sea-level-static test conditions).

pressure rate of 1.8. This error is the result of the flow angle just outside of the nozzle boundary layer being affected by the expansion to the lower ambient pressure; therefore the flow angle is something less than the measured wall angle and is not accurately defined. This error manifests itself in the flowfield calculations and grid spacing. With proper attention, the continuity balance will give measurement uncertainties comparable to the engine bellmouth system. For the afterburning case, a direct measurement of the tail-pipe temperature profiles is generally not possible, and bulk values are estimated based on a tail-pipe heat and momentum balance or a value obtained from the engine math-model simulation for the particular test conditions. Obtaining a value in this manner means that the pressure, temperature, and gas property errors are larger and that profile effects are not correctable.

*III.2.4.2  Gross Thrust*

A thrust error chart for the scale-force and momentum balance thrust measurement methods is presented in Fig. III.15. The magnitude of the individual terms in these methods is a function of the engine cycle and operating conditions. The chart in Fig. III.15a corresponds to results from a nonafterburning turbojet engine operating at military power and Fig. III.15b to an afterburning condition.

The largest error sources in the momentum method are the drag of the survey rake (although this is greatly dependent upon the particular rake design) and errors from nonuniform flow profiles of total pressure, total temperature, static pressure, and angularity in the flow. These types of errors, however, can be corrected using theoretical and experimental treatment. Other error sources such as flow swirl and viscous effects, gas properties, and nozzle leakage can be controlled by careful design of the experiment. Determination of the flow angle at the nozzle wall is not an error source (as was the case for the continuity balance) since the engine mass flow rate as determined by the facility flow-metering nozzles can be used to evaluate this angle. For the afterburning case, again the need to approximate the tail-pipe temperature and gas properties adds significantly to the measurement error. Note also that there is no rake effect for the afterburning case.

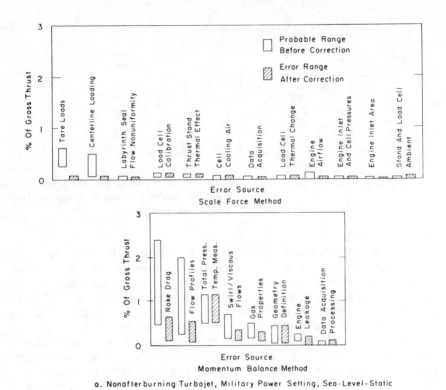

Fig. III.15    Thrust measurement elemental error charts.

The largest correctable error for the scale-force method nonafterburning (i.e., low-altitude) case is the tare load effect (Fig. III.15a). The tare load shown reflects the stiffness of the thrust stand support systems and service systems. The centerline loading error quantifies the effect of the bending of the thrust stand and engine-mounting system if the data load cell is not located on the engine thrust centerline. However, the tare loads and centerline loads can be established with such a high degree of precision that the overall effect of each of these errors is less than ±0.1% of the measured gross thrust. Similarly, with additional effort, the probable error range for the other primary sources of scale-force error can also be established with a high degree of precision.

The largest correctable error for the scale-force method for the afterburning (i.e., high-altitude) case is the effect of varying test cell ambient pressure on the data load cell and thrust stand (Fig. III.15b). For present-day equipment the magnitude of this effect typically lies in the range of 0.5–1.5% of full scale of the scale-force measuring system. However, the load cell and thrust stand response to changing pressures can be established with such a high degree of precision that the overall effect of this error is less than ±0.1% of the measured force.

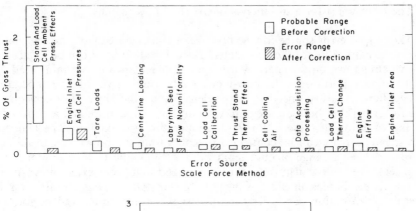

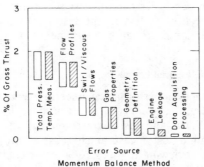

b. Afterburning Turbojet, Military Power Setting, Altitude = 40,000Ft, Mach No = 1.2

**Fig. III.15 (continued)  Thrust measurement elemental error charts.**

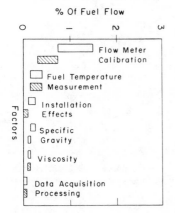

**Fig. III.16    Fuel-flow error chart (all engine systems).**

*III.2.4.3   Fuel Flow*

The error chart for a turbine volumetric fuel flow meter is presented in Fig. III.16.

The largest error source for the flow meter is associated with the repeatability of the calibration of the flow meter. The calibration error can be reduced to the values shown on the chart by using multiple meters and confining readings to the linear portion of the flow-meter calibration curve. Fluid temperature affects fluid viscosity, specific gravity, density, and the physical dimensions of the meter, so that both an accurate value of the temperature and a knowledge of the variation of these factors with temperature are required. The temperature-related errors can be controlled by calibrating the flow meter in the test fluid and over a range of fluid temperatures. Since the turbine flow meter is also sensitive to flow disturbances upstream and to a lesser extent downstream of the meter, there is an effect of the plumbing installation configuration on the fuel-flow calibration curve. By careful design considerations and performing calibrations with the flow meters in the testing configuration, the plumbing configuration installation error can be made negligible.          .

### III.2.5   Measurement Resource Estimating

Once the test program performance uncertainty requirements have been defined and the measurement uncertainty error charts have been generated, the tester can use the charts and test requirements to arrive at a measurement system selection and a test measurement cost estimate. The measurement selection is made by reviewing the individual error charts to determine which systems can meet the test requirements. Having defined the measurement system, the hardware and software requirements can be defined and the work efforts required to address the correctable error sources can be estimated. An example of an engine thrust measurement resource curve showing the resource requirements for different approaches is shown in Fig. III.17. It can be seen

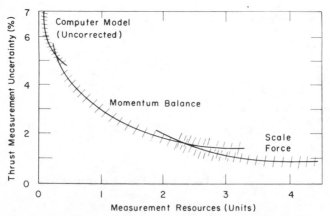

Fig. III.17   Impact of thrust measurement uncertainty on program measurement resources (nonafterburning turbojet engine).

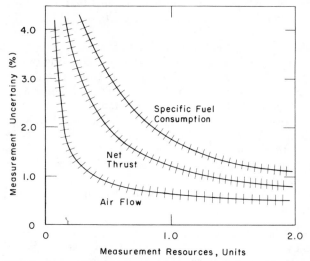

**Fig. III.18   Relative impact of uncertainties for various engine performance parameters on measurement resources (nonafterburning turbojet engine).**

that as the thrust measurement uncertainty requirements are reduced, the proposed measurement method changes and resource requirements increase exponentially approaching the current state-of-the-art level. Similar plots can be made for airflow, fuel flow, and so on. Figure III.18 shows a relative comparison of several engine performance parameters and their measurement uncertainty requirements on measurement resources.

### III.2.6   Measurement Validation

Propulsion facility measurement validation procedures are presented in Ref. 12. The validation process consists of: 1) establishing a thrust measurement calibration hierarchy with traceability to the National Bureau of Standards, 2) quantifying and categorizing elemental measurement errors into precision and bias terms, 3) propagating the elemental errors to obtain measurement uncertainty values, and 4) establishing an influence coefficient matrix to propagate element measurement uncertainties to a final in-flight thrust measurement uncertainty value. These topics are discussed in more detail in Chap. VI. Execution of this process is a difficult task, requiring a well-planned and coordinated effort among the engine manufacturer, ground- and flight-test personnel, and the airframe manufacturer to ensure that all systematic errors contributing to the final measurement uncertainty are identified and properly managed. The errors that are particularly difficult to isolate are:

1) Errors caused by unanticipated instrumentation problems (example: instrumentation bias shifts subsequent to calibration or flight-test maneuver/instrumentation response incompatibilities).

2) Uncertainty analysis error propagation items that are either erroneously excluded, improperly modeled, or inaccurately bounded (example: incorporation in the error propagation analysis error bands for instrumentation equipment without a sufficient historical data base).

3) Errors attributable to ground-test-facility limitations or test techniques (examples: basis for assessment of transient engine operation in flight; test envelope assessment at cold day conditions or airflow; temperature limitations for low-altitude high Mach number test conditions).

4) Errors attributable to changes in engine operating characteristics (examples: "rematching" of multispool engines caused by differences in installation effects between the ground-test facilities and flight environment; engine control system operational differences between flight and that defined for the engine performance data base).

5) Errors caused by what are classified as "mistakes" (examples: data reduction and programming errors).

To preclude gross error omissions of the type discussed above, measurement self-consistency checks should be included as part of the ground and flight measurement evaluation process. Maximum confidence of a measurement is only achieved when two or more methods with relatively weak coupling between the inputs and outputs are employed or compared simultaneously and when data comparisons are made with other facilities. Disagreement outside the measurement uncertainty bands of comparative methods indicates the presence of an unaudited error source or flaw in the uncertainty analysis. Measurement consistency checks also provide a substantial diagnosis and understanding of the individual factors that make up the final in-flight thrust of a vehicle. References 12 and 13 contain an in-depth discussion of the measurement validation process used in ground-test facilities.

*III.2.6.1   Measurement Self-Consistency Checks*

Basically four types of measurement checks are performed for engine airflow measurements. The airflow validation methods are, for convenience, called: duct checks, tail-pipe checks, turbine nozzle checks, and nozzle coefficient checks. Duct checks consist of comparing engine airflow from the facility primary airflow measurement system with calculated airflow using the engine bellmouth and/or measurements at the engine inlet (face) station. Tail-pipe checks refer to comparison of the primary airflow measurement value with the value obtained using the tail-pipe continuity balance described in Sec. III.2.3.2. Turbine nozzle checks are based on a comparison of the high-pressure turbine nozzle flow function over the engine operating envelope. For choked turbine flow, the turbine nozzle flow function should be relatively constant independent of test conditions. The final check, nozzle coefficients, consists of comparing the as-tested nozzle discharge coefficients with the engine manufacturer's predicted values from test-rig and scale-model tests. Extreme care must be exercised, however, when comparing cold flow model and full scale hot engine coefficients to ensure that all the differences in the data are properly assessed. Additional discussion on model–to–full scale comparisons is presented in Sec. III.2.7.

There are two types of measurement checks commonly used to validate engine scale-force thrust values: tail-pipe momentum checks and nozzle thrust coefficient checks. The tail-pipe momentum check refers to comparison of gross thrust as determined using the scale-force method with the value of gross thrust obtained using the tail-pipe momentum balance described in Sec. III.2.3.2. The nozzle coefficient checks again consist of comparing the as-tested nozzle thrust coefficients with the predicted coefficient values from test-rig and model data and are discussed more fully in Sec. III.2.7.

With respect to fuel-flow checks, the type of checks made are comparison of the facility fuel-flow-meter values with the engine meter (if so equipped) and the fuel pressure drop characteristic evaluation across the engine fuel injector, and comparison of the combustor burner efficiency with the engine manufacturer's predicted values.

### III.2.6.2 Facility-to-Facility Comparisons

On nearly every engine performance test program, the engine will have been tested on a sea-level test stand prior to being tested in an altitude test facility. It is, therefore, standard practice to compare engine sea-level stand data with the corresponding sea-level data from the altitude test facility test installation.

In order to conduct a sea-level check, it is necessary to first complete an engine cycle matching assessment which includes a comparison of: rotor speed ratio, compressor and overall engine pressure and temperature ratios, component efficiencies, differences in engine face flow profiles, and engine ram ratios. Some of the factors that have resulted in engine "rematching" in the past are: 1) engine health and performance degradation, 2) nonuniform inlet flow conditions at the sea-level stand inlet bellmouth, 3) large secondary flow effects in the vicinity of the nozzle aftbody on sea-level stands and capture diffuser talkback, 4) differences in instrument sampling rates, 5) differences in engine face boundary-layer displacement, and 6) cell heating and cell flow recirculation effects in altitude test cells.

Any differences in engine operational behavior must be taken into account, presumably using an engine computational simulation model, before comparing data. After all engine assessment checks have been completed and engine operational differences properly accounted for with respect to changes in engine performance, the measurement data should agree (with overlapping) within the sea-level test stand and altitude test facility measurement uncertainty bands.

Periodically interfacility engine performance correlations should be conducted to determine whether significant differences in measured engine performance exist. Results from one such correlation test using a low-bypass turbofan engine and involving three different ground-test facilities are presented in Fig. III.19.[17] Agreement between Facilities B and C net thrust and specific fuel-consumption values was within 1%, which was within the quoted interfacility measurement uncertainty band of 1.5%. The agreement with Facility A was less (approximately 2–2.5%) and was tracked to an inaccuracy in Facility A's bellmouth flow coefficient and omission of an inlet momentum term in the calculation of gross thrust, thereby enabling their measurement system to be improved.

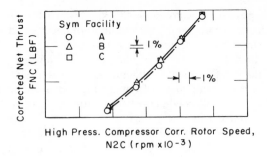

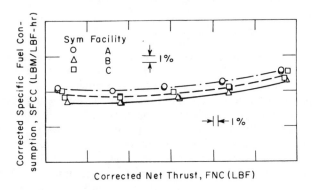

**Fig. III.19   Interfacility data comparison.**

One of the most ambitious undertakings in interfacility propulsion data comparisons is the Uniform Engine Test Program (UETP) initiated by the Propulsion and Energetics Panel of the Advisory Group for Aerospace Research and Development (AGARD), which is a subgroup of the North Atlantic Treaty Organization (NATO). The intent of the UETP is to enhance measurement practices used at government-operated engine test facilities and to establish a common basis for data comparison among the various test facilities. The approach adopted was to have each program participant measure and compare the internal, aerothermodynamic, steady-state performance of two J57-P-19W turbojet engines supplied by the United States Air Force. Each test participant is to document the performance of the J57 engines at specified environmental test conditions and engine power settings. The program participants are:

NASA Lewis Research Center, Cleveland, Ohio, USA
Arnold Engineering Development Center, Tennessee, USA
National Gas Turbine Establishment, Pyestock, Farnborough, England
Centre d'Essais des Propulseurs, Saclay, France
National Research Council, Engine Laboratory, Ottawa, Ontario, Canada
Turkish Air Force Air Materiel Logistics Center, Ankara, Turkey

The results from this program should be available in the 1986–1987 time frame.

### III.2.7  Scale-Model Test Comparisons

Scale-model tests have long been used for the performance prediction of exhaust system hardware concepts and for providing design guidance information early in the engine development cycle. Scale-model tests are conducted to develop nozzle coefficients to characterize the engine exhaust system performance. Model test results, when used in conjunction with full scale engine ground-test results, provide data for: 1) determination of the magnitude and variation of nozzle performance with nozzle pressure ratio, 2) extrapolation of full scale nozzle coefficients to nozzle pressure ratios not capable of being obtained in full scale engine test facilities, and 3) adjustment of propulsion facility uninstalled nozzle coefficients for flow effects. Many of the details involved in scale-model testing are discussed in Chaps. IV and V of this document, and a thorough presentation on the subject is presented in Ref. 12. The present discussion is limited to the problems of using scale-model nozzle coefficients to predict full scale engine thrust performance. A list of flowfield considerations that must be quantified before using model results to predict full scale engine performance are:

1) The three-dimensional nature of flow in the nozzle.
2) Corrections for real-gas effects which may arise in applying some model nozzle test data to full scale nozzles (particularly at high-pressure and low-temperature conditions).
3) Nonuniformity of pressure and temperature profiles across the exhaust duct at the nozzle entry measurement plane.
4) The coverage of the pressure and temperature probes, which will not, in general, give representative mean values.
5) Local flow direction, including swirl, in the plane of measurement.
6) Value of $\gamma$ used for isentropic groups (if ideal-gas thermodynamics are not used).
7) Dissociation of real gases at high temperature, and energy-mode fixation during rapid nozzle expansion.
8) Pressure losses between plane of measurement and nozzle entry, particularly with reheat.
9) Mass flow leakage from the tail pipe and nozzle.

To make these types of data assessments requires information pertaining to the full scale engine tail-pipe flowfield—information such as the extent of blade wakes, flow swirl angles, streamline curvature effects, viscous mixing effects, flow dynamics, magnitude of induced waveforms from rotating machinery, and real-gas properties with afterburner operation.

Another very difficult problem is relating model nozzle performance coefficients obtained using uniform cold flowfields to the nonuniform flow that typifies engine tail-pipe flowfields. Table III.6 shows a comparison using nozzle flow-averaging methods which do not account for the full scale engine flow effects. The results in this table were derived using actual nozzle data shown in Fig. III.20.

Table III.6   Comparison of flow-averaging methods

| Averaging method[a] | $\bar{P}$, psia | $\bar{T}$, °R | Discharge coefficient $C_D$ | Thrust coefficient $C_G$ |
|---|---|---|---|---|
| Uniform flow (scale-model test data) | 14.7 | 520 | 0.92 | 0.995 |
| Area-weighted | 14.83 | 1122 | 0.98 | 0.97 |
| Mass-flow-weighted | 14.84 | 1122 | 0.98 | 0.97 |
| Mass impulse (Kuchar/Tabakoff) | 14.83 | 1050 | 0.94 | 1.00 |
| Momentum | 14.83 | 1099 | 0.97 | 0.98 |
| Adjusted coefficient | 14.83 | 982 | 0.92 | 1.04 |
| Mass energy (Pianko) | 14.47 | 1120 | 1.00 | 0.98 |
| Mass, momentum, energy (Dzung) | 14.82 | 1127 | 0.99 | 0.96 |

[a] The details of each of the flow-averaging methods are contained in Ref. 3.

A review of the nozzle performance coefficients in the table shows that, depending on the flow-averaging method selected, the indicated mass flow efficiency of the nozzle lies between 92 and 100%. Correspondingly the nozzle thrust efficiency can vary from 96 to 104% efficiency. For the example shown in Table III.6, the error from using uncorrected model data could be as much as $-8\%$ in mass flow and $\pm 4\%$ in thrust, depending on the method used to average the full scale data.

Meaningful model–to–full scale engine performance comparisons can be accomplished if sufficient effort and resources are applied. Figure III.21 is another example of a scale-model–to–full scale engine data comparison using a simplified one-dimensional nozzle coefficient approach. In this comparison, the gas properties at the full scale exhaust nozzle inlet were measured experimentally during test operations. The values of the gas properties were used in combination with scale-model thrust coefficient data to obtain a gross thrust value independent of the scale-force measurements made for the full scale engine. At the higher levels of thrust, the scale-model coefficient thrust value was 4% larger than the value obtained by the scale-force method, and this difference increased to 5.5% at the lower values of gross thrust. This level of difference in thrust measurement is obviously unacceptable for use in an engine development and qualification program. To explain this difference, the thrust from both methods was analyzed and compared in an attempt to quantify the term-by-term differences and to analyze and resolve those differences. The evaluation used both inviscid and viscid complex computer programs to quantify flowfield differences, in addition to available experimental data to estimate duct pressure losses and nozzle leakages. The results of this effort are shown in Fig. III.22. The initial values of the data comparison are repeated along the top line shown in Fig. III.22. The detailed analysis of the thrust information showed that 2–3% of the differences in the thrust values were caused by improper accounting for measurement rake drag and wall drag in the engine tail pipe. Another 1% of the difference was related to improper

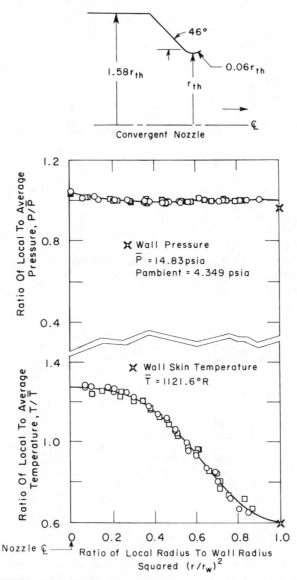

**Fig. III.20 Measured nozzle total pressure and temperature profiles.**

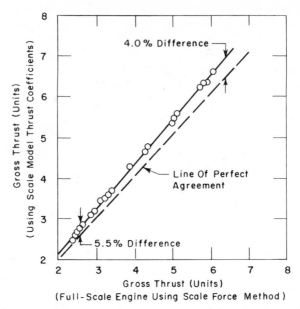

**Fig. III.21   Gross thrust comparison using measured full scale engine data and scale-model predictions—initial results (nonafterburning turbojet).**

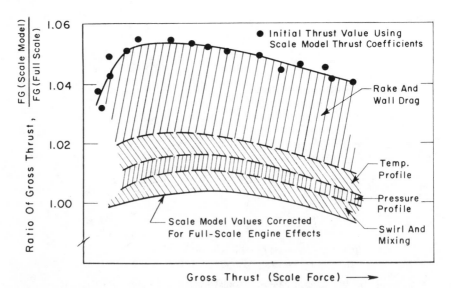

**Fig. III.22   Resolution of scale-model and full scale engine thrust data (nonafterburning turbojet).**

accounting for the substantial profile variations of total pressure and total temperature at the nozzle inlet. And, finally, another 1% of the difference was related to the gas swirl and mixing losses produced by the turbine and exhaust sections of this particular engine. In this example, the nozzle in/out leakages were about equal. When all these terms were properly placed into the scale-model predictions and scale-force measurements, the results from the two methods agreed within approximately ±0.5%, as shown by the bottom line in Fig. III.22. Thus, resolution of what was initially an unacceptable variation in two experimental measurements has become a single value of measured thrust *and* a series of explicit values of second-order engine flow effects. Thus, success in understanding and resolving scale-model and full scale engine data goes beyond a comparison of one-dimensional model nozzle coefficients and must include the details of the thermodynamic and aerodynamic parameters affecting engine airflow and thrust prediction.

### III.2.8 Short-Duration Turbine Engine Testing

A relatively recent development in ground testing for engine performance is short-duration turbine engine testing (SDTET).[2] As shown in Fig. III.23, the method consists of obtaining engine performance data while slowly sweeping the power lever across the range of interest instead of discretely acquiring data in the conventional steady-state manner. The potential advantages of SDTET are acquisition of data in shorter time periods, thus at lower cost, the definition of functional relationships among engine variables on a continuous, rather than separated, distinct point basis. SDTET therefore offers the promise of more and better-defined performance data at lower cost. Few benefits are free, however, and the price paid for SDTET is increased data uncertainty, the requirement to assess and correct for transient thermal, mechanical, and engine cycle mismatch nonequilibrium effects, and the need to accurately record and process transient rather than steady-state data. Thermal nonequilibrium effects vary for any given engine as a function of flight condition. Figure III.24 shows the variation of engine thermal time constant with flight condition relative to the sea-level-static value. Thus, any successful SDTET technique must account for these effects.

Short-duration testing has been explored and feasibility demonstrated as described in Ref. 2. A medium-bypass ratio engine in an altitude test cell was used as the test vehicle. Methods for transiently measuring and processing pressure, temperature, airflow rate, fuel-flow rate, and thrust data (via scale force) were developed and demonstrated. Because slow, smooth power lever sweeps, on the order of 1 min, were selected as the test technique, measuring systems with frequency response of only about 5 Hz were required. Response compensation was applied to enhance thermocouple transient response. All other variables possessed sufficient frequency response so that compensation was not necessary. All data were processed through a low-pass digital filter for high-frequency noise rejection and to assure matched transient response among the independent variables from which key dependent variables such as thrust and airflow were calculated.

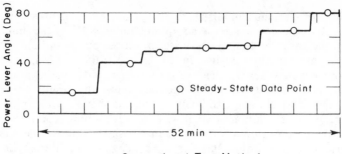

a. Conventional Test Method

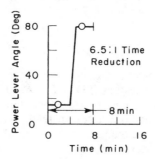

b. One-Min Sweep (SDTET Method)

**Fig. III.23  Comparisons of conventional steady-state test with sweep test time requirements.**

Analytical approximate methods for adjustment of the transient engine data to eliminate nonequilibrium effects were also developed. Application of the corrections thus yield equivalent equilibrium running, or quasi-steady-state data from the transient data recorded during the slow, smooth sweep. The effects that must be adjusted appear principally in the energy input (fuel flow) to the engine. The effects consist of transient heat soak into the metal parts of the engine, transient heat storage in the fluid contained by the engine at any instant, energy to overcome rotor inertia (to accelerate the engine), and extra energy to rematch the compressors during acceleration relative to the equilibrium running operating lines. Figure III.25 shows the size of the fuel-flow correction for a 100-s power transient at Mach 0.26 and 5000-ft altitude. The total correction is 7% with the metal heat soak predominating.

An approximate expression for the metal heat-soak term is shown in Ref. 2 to be

$$\Delta WF = K\frac{\partial T_{\text{turbine}}}{\partial t} + \tau_E\frac{\partial^2 T_m}{\partial t^2}$$

where $WF$ is the energy at any instant tied up in heat transiently stored in the

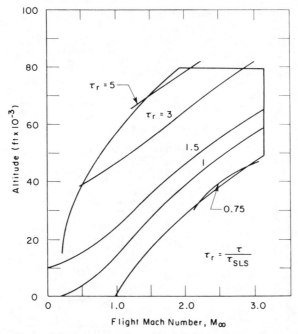

**Fig. III.24    Effect of flight condition on thermal time constant for a turbojet engine.**

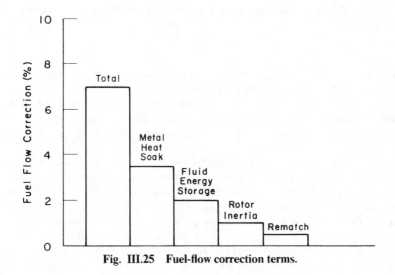

**Fig. III.25    Fuel-flow correction terms.**

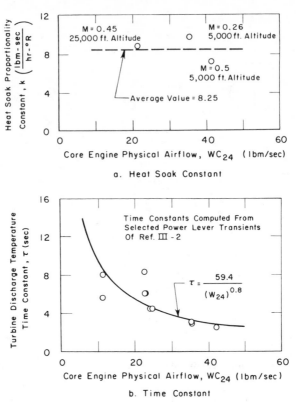

Fig. III.26   **Empirical method for estimataion of transient energy storage in engine metal and fluid.**

metal parts in terms of fuel, $K$ is the bulk heat-soak proportionality constant, and $\tau_E$ is an overall engine thermal time constant.

$T_{\text{turbine}}$ is the engine hot discharge section control temperature. The constants $K$ and $\tau_E$ were determined experimentally by comparing a few transients to steady-state data. Figure III.26 is an example of engine "thermal calibration."

Several observations are of particular importance. First, if the test transients are smooth, the second derivative of the adjustment equation goes to zero, thus the first derivative term predominates. Second, the variation of $K$ is shown in Ref. 2 to vary little with flight condition, therefore $K$ need not be evaluated all over the flight map but only at a few points. Third, the effects of tip and seal clearance transient changes are implicitly included in the heat-soak term because the constants are derived from comparisons of actual experimental steady-state and transient data. Finally, the transients are performed slowly as well as smoothly, therefore, the magnitude of the correction is less than 10%; thus, if the adjustment procedure is only 5% accurate, the uncertainty in the

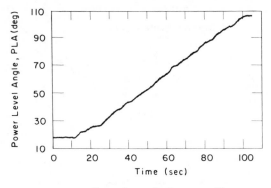

a. Power Lever Angle versus Time

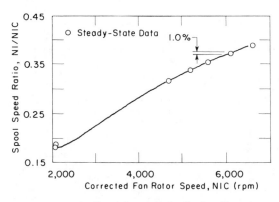

b. Spool Speed Ratio During Transient

**Fig. III.27  Test conditions, transient data, and comparisons with steady-state results for SDTET power lever transient at Mach 0.26, 5,000-ft altitude.**

final result from the adjustment is on the order of 0.5%. The rate may be even slower, for example, 5-min transients instead of 1 min, and still provide a significant savings over conventional steady-state testing. In this case, the adjustment practically vanishes. Also, the comparisons of data from power decelerations as well as accelerations will permit estimation of tip and seal clearance effects from bulk heat storage.

Engine cycle rematch and spool-adjustment methods based on rotor inertias and readily known compressor map characteristics were derived and applied. The inertial and cycle rematch adjustments as well as the thermal adjustments require determination of various derivatives from experimental data. A method of extracting smooth derivatives is also provided in Ref. 2.

Figure III.27 shows example results for a 100-s power transient at Mach 0.26 at 5000-ft altitude. Figure III.27a shows the manual but smooth power lever

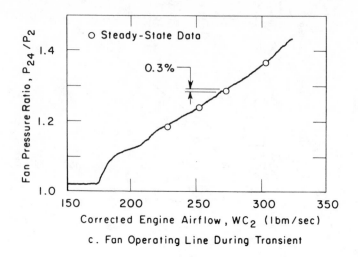

c. Fan Operating Line During Transient

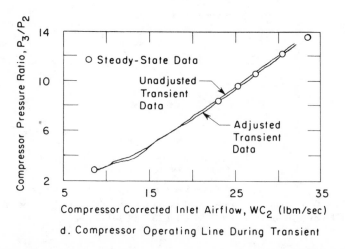

d. Compressor Operating Line During Transient

**Fig. III.27 (continued) Test conditions, transient data, and comparisons with steady-state results for SDTET power lever transient at Mach 0.26, 5,000-ft altitude.**

transient. Figure III.27b shows spool-speed ratio and indicates excellent agreement between steady-state and transient data verifying that for slow transients, mechanical nonequilibrium effects can virtually be eliminated. Figures III.27c and III.27d indicate that the fan and adjusted compressor operating lines agreed well with steady-state results. Finally, Figs. III.27e and III.27f show net thrust and specific fuel-consumption agreement. The agreement on thrust was even better on cases where the test conditions are better maintained during the transient.

The experiments described above were conducted manually with little smoothing applied to the transient data. Subsequent experiments employing

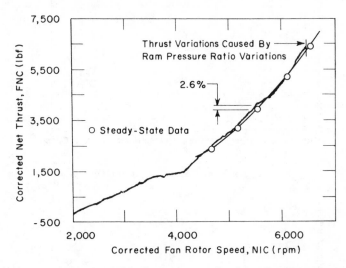

e. Corrected Net Thrust Versus Corrected Fan Rotor Speed

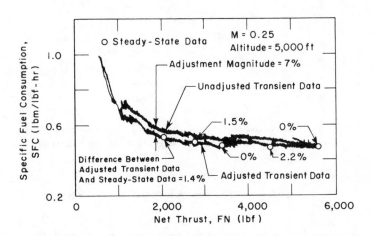

f. Specific Fuel Consumption Versus Net Thrust

**Fig. III.27 (continued) Test conditions, transient data, and comparisons with steady-state results for SDTET power lever transient at Mach 0.26, 5,000-ft altitude.**

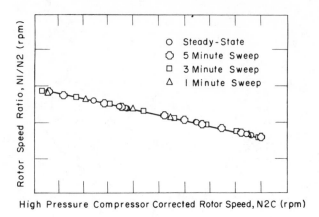

a. Spool Speed Match Comparison

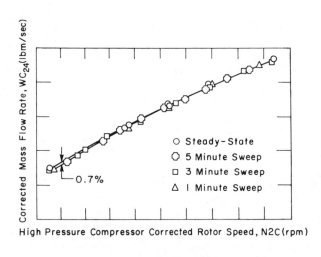

b. Airflow Comparison

**Fig. III.28   SDTET results for small turbofan engine.**

computer-controlled power lever and flight condition (test facility) control on a smaller low-bypass ratio turbofan engine showed even better results than those on Fig. III.27. Figure III.28 is an example. For this case, no adjustments were required because of the slower, smoother rates and smaller thermal and mechanical constants of the small-size engine.

In summary, SDTET promises significantly more performance data in less test time, better definitions of performance, and insight into transient effects not available from conventional testing. It requires more complex transient data measurement systems and processes, and produces data with some increase in uncertainty. Tables III.7 and III.8 present steady-state and SDTET

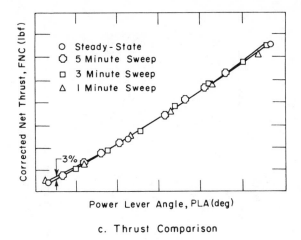

c. Thrust Comparison

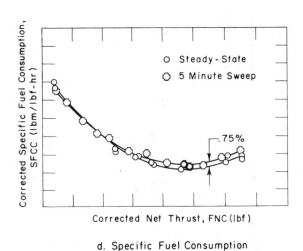

d. Specific Fuel Consumption

**Fig. III.28 (continued)  SDTET results for small turbofan engine.**

uncertainties. As may be noted, in this particular experiment the thrust and SFC uncertainties were approximately 20% higher than the steady-state values at the same test conditions.

### III.3    Engine Performance Estimation by Flight Testing

Flight testing is the process of evaluating the integrated performance of full scale airframe and propulsion hardware and software systems under actual environmental conditions. Flight testing can include climb tests, turning performance evaluations, takeoff/landing tests, and cruise performance evaluation. While flight testing requires no environmental simulation as required in

Table III.7   Steady-state data uncertainty; net thrust and specific fuel consumption

| Test conditions | | | Uncertainty | | | | | | | |
|---|---|---|---|---|---|---|---|---|---|---|
| | | | Net thrust | | | | Specific fuel consumption | | | |
| Altitude, ft | Mach number | Power setting | Precision index, % of level | Degree of freedom | Bias, % of level | Total uncertainty, % of level | Precision index, % of level | Degree of freedom | Bias, % of level | Total uncertainty, % of level |
| 5,000 | 0.26 | 95% maximum continuous | 0.17 | > 30 | 0.19 | ±0.53 | 0.26 | > 30 | 0.54 | ±1.06 |
| 5,000 | 0.26 | 45% maximum continuous | 0.25 | > 30 | 0.19 | ±0.69 | 0.32 | > 30 | 1.02 | ±1.66 |
| 5,000 | 0.50 | Maximum | 0.18 | > 30 | 0.20 | ±0.56 | 0.26 | > 30 | 0.44 | ±0.96 |
| 25,000 | 0.47 | 90% maximum continuous | 0.24 | > 30 | 0.17 | ±0.65 | 0.31 | > 30 | 0.93 | +1.55 |

Table III.8 SDTET data uncertainty; net thrust and specific fuel consumption

| Test conditions | | | Uncertainty | | | | | | | |
|---|---|---|---|---|---|---|---|---|---|---|
| | | | Net thrust | | | | Specific fuel consumption | | | |
| Altitude, ft | Mach number | Power setting | Precision index, % of level | Degree of freedom | Bias, % of level | Total uncertainty, % of level | Precision index, % of level | Degree of freedom | Bias, % of level | Total uncertainty, % of level |
| 5,000 | 0.26 | 95% maximum continuous | 0.20 | > 30 | 0.22 | ±0.62 | 0.30 | > 30 | 0.63 | ±1.23 |
| 5,000 | 0.26 | 45% maximum continuous | 0.29 | > 30 | 0.22 | ±0.60 | 0.37 | > 30 | 1.10 | +1.92 |
| 5,000 | 0.50 | Maximum | 0.21 | > 30 | 0.23 | ±0.65 | 0.30 | > 30 | 0.51 | ±1.11 |
| 25,000 | 0.47 | 90% maximum continuous | 0.20 | > 30 | 0.20 | ±0.76 | 0.36 | > 30 | 1.06 | ±1.00 |

ground tests, a flight environment is not easily controlled for systematic test evaluations. Also, aircraft volume and weight limitations reduce the number of flight measurement options available to the tester. In-flight performance evaluation schemes consist of aircraft energy change and excess thrust methods[18] and propulsion system methods based on indirect measurement methodology using ground-test data. Some of the more difficult problems associated with flight testing involve the methods used to monitor and maintain the aircraft and environmental test conditions, i.e., aircraft attitude, ambient pressure, airspeed, and ambient air temperature. In this section the procedures and problems associated with evaluation of engine thrust performance in steady-state, level flight are addressed. The discussion is limited to engine *modified net thrust*, which includes installation effects of inlet internal performance, nozzle internal performance, bleed-air extraction, shaft power extraction, and axial forces on plug and pylon surfaces buried in the exhaust jet. Evaluation of *installed net propulsive thrust*, which includes external force increments due to the inlet and exhaust system and external control surface trim forces, is discussed in Chap. V of this document. Much of the material in this section is based on Ref. 12 to which the reader is referred for a detailed treatment of in-flight thrust determination methods.

### III.3.1  Preflight Test Considerations

An important preflight activity is the liaison work between the flight planning personnel and the ground-test conductors. A continuous cross flow of information should be maintained concerning data-handling procedures, calculational routines, and engine anomalies that arise during the ground tests. Other program safeguards are a preflight ground-to-flight measurement uncertainty analysis and an early review of the flight-test data by both flight- and ground-test personnel to ensure that the engine flight behavior is representative of ground-test results. The preflight uncertainty analysis is essential since it is only through this type of effort that shortcomings in proposed flight instrumentation, test matrices, and calculational routines can be identified.

### III.3.2  Engine Installation Effects

Flight-test data include effects on engine operation as a result of the aircraft installation. The more prominent effects are: 1) engine accessory horsepower extraction, 2) engine compressor bleed airflow, 3) inlet airflow pressure recovery, and 4) inlet airflow distortion. As a part of the propulsion ground-test program, the sensitivity of the engine performance to each of the installation effects should be evaluated.

Typically air-vehicle horsepower extraction is less than 1% of engine power and is best accounted for using analytical means (such as validated math models) as opposed to in-flight measurements.

Compressor bleed effects are related to the bleed airflow rate so that flight-test results may be evaluated using ground-test data with the same bleed flow extraction. In-flight, compressor bleed flow rates are measured using a flow coefficient determined during the ground-test program and onboard pressure and temperature instrumentation. Because of the low velocities for bleed flows, an accurate measurement (better than 3%) of bleed flow rates is difficult.

The inlet airflow recovery factor is the measure of the freestream pressure recovered in the inlet. Inlet recovery factors are usually defined from wind-tunnel scale-model tests and are used in flight to define the average engine face pressure based on flight conditions and the engine flow rates. The effect of inlet recovery factors on engine performance is defined in propulsion ground-test facilities by varying engine inlet pressure and ram ratio ($P_2/P_0$) and recording the change in engine performance.

With respect to inlet airflow distortion, the flow quality entering the engine through the aircraft inlet is a function of inlet configuration and the aircraft flight velocity and attitude. Since cruise performance testing is done at level, steady flight conditions, inlet flow distortions should be minimal. However, in cases where inlet flow distortions are more than several percent of the bulk averaged pressure value, it may be necessary to equip the flight engine with a full set of inlet pressure rakes. Engine performance adjustments can be estimated using the ground-test distorted flow performance data. In propulsion ground-test facilities, flow distortions are simulated using screens or airjets[19] to match wind-tunnel derived pressure patterns. For the purpose of analysis, various analytically defined distortion indices[20] are used to characterize the different types of flow distortions.

### III.3.3   Engine Evaluation Methods

In this subsection the primary methods used to determine in-flight engine thrust are reviewed. Selecting an in-flight thrust measurement method depends on the flight-test objectives, test matrix, available instrumentation, and accuracy requirements. Every flight-test program should include at least one to two alternate thrust prediction schemes (preferably with unlinked methodology) in the event of loss of instrumentation which would negate the selected method and also to serve as a verification cross-check.

*III.3.3.1   Engine Computer Math Model*

An engine math model is an extremely effective diagnostic tool for flight-test data analyses since it provides a capability for component evaluation and can be used to adjust engine data for cycle changes that may occur between the ground and flight tests. The engine math model also serves to monitor the engine health and evaluate engine operational performance deterioration. A complete discussion of engine math models is presented in Sec. III.1.6.

To obtain in-flight thrust, the inlet performance, nacelle secondary airflows, and installation losses (shaft horsepower and bleed flow) must be specified. The output performance may be limited to the overall net thrust and fuel flow or may include detailed component performance at each flight-test condition. The accuracy of the simulation is dependent upon the quality of the component and overall engine performance data available to make the computer simulation. For flight testing, the engine math model should be adjusted with engine data that corresponds to the "bill-of-material" flight engine hardware if accurate flight-test results are to be achieved. Particularly accurate results (2–3%) are possible if a model is "calibrated" to a particular flight-test engine.

*III.3.3.2   Gas-Path / Nozzle Methods*

Another common method of evaluating in-flight engine performance is with the use of gas-path and nozzle methods. Gas-path/nozzle methods utilize calibrated measurements of gas-generator flow properties at various stations within the engine or station properties calculated from calibrated measurements made at other stations within the engine.

All gas-path/nozzle methods require determination of ambient pressure, $P_0$; freestream total pressure, $P_1$; and inlet total temperature, $T_2$. Other variables that may require determination include:

1) Inlet total pressure, $P_2$.
2) Nozzle mass flows, $W_9$ and/or $W_{19}$.
3) Engine inlet mass flow, $W_2$.
4) Nozzle areas, $A_8$ and/or $A_9$, and $A_{18}$ and/or $A_{19}$.
5) Nozzle entry temperatures, $T_7$ and $T_{17}$.
6) Nozzle entry pressures, $P_7$ or $P_{s7}$, and $P_{17}$ or $P_{s17}$.

The measured or calculated local variables will depend on the gas-path/nozzle method option adopted. In many cases, aircraft and calibrated flight development engines are instrumented to provide more information than may be required to meet the need for in-flight thrust determination. Notable examples are: accurate fuel-flow measurement to determine aircraft specific range and engine specific fuel consumption, and data to compare engine component behavior with projected component performance or to confirm control functions. Various stations throughout the engine can be used for evaluating the flow quantities needed for in-flight performance determination. A common method used for mass flow determination is a low-pressure (fan) compressor calibration with corrected rotor speed. The fan pressure ratio and referred speed are measured in flight and the flow determined from a ground-test calibration curve. If the compressor is equipped with variable stators, a vane-angle flow correction correlation is also required. The in-flight spatial flow distributions at the compressor inlet and exit face stations should be comparable with the measurements made in ground tests. While it is realized that changes in circumferential and radial profiles, radial tip clearance, and humidity will affect the flow evaluation process, in practice these factors are very difficult to take into account.

The nozzle method is commonly used to evaluate both in-flight engine airflow and thrust performance. The advantage of nozzle coefficients is that they provide a direct indication of the engine performance independent of other gas-path methods and therefore serve as a credibility check. The general procedure is to relate real nozzle performance to an ideal nozzle through the use of empirically established coefficients. The coefficients may be derived from full scale engine ground tests or from scale models of the full scale nozzle configuration. The nozzle coefficients are based on one-dimensional isentropic-expansion assumptions for the nozzle flow; the problem, however, is that real engine nozzle flows are neither one-dimensional nor isentropic. For full scale engine ground testing, this generally does not present a problem since the nozzle coefficients to be used in flight are obtained from a matching flowfield. For full scale testing, the concerns with using nozzle coefficients to predict internal engine in-flight performance are: 1) engine-to-engine performance variations (this is estimated if the actual flight engine is calibrated in the ground-test facility), 2) differences in engine tail-pipe instrumentation, 3) engine "rematch" in flight, 4) engine operational deterioration, 5) tail-pipe geometry changes caused by dynamic flow conditions, and 6) insufficient ground-test data to cover the flight operational envelope.

For engines that are only tested on a sea-level stand, the performance coefficients must be extrapolated to the flight conditions, and this is generally accomplished using scale-model data. This then imposes additional problems, such as: 1) flight altitude, Mach number aerothermodynamic effects on engine performance, and 2) control schedule effects for the flight-test conditions.

For afterburning engines, direct in-flight measurement of the exhaust nozzle entry flow conditions is generally not feasible. Instead, inlet pressure and temperature to the afterburner are measured and afterburner pressure loss and heat balance procedures are used to infer the nozzle entry flow properties. For some afterburning applications, ground-test tail-pipe static-pressure correlations can be used to provide the nozzle inlet flow properties.

### III.3.3.3   Calibrated Flight Engines

The most accurate means of determining in-flight engine performance is the direct calibration of the flight engine in an altitude ground-test facility. This was the procedure used in the fly-off competition between the General Dynamics and Boeing Air Launch Cruise Missile (ALCM) candidates.[21] The test philosophy used for the ground calibration of the ALCM flight engines was:

1) Actual flight engines.
2) Actual in-flight engine instrumentation and signal conditioning equipment.
3) Three calibration test conditions to cover the flight-test envelope.
4) Engines calibrated between cruise and military power settings using four-point power hooks.
5) Inlet flow distortion calibration data generated using screen patterns based on wind-tunnel measured pressure profiles.
6) Generation of compressor bleed flows and power extraction engine performance calibration data.

The objectives of the engine ground calibration tests were to:

1) Develop engine airflow and gross thrust calibration coefficients/curves.
2) Trim the flight engine status deck.
3) Provide a preflight ground-through-flight measurement uncertainty analysis to identify measurement problems prior to the actual flight tests.

A significant point to be made with respect to flight engine calibration is the reduction in required engine instrumentation. For the ALCM tests a total of only 16 engine performance measurements were required.

The ALCM in-flight engine thrust comparisons are shown in Fig. III.29. Three methods were used to evaluate the engine in-flight thrust and all of the methods agreed within 1%. In addition, the in-flight measured gross thrust agreed within 0.5% of the preflight engine ground-test data, indicating that for this test the engine cycle and schedule did not rematch from the uninstalled engine (i.e., ground-test configuration) to the installed engine (i.e., flight configuration).

Preflight engine calibration offers several advantages that greatly simplify the in-flight analysis effort:

1) Simple, accurate approach. The in-flight thrust measurement uncertainty using this approach is 2–3% (if necessary, the engines can also be calibrated postflight test for further improved accuracy) as compared to 5–10% for uncalibrated engines.

2) Significantly reduced engine measurement requirements.

3) Altitude preflight functional checkout of the engine.

4) The basis, through measurement uncertainty analysis, for evaluating proposed ground-through-flight prediction methods and thus identify measurement problems prior to testing.

The calibration process is more difficult for engines with extensive variable geometry features. In those cases, a calibrated math model is almost mandatory as an in-flight performance adjustment tool if the number of flight engine calibration hours are to be kept to an acceptable level.

### III.3.4  Flight Instrumentation Considerations

Methods of acquiring data in flight engine tests are considerably different from those used for ground propulsion tests. Space and weight limitations and installation problems frequently prevent the use of extensive instrument packages, signal conditioning equipment, and in-place calibration equipment. As a result, one should expect higher measurement uncertainty in flight than achievable in propulsion ground-test facilities. How much higher will depend on the particular in-flight performance prediction method selected and the instrument requirements associated with that method.

The choice of an in-flight measurement system depends on several factors, such as:

1) Results from a preflight measurement uncertainty assessment.

2) Instrument experience background of personnel working the test.

3) Availability of instrumentation hardware and software and instrument support systems.

4) Test program resource and schedule requirements.

5) Type of data generated in the ground calibration tests.

Nearly all of the in-flight thrust prediction schemes, however, require some measurement of engine internal pressures and temperatures, rotor spool speeds,

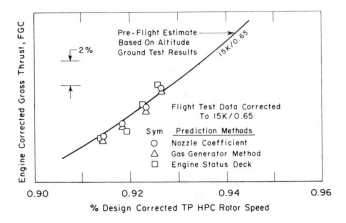

**Fig. III.29   ALCM inflight thrust results.**

fuel flow, and engine flow areas. The usual aircraft measurements are flight pressure altitude, ambient temperature, airspeed, attitude, and acceleration for nonsteady-state flight conditions. Some of the engine performance prediction methods require only a few inputs in order to make an engine performance estimate (the engine math model only needs a specified aircraft flight condition and engine power setting); but it is still necessary to make sufficient additional measurements to ensure that the measured profiles have not changed between the ground calibrations and the flight tests. Profile changes must be identified and quantified with respect to the flow-averaged values if accurate in-flight test results are to be achieved. The importance of flow profile changes on engine performance is discussed in Sec. III.1 and quantified in Ref. 3 and can amount to several percentage points change in the calculated engine component efficiency.

In all instances, individual instrument readouts should be corrected for recovery factors, flow misalignment, thermal corrections, flowfield interference effects, and time-unsteady corrections before being used in computations. These corrections, while difficult to assess, are nonetheless very important. For instance, the measured temperature in the hot-gas, low-velocity sections of the engine can be in error by as much as 5–15% from the true gas temperature if the necessary corrections are not made to the thermocouple temperature probes.[22] It is also important to ensure that the type of measuring instrument and instrument corrections used in flight and for the ground-test calibrations be consistent. For the case where the actual flight engines and flight instrumentation package are calibrated directly in a propulsion altitude test facility, an instrument correction need not be made to the engine internal measurements since the correction becomes redundant when the calibrated data are used to back out the engine in-flight performance information.

Instrument corrections are also important to aircraft attitude measurements. A list of angle-of-attack corrections for a noseboom-mounted vane system consists of:

1) Misalignment vane to boom.
2) Mechanical misalignment boom to fuselage reference line.
3) Boom bending due to inertial loads.
4) Fuselage bending due to inertial loads.
5) Boom bending due to airloads.
6) Fuselage bending due to airloads.
7) Upwash correction (boom, fuselage, wind).
8) Pitching velocity corrections.

Many of these same corrections apply to boom-mounted pitot-static-pressure probes and attitude cone probes which use an array of surface static-pressure taps to indicate aircraft attitudinal position.

Engine exhaust nozzle flow area can be a difficult in-flight measurement. For fixed nozzle areas, only a nozzle material temperature expansion/contraction relationship is required to calculate the flow area in flight. Variable area exhaust nozzles, however, are much more difficult to evaluate. Most variable area nozzle flow areas are measured indirectly by statically calibrating the displacement of the nozzle positioning gear or linkage displacement. Because of the indirect nature of this method and the mechanical dynamics experienced in flight, this measurement method can result in large errors in the calculated flow area. If possible, when using variable area nozzles, it is desirable to select a thrust measurement method that does not require, or at least is not greatly influenced by, measurement error in the nozzle flow area.

All flight instrumentation systems should be tailored to correspond to the operating range over the specified test conditions. In general, the narrower the instrument operating range, the easier it is to devise a system to meet the measurement requirements. For a wide range of operating test conditions, two or more measurement systems may be required to meet the flight program measurement accuracy requirements. This type of information should fall out from the preflight measurement uncertainty analysis.

Finally, test procedures should include pre- and postflight checks as well as instrumentation monitoring during a flight. Preflight checks should include an inspection of the mechanical condition of transducers, connectors, cabling, tubing, recorders, etc. Each channel should be sampled and compared with known levels—e.g., differential transducers at zero and absolute transducers at barometric pressure—and, where appropriate, leak-checked. Stability and noise levels of all outputs, particularly with regard to any adverse interaction with aircraft systems, should also be checked.

### III.3.5   Unsteady Influences on Thrust Determination in Flight

The in-flight thrust measurement methodology commonly used is based on steady-state relationships. Indeed, the fundamental definition of thrust asserts that the thrust axis is parallel to the flight path and that thrust and drag are equal. Figure III.30 shows an actual in-flight thrust time history from a

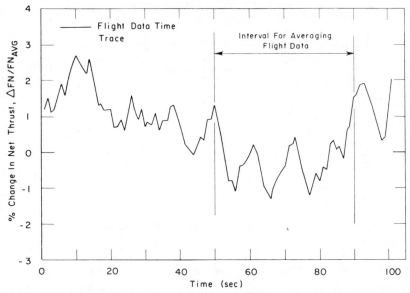

Fig. III.30    Thrust history for steady-state test conditions (data averaged over 1-s time intervals).

low-bypass turbofan engine cruise test. Even with averaging the data over a 1-s time interval, unsteady effects can be identified in the engine thrust data. Unsteady effects can influence thrust determination in flight and the magnitudes of these effects must be estimated. Unsteady effects come from combinations of three sources:

1) Unsteady terms in the basic force equations which are most ordinarily ignored when deriving the steady-flow thrust equations.

2) Departure from expected steady-state engine performance because of engine transient effects present in flight or power transients.

3) Measurement errors because of inadequate transient response of various measurement systems.

The purpose of this discussion is to present information as guiding principles for estimating the magnitude of the unsteady influences on in-flight thrust. In most cases, unsteady thrust effects will be relatively small and will not significantly increase the thrust uncertainty determination, providing the unsteady or transient process is performed in a slow and steady manner. In the following, the unsteady form of the in-flight force equations, including a method for estimating the size of the unsteady terms relative to the steady terms, is presented, and gas-generator thermal transient terms and transient measurement considerations are discussed. A more detailed discussion is given in Ref. 12.

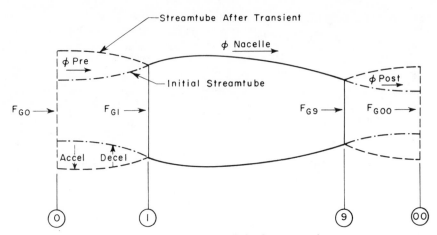

**Fig. III.31   Unsteady inflight thrust equation.**

*III.3.5.1   Unsteady In-Flight Thrust Equations*

Excess thrust is defined as installed thrust minus aircraft drag and will result in aircraft acceleration or climbing. Measured acceleration or climb rate may be used to assess drag from in-flight thrust in unsteady flight. The relationships governing such aircraft dynamics are presented in other parts of this document. In addition to aircraft dynamics, additional unsteady terms appear in the thrust equation when the flow through and around the propulsion system is not steady as is assumed in the development of conventional in-flight thrust equations. To identify the unsteady terms, consider the simple propulsion nacelle shown in Fig. III.37. Under actual test conditions, all of the force terms identified in Fig. III.37 are functions of time. The time dependence may result from the aircraft climbs, descents, power changes, or combinations of power and flight changes. Using the simple nacelle control volume, unsteady overall net thrust $F_N$ is given in two major parts: a quasi-steady part and a part made up of transient terms as shown in Fig. III.31. The unsteady thrust equation is written relative to the airplane frame of reference, the same as the preceding steady-state thrust equations. The additional transient terms from translational, centripetal, and Coriolis accelerations which arise from axis transformation from the stationary Earth to the moving aircraft frame of reference are not included, since these terms are included within the aircraft acceleration term or are implicit in the unsteady thrust equation or are negligible when compared to the other transient terms. The quasi-steady terms are calculated just as in the steady-state case except that all terms are computed using instantaneous values. The transient terms account for the instantaneous change of momentum within the control volume and for the expansion or contraction of the pre- and poststream tubes that occur during transients.

*III.3.5.2   Magnitude of Time-Dependent Terms*

Transient terms are represented as additional terms or corrections to the standard steady-state thrust equations. Exact computation of the transient terms is very difficult and usually not necessary. A generally satisfactory approach is to approximate the correction terms, note the conditions that cause them to be significant, and avoid testing at those conditions. In this manner, the simpler steady-state equations may be used through mild, well-controlled transients and yield satisfactory results.

Approximate methods of estimating the relative magnitude of the transient terms are derived in Ref. 2. The combined size of the transient terms depends on the time rate of change of airflow $\partial W/\partial t$ during the transient and may be estimated as

$$\frac{\text{transient portion of thrust}}{\text{quasi-steady-state thrust}} = ( L + D_{\text{NAC}} ) \frac{\overline{\partial W}}{\partial t} \bigg/ FN_{\text{SS}} \qquad \text{(III.8)}$$

where $L$ and $D_{\text{NAC}}$ are the nacelle length and entry diameter, respectively. Evaluation of this term for a large turbojet produced the following results:

| Type transient | Flow derivative range, percent change in engine airflow | Transient term in Eq. (III.8), percent of steady state |
|---|---|---|
| Smooth flight maneuvers representative of transient or dynamic aircraft tests at constant power, $M = 0.9$ at 20,000 ft | 0–1 | 0–0.3 |
| Smooth in-flight power transients 3–60 s from mid to maximum power setting, $M = 0.9$ at 20,000 ft | 5–35 | 0.5–3.0 |
| Smooth 30-s transient idle to maximum power in an ATF, $M = 0$ at 0 ft | 5 | 0.1 |

The values in the table are based on airflow and thrust values at a midpoint during the transients and give only an order of magnitude indication of the size of the terms during the transients. Both airflow and thrust change as a function of time during the transient, thus the ratio must also change.

The above examples are for smooth transients. If erratic flight maneuvers or power changes are executed, large instantaneous values of airflow derivative can occur. Therefore, a "smooth" test pilot will produce results with less apparent scatter than a nervous one on a turbulent day.

For most applications it is sufficient simply to compute the approximation to justify ignoring the transient terms and the assessment of the additional uncertainty incurred. For rapid or erratic flight or throttle transients, errors of appreciable size are produced, but monitoring of the approximation through the transient will serve to isolate regions of excessive unsteady effect. In general, smooth aircraft maneuvers with a fixed power setting will produce little influence because the propulsion flow derivatives are minimal. Power transients, however, must be monitored closely and executed smoothly and slowly to keep transient effects small.

### III.3.5.3   Engine Transient Thermal Effects in Flight

Thrust delivered by a gas turbine engine transiently passing through a given power setting differs from that produced under steady-state conditions at the same power setting. For transient rates produced by flight condition or power lever transients, a principal cause of this difference is transient thermal effects as discussed earlier in this chapter.

Principal energy input sources are engine fuel flow and the enthalpy of the inlet airstream. For power transients, both sources change because the engine accelerates as a result of increased fuel flow. For flight condition transients with fixed power lever setting, the enthalpy of the inlet air changes, and only the fuel-flow rate is changed to accommodate the modified inlet conditions.

For a typical power lever change from a low to a high power setting, the energy added to the engine must be sufficient to:

1) Provide the basic steady-state required-to-run energy at the instantaneous operating point.

2) Accelerate the rotors by overcoming rotor inertia.

3) Make up for transient energy losses through higher heat transfer to the metal parts.

4) Make up for additional transient compressor work required during the acceleration because of higher back pressure on the compressor.

5) Make up for changed efficiencies because of transient tip and seal clearance changes.

### III.3.5.4   Accounting Methods for Thermal Transient Effects

Transient thermal effects may be treated either by avoidance or compensation. In avoidance, flight and power maneuvers are restricted to rates that hold the effects to insignificant magnitudes. In compensation, provisions are made to correct for the effects. The best approach depends on the selected method for thrust determination.

### III.3.5.5   Math-Model Method

The overall performance method is usually based on steady-state operating characteristics of the engine. Therefore, avoidance should be given strong consideration. Experimental results from altitude test facility tests and transient math-model calculations indicate that a correlation may be formed for assessment of transient effects, as shown in Fig. III.32.

The correlation for a specific engine may be constructed from comparisons of steady-state and transient test results or from a transient engine math model which accounts for the thermal transient effects previously discussed. The correlation may then serve as a guide for avoidance of transient rates producing excessive transient thrust effects.

Mathematical models or computer simulations of engine performance are sometimes used along with a few key engine parameters as the in-flight thrust overall performance model. If transient models incorporating the thermal effects are used, compensation may be achieved. Care must be taken to assure

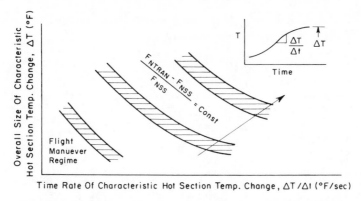

Fig. III.32   Correlation of transient temperature effects on thrust.

that sufficient measurements are made to describe the transient variations to the model. Typical transient math-model inputs include engine fuel-flow rate, engine entry temperature and pressure, ambient pressure, variable geometry positions, and redundant engine cycle parameters for checking test results.

### III.3.5.6   Gas-Path / Nozzle Methods

Gas-path/nozzle methods use engine and nozzle entry measurements along with nozzle coefficients to compute thrust. Over the range of power and flight transients normally encountered, this method is nearly self-compensating for thermal transient effects. The instrumentation used must be of sufficiently high and matched transient response and must be properly sampled to follow the transient.

Errors can be introduced by flow temperature and pressure profiles entering the nozzles that differ from the steady state. Such effects are usually second-order and can be assessed using internal nozzle flow analysis methods. Care must be taken to assure that the specific method used accounts for airflow split between fan and gas-generator flow paths on a transient basis and for transient engine variable geometry features.

### III.3.5.7   Measurement Error from Inadequate Transient Response

All considerations described earlier in this chapter in the discussion on short-duration testing apply to transient measurements for the assessment of unsteady thrust effects in flight. Special attention must be given to assuring: adequate transient, dynamic, or frequency response; matched frequency response; control of extraneous noise; and proper spatial sampling. A representative approach for making a set of ground-test measurements to define engine transient performance is given in Ref. 2.

### III.3.6   Calculational Considerations

This final subsection presents a simplified example for both a low- and high-bypass turbofan engine that illustrates how ground-test data are used to

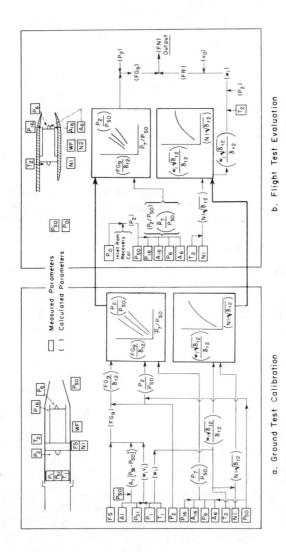

Fig. III.33  Ground-to-flight gross thrust/nozzle pressure ratio correlation process (low-bypass turbofan engine).

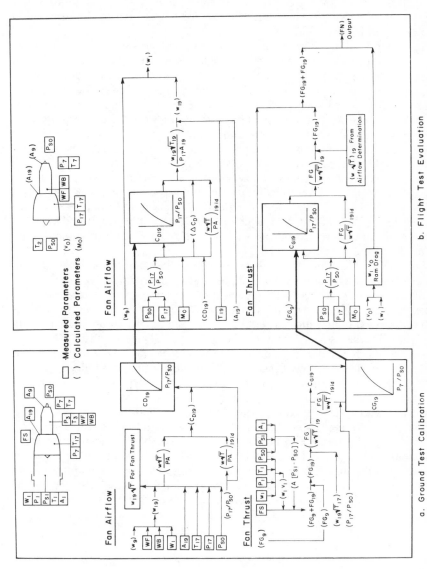

**Fig. III.34  Ground-to-flight gross nozzle method correlation process (high-bypass turbofan engine).**

derive engine performance correlating parameters for in-flight thrust evaluations. For clarity, only the prime parameters used in the ground-to-flight thrust correlation are presented. It is recognized, however, that many additional parameters will be measured for the purpose of cross-checks and to verify engine control limits to schedules. The examples presented in this subsection were obtained from Ref. 12.

Figure III.33 illustrates the gross thrust-nozzle pressure ratio calibration measurement process for a low-bypass, single-exhaust turbofan engine. The fan and core discharge pressures and test cell ambient pressure are used to calculate the nozzle pressure ratio. The engine gross thrust, fuel flow, and airflow are evaluated using the facility measurement techniques described in Sec. III.2. Therefore, for this example, the ground-test facility output is an engine airflow calibration as a function of corrected engine rotor speed, and engine thrust as a function of engine ram pressure ratio and nozzle pressure ratio. In flight, the measurement process is very similar to the ground process except that now the measured inputs are used to back out engine gross thrust and airflow from the ground calibration curves. The engine gross thrust and airflow are then used in turns to calculate the in-flight modified net thrust. Two other pieces of ground-test data, which may or may not be required depending on the application, are a wind-tunnel inlet ram recovery calibration and wind on/off correction factor.

Figure III.34 illustrates the gross thrust and airflow calibration measurement process for a high-bypass, dual-exhaust turbofan engine. The fan nozzle discharge pressure and test cell ambient pressure are used to calculate the fan nozzle pressure ratio. The engine thrust and airflow values are used to derive a fan nozzle velocity and discharge coefficient, respectively. In order to simplify the figure, it was assumed that the core airflow ($W_9$) and thrust ($F_{G9}$) are known inputs. Normally the core airflow would be obtained using a choked high-pressure turbine flow function and the core thrust obtained from a tail-pipe momentum calculation or from corrected model coefficient data. The fan nozzle thrust and mass flow are equal to the difference between the total measured values and the core stream values. In flight, the fan thrust and airflow are determined using the fan nozzle pressure and temperature measurements and nozzle thrust and flow coefficients from the ground tests. The fan coefficients are adjusted for the effect of freestream suppression as required. Ram drag is calculated from inlet total airflow determined by adding the fan and primary nozzle gas flows and adjusting for fuel flow and air bleeds. The core airflow and thrust are calculated in the same manner as in the ground tests.

## References for Chapter III

[1]Oates, G. C., ed., "The Aerothermodynamics of Aircraft Gas Turbine Engines," AFAPL-TR-78-52, July 1978.

[2]Kimzey, W. F. and Wantland, E. C. Jr., "Short Duration Turbine Engine Testing for Energy Conservation," AIAA Paper 77-991, July 1977.

[3]Wazelt, F., ed., "Suitable Averaging Techniques in Non-Uniform Internal Flows,"

AGARD Advisory Rept. 182, June 1983.

[4] McKinney, J. S., "Simulation of a Turbofan Engine," AFAPL-TR-67-125 (Project 3066, Task 306603), Nov. 1967.

[5] Hutcheson, L. C., "Turbine Engine Mathematical Model Validation," AEDC TR-76-90, Dec. 1976.

[6] In-Flight Thrust Measurement Uncertainty, Society of Automotive Engineers, Inc., AIR 1678, July 1985.

[7] Lazalier, G. R., Reynolds, E. C., and Jacox, J. O., "A Gas Path Performance Diagnostic System to Reduce J75-P-17 Engine Overhaul Costs," ASME Paper 78-GT-116, April 1978.

[8] Pryzbylko, S. J., "Control Systems and Transient Simulations," in "A Review of Current and Projected Aspects of Turbine Engine Performance Evaluation," AFAPL-TR-71-34, Feb. 1972.

[9] Sellers, J. F. and Daniele, C. J., "DYNGEN—A Program for Calculating Steady-State and Transient Performance of Turbojet and Turbofan Engines," NASA TN D-7901, April 1975.

[10] Hutcheson, L. C., "AEDC Development and Validation of a J79 Engine Status Deck for the AFFTC Dymotech Program," presented at the 2nd Annual Dynamic Workshop held at AEDC, May 8–9, 1979.

[11] Smith, R. E. Jr. and Wehofer, S., "Measurement of Engine Thrust in Altitude Ground Test Facilities," AIAA Paper 82-0572, presented at the AIAA 12th Aerodynamic Testing Conference, March 22–24, 1982.

[12] In-Flight Thrust Determination, Society of Automotive Engineers, Inc., AIR 1703, Aug. 1985.

[13] Smith, R. E. Jr. and Wehofer, S., "Uncertainty of Turbine Engine Performance Measurements in Altitude Ground Test Facilities," Proceedings of the 29th International Instrumentation Symposium, Vol. 29, 1983.

[14] Kimzey, W. F., "Some Insights into High-Bypass Ratio Turbofan Engine Nozzle Performance from Full-Scale Tests in an Altitude Test Cell," Paper from the AIAA Sixth Propulsion Joint Specialist Conference, San Diego, Calif., June 15–19, 1970. Published by Johns Hopkins University, Applied Physics Laboratory, Silver Springs, Md., June 1970.

[15] Poland, D. T. and Schwanebeck, J. C., "Turbofan Thrust Determination for the C-5A," AIAA Paper 70-611, presented at the AIAA 6th Propulsion Joint Specialist Conference, June 15–19, 1970.

[16] Discussions on unpublished High Bypass Performance Data with C. R. Bartlett, E. Matchett, and M. W. McIlveen, Sverdrup Technology, Inc., AEDC Group, Nov.–Dec. 1983.

[17] Couch, B. D. and Hinds, H. S., "Turbine Engine Performance Correlation Between Three Ground Test Facilities," AEDC-TR-79-70, June 1980.

[18] Godwin, O. D., Frazier, R. D., and Durnin, R. E., "Performance Flight Test Techniques," Edwards Air Force Base, FTC-TIH, 64-2006.

[19] Overall, B. W., "Distortion Screen and Airjet Distortion Generator," AEDC-TR.

[20] Aerospace Recommended Practice, "Gas Turbine Engine Inlet Flow Distortion Guidelines," ARP-1420.

[21] Wehofer, S. et al., "Flight Engine Performance Evaluation for the Air-Launched Cruise Missile," AIAA Paper 77-120, presented at the AIAA Aircraft Systems and Technology Meeting, Aug. 22–24, 1977.

[22] Glawe, G. E., Holanda, R., and Krurese, L. N., "Recovery and Radiation Correction and Time Constants of Several Sizes of Shielded and Unshielded Thermocouple Probes for Measuring Gas Temperature," NASA TP 1099, Jan. 1978.

# Prediction and Verification of Aerodynamic Drag
# Part I:   Prediction

Charles E. Jobe*
*Air Force Wright Aeronautical Laboratories,
Wright Patterson Air Force Base, Ohio*

## IV.1   Introduction

$T$HE state of the art in aircraft drag prediction was defined as of 1973 by an AGARD conference entitled "Aerodynamic Drag."[20] This chapter draws heavily on that work and attempts to update, to the extent possible, those portions relevant to aircraft drag prediction at subsonic, transonic, and supersonic speeds. Special drag prediction problems peculiar to short takeoff and landing (STOL) designs, such as jet flap thrust recovery, and energy-efficient aircraft concepts, such as laminar flow control, are beyond the scope of this chapter, as is drag due to speed brakes and fighter aircraft weapons carriage. Certainly a sound physical understanding of drag will be required before drag prediction becomes a science.

The assessments, projections, and conclusions in this chapter agree, in general, with those of Wood,[161] who briefly addressed four relevant questions:

1) How well do ground-based estimates for drag polars and engine characteristics correlate to flight-test results?
2) In what areas do performance prediction techniques work best or worst, and why?
3) What are the differences in the way various manufacturers (airframe or engine) predict drag polars and engine characteristics?
4) Are there portions of various performance prediction techniques that could or should be combined to form a better or best method?

Wood's answer to question 1 is, "very well for up-and-away flight"; however, his survey of 12 commercial transport aircraft showed that 6 predictions

---

*Technical Manager; currently, Plans Engineer, Aeronautical Systems Division (AFSC), Integrated Plans Group.

were low (by as much as 22%), 4 were high (by up to 10%), and 2 were right on. The second question received a similar answer: predictions are best for up-and-away flight in the clean aircraft configuration. An entire symposium would be needed to address questions 3 and 4 unless the simple answer that no single methodology is best for every situation is accepted. Wood also observed that, "the quality of the estimate is more a function of the time and care taken to include all the details and higher-order terms than it is of the particular equations used."

Williams'[158] synthesis of responses to the AGARD-FMP questionnaire on prediction techniques and flight correlation is also worthy of detailed study. The five questions asked of experts in six NATO countries were:

1) What are the advantages/disadvantages of different prediction techniques?
2) What portions of the flight regime cannot/should not be addressed by ground-based techniques?
3) Are there areas where analytical prediction can be better than wind tunnel and/or simulation results; or vice-versa?
4) Are there methods of reducing differences between prediction and flight-test results?
5) Are there any new prediction techniques that should be emphasized?

Williams' section on background to prediction/design needs and discussion of increasing technical demands is also especially relevant to the understanding of drag prediction.

The initial sections of this chapter describe drag prediction methods typically used to define the most promising configuration for further detailed analysis and wind-tunnel testing. These methods are mainly empirical in origin, although numerical aerodynamics is presently providing timely data for this phase of design in some organizations. The later sections of the chapter describe drag prediction methods based on wind-tunnel results from models specifically constructed for the particular design project and guided by aerodynamic theory, known as wind-tunnel–to–flight correlations. Rooney, Craig, and Lauer[120] and Craig and Reich[34] have recently published detailed accounts of the meticulous testing performed to correlate the wind-tunnel and flight-measured aerodynamic drag of the Tomahawk cruise missile. The size of this missile permitted wind-tunnel testing of the full scale vehicle—yet unexplained discrepancies remained. These correlations and many others for all types of aircraft and missiles form the basis of evaluation for a new design.

Despite many recent advances, it is generally conceded that accurate drag predictions, based entirely on the solution of the equations of motion, or some reduced form thereof, are at present limited to the prediction of drag due to lift in the linear range and supersonic wave drag for a limited class of slender configurations. All other drag predictions ultimately depend on empirical correlations. Progress in computational fluid dynamics or, more generally, numerical aerodynamics, has been tremendous in the last decade and will

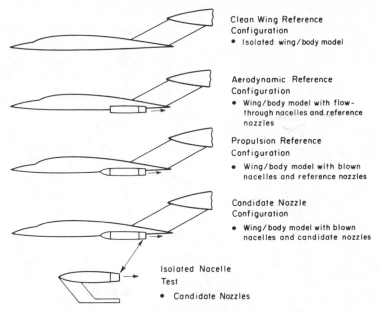

Clean Wing Reference
Configuration
• Isolated wing/body model

Aerodynamic Reference
Configuration
• Wing/body model with flow-
through nacelles and reference
nozzles

Propulsion Reference
Configuration
• Wing/body model with blown
nacelles and reference nozzles

Candidate Nozzle
Configuration
• Wing/body model with blown
nacelles and candidate nozzles

Isolated Nacelle
Test
• Candidate Nozzles

Fig. IV.1  Test approach.

continue into the future due to the introduction of new computers, faster, more accurate solution algorithms, improved resolution of grids, and new turbulence models.[73] However, except for the isolated cases of drag due to lift at small angle of attack and supersonic wave drag for smooth, slender bodies, drag prediction is beyond the capability of current numerical aerodynamic methods.

The aerodynamics of the aerodynamic reference model,[147] usually a subscale wind-tunnel wing-body model with flow-through nacelles and reference nozzles, is considered initially. This is one of many models tested during an aircraft development program (Fig. IV.1) and serves as the basic model for drag prediction. The uncorrected airplane lift, drag, and pitching moment characteristics are derived from tests of this model. Drag corrections to account for the differences between the data from this model and the actual flight vehicle are discussed in the last sections of this chapter. The methods of properly combining these drag data, including thrust effects, are the subject of this book.

### IV.1.1  Drag

The resultant aerodynamic force caused by a flight vehicle's motion with respect to the atmosphere is the summation of the pressure, or normal forces, and the tangential, or skin-friction forces, acting on the vehicle's surface. This resultant force is conventionally resolved into a lift component and a drag component in the vehicle's plane of symmetry. The lift force is the aerodynamic

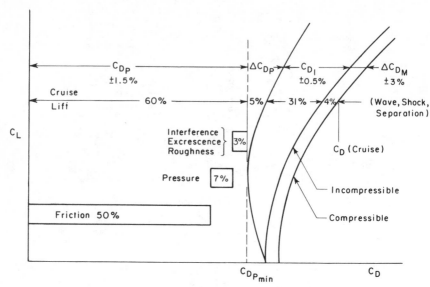

**Fig. IV.2   Transport aircraft drag buildup.**

reaction perpendicular to the freestream velocity direction. Lift is not directly
a subject of this chapter; however, drag, the component of the total force that
opposes motion in the flight-path direction, follows approximately a parabolic
variation with lift for heavier-than-air flight as shown in Fig. IV.2. Thus, the
relative state of motion of the vehicle (equilibrium, or accelerated flight) is
determined by the lift force, the drag force, the vehicle weight, and the
throttle-dependent thrust force.

Most aircraft normally spend the majority of their flight time on nominally
straight, unaccelerated flight paths where all the forces are in static equilibrium. This is the cruise condition that is usually considered a critical
condition for the design of an airplane. This chapter is mainly concerned with
cruise drag prediction (Fig. IV.2) because of the importance of this flight
condition. Drag prediction for accelerated flight (takeoff, landing, maneuver)
will be treated as an important, although ancillary, issue.

The parabolic variation of drag with lift, or angle of attack, appropriately
additive to the zero lift drag, is shown in Fig. IV.2. Drag is produced by the
tangential (skin-friction) and normal (pressure) forces acting on the vehicle's
surface due to relative fluid motion from a basic fluid-mechanics standpoint.
Total drag has been divided into a multitude of various components, depending on the phase of the design process, the aircraft mission and configuration,
the experimental data available, and the persuasion of the drag analyst. A
properly defined, consistent reference condition for the entire aircraft and
propulsion system is crucial to the correct final comparison of theoretical,
empirical, and ground- and flight-test data in the latter design stages for

performance guarantees. The operating reference condition is straightforward for commercial transports with a single known cruise design point; however, military fighter aircraft with multiple design points pose severe additional problems. It is often difficult to define an operating reference condition in this case.

### IV.1.2  Level of Drag Prediction Detail

Three somewhat distinct levels of drag prediction sophistication and reliability are usually described by authors on this subject.[20,106,110,125,167] In reality, drag predictions are updated continuously as data from numerical methods and wind-tunnel tests become available. However, decisions that affect the final design are made based on the best available information at many stages of the design process.

*IV.1.2.1  The Preliminary Design Phase*

This is the beginning of the design process. It has also been called the feasibility[20] and conceptual design phase.[167] Primarily, empirical methods are used to assess the relative merits of many design concepts against the mission specifications generated by an apparent market opportunity or military statement of operational need. Drag prediction error, as measured with respect to future flight-test data, is highest due to method error and geometric uncertainty in the configuration definition[13] (Fig. IV.3). Relative accuracy in the methodology is necessary at this stage in order to select the most efficient configuration to meet the design objectives. Absolute accuracy is desirable in order to give equal consideration to all new concepts and areas of the design space or mission profile. Promising new concepts could be excluded from further consideration at this stage if drag methodology incorrectly predicts that the mission cannot be achieved. An overall drag target or maximum drag level is determined during this design phase. Several compilations of drag prediction methodology are available for rapid performance assessment.[4,63,101,107,122,135] Each has its own peculiarities and limitations. Additionally each airframe manufacturer has compiled drag handbooks that are highly prized and extremely proprietary.

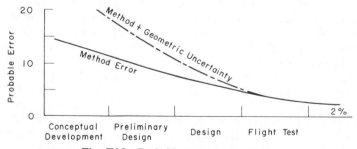

Fig. IV.3  Probable error during design.

*IV.1.2.2   The Detailed Design Phase*

Drag validation becomes a more specialized and detailed continuous process. This phase has also been called the development,[20] preliminary,[124] and project definition[109] phase. Estimates and assumptions are gradually replaced by predictions supported by intensive wind-tunnel test data on the determinate aircraft design, as derived from the clean wing reference configuration and aerodynamic reference configuration models (Fig. IV.1). Drag subtargets for each component of the configuration, consistent with the overall target, are assessed and synthesized. In some development programs, prototype flight-test data provide the basis for diagnostics, reassessments, and analysis if drag improvements are found to be necessary.

*IV.1.2.3   The Final Design Phase*

Production aircraft performance guarantees with error bounds are made. This is also the preproduction[20] or wind-tunnel–to–flight-correlation[104] phase. A detailed analysis and interpretation of the prototype test performance is conducted. The drag predictive model is calibrated by reference to the prototype data. An excellent description of this process and its difficulties has been published by Rooney and Craig[119] and Craig and Reich[34] for the Tomahawk missile.

Theory, empirical methods, and ground- and flight-test measurements are continuously intertwined throughout the design, development, and production phases of an aircraft's life. Each drag prediction method contains its unique strengths and weaknesses.

Strictly empirical methods, conventionally based on flight-test data, include the real aircraft effects of flow separations, Reynolds number, gaps, steps, protuberances, rigging, etc., but are not readily generalized. Problems have even arisen when these methods are applied to aircraft of the same family from which the data base was generated.[158]

Current theoretical methods that utilize large computers are capable of immediate generalization to almost any configuration but are limited in accuracy because of the lack of real aircraft effects. Prediction of the drag due to lift is usually quite accurate up to drag levels where flow separations become significant. Prediction of the zero lift drag remains empirical.

Progressive wind-tunnel development programs contain a greater degree of reality than theoretical methods, lacking some real aircraft effects, but are costly and time-consuming. Correct interpretation of the data remains an art, particularly when wall effects, Reynolds number scaling, and propulsion system model data are included. Systematic drag measurement errors greater than 1% cannot be tolerated if drag prediction methods are to be validated.[19]

A typical breakdown of flight-test data is shown in Fig. IV.2. From these data, drag correlations for each drag component are made and drag design charts are constructed as a basis for the preliminary design empirical method. The preliminary design study is considered as the reference point for judgment on program success. The initial decisions on wing area, aspect ratio, thickness ratio and matching engine cycle and thrust requirements are based on early lift-to-drag ($L/D$) estimates. A lower bound for $L/D$ must be estimated in

Table IV.1   An aerodynamicist's rating

| $L/D$ level achieved | $M_{DD}$ | Rating |
|---|---|---|
| ± 3% | ± 0.002 | Amazing |
| ± 5% | ± 0.004 | Very good |
| ± 7% | ± 0.006 | Average |
| ±10% | ± 0.010 | Below average |

Table IV.2   Drag scatter

| Aircraft | Flight-test data scatter | Ref. |
|---|---|---|
| B-747 | ± 1% (drag) | 13 |
| C-5A | ± 3.5% ($C_{D_p}$)±2% (drag) | 86 |
| C-141 | ± 3.3% (drag) | 86 |
| B-707 | ± 5% (drag) | 13 |
| DC-10-30 | ± 3.5% (range factor) | 44 |
| F-14 | ± 1% ($C_{D_0}$) | 118 |
| YF-16 | ± 5% (drag) | 18 |
| F-16 | ±12% ($C_{D_{min}}$) | 152 |
| F-18 | ± 5% (drag) | * |
| Tomahawk | ± 1.5% (drag) | 34 |
| Tornado | ± 3% (drag)±1% ($C_{D_0}$) | 71 |
| Alpha-Jet | ± 5% (drag) | 124 |

*See afterburning turbofan engine example, Chap. 7 of R. B. Abernethy et al., "In-flight Thrust Determination," Society of Automotive Engineers, AIR 1703, 1985.

order to assure that the propulsion system will provide the power necessary at the airplane critical flight condition. Conservatism at this design stage will result in an inefficient engine and a noncompetitive airplane.

The designer encounters errors from two sources at this stage: errors due to lack of configuration, geometric definition, and errors due to inaccurate methods. These errors have been estimated by Bowes[13] for long-range subsonic transports and are summarized in Fig. IV.3. He also suggests an aerodynamicist's rating scale (see Table IV.1), where $M_{DD}$ is the drag divergence Mach number. The need for an accurate force and moment model as early as possible in the design cycle is clearly evident from the table. Most commercial aircraft flying today fit within this table; however, military programs have been far less successful in achieving these accuracy levels. This may be due in part to the less precise flight-test data recorded from the older military aircraft as shown in Table IV.2.

Table IV.2 should not imply that the data scatter is constant for a given aircraft throughout its flight envelope. The scatter stated is representative of the cruise condition. Figure IV.4, from Arnaiz'[3] study of the XB-70 flight-test data, clearly demonstrates this point by showing the actual number of

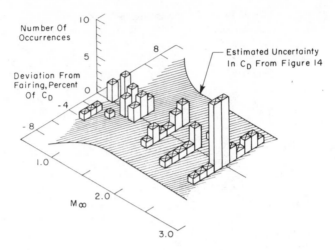

**Fig. IV.4    Drag data scatter about drag polar fairings with Mach number.**

occurrences of scatter and the estimated scatter plotted against Mach number. The drag uncertainty estimates were obtained by combining many individual errors in weight, Mach number, static pressure, angle of attack, thrust, acceleration, etc., in a root-sum-square calculation.

Performance estimates for military aircraft are more accurate than Table IV.2 indicates since the major problems occur when one attempts to isolate the thrust component and the drag component from the measured performance data. Current steady-state, level flight-test data indicate that a repeatability of $\pm 1\%$ can be achieved if meticulous care is exercised during the engine calibration and flight-test processes. This level of accuracy must be attained[159] in order to obtain useful correlations for the smaller drag components shown in Fig. IV.2. These component correlations, necessary to develop accurate prediction methods, are based on convention, tradition, and aerodynamic theory.

Flight-test data are, however, not presently capable of resolving drag into components. The accuracy of each component cannot be checked since only total drag, deduced from performance measurements, is available. Additional instrumentation that could resolve drag into components is superfluous to aircraft performance guarantees and is usually not included on the flight-test aircraft because of the additional cost to the development program.

The relative size of each drag component varies with aircraft type and flight condition. Representative examples of the cruise drag and takeoff drag breakdowns for four different aircraft types are shown in Fig. IV.5 from Butler.[19] The drag breakdown also varies with aircraft range and purpose (military or civilian transport) within each aircraft type.

The conventional approach to drag estimation is to estimate and sum the zero lift drag of each major physical component of the aircraft, with allowances for interference effects and other small contributions, and then add

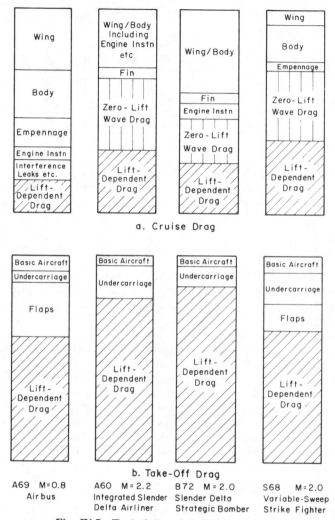

a. Cruise Drag

b. Take-Off Drag

| A69  M=0.8 | A60  M=2.2 | B72  M=2.0 | S68  M=2.0 |
| Airbus | Integrated Slender | Slender Delta | Variable-Sweep |
| | Delta Airliner | Strategic Bomber | Strike Fighter |

Fig. IV.5   Typical aircraft drag breakdowns.

the drag due to lift. Their summation is intrinsically inaccurate since aircraft components are mounted at incidence to each other so that each is not at zero lift simultaneously. Butler[19] suggests that this classification of aircraft drag into components independent of, and dependent on, lift should be rejected and drag could be estimated by compounding elements arising from different basic causes associated with fluid dynamics as shown in Fig. IV.6.

Currently each manufacturer has a methodology, or drag handbook, by which a thorough drag estimate is made. There is no single industrywide drag handbook. Manufacturers may even use several drag accounting breakdowns (depending on the customer) for the same airplane. There exist several basic

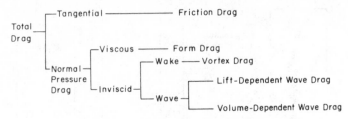

Fig. IV.6   Components of aircraft drag.

sources for fundamental drag level prediction; however, total drag comparisons, between prediction and test, require accurate flight data and expert analysis and are highly proprietary to each manufacturer.[161]

## IV.2   Subsonic Drag

The aircraft drag at subsonic speeds, here taken to be less than the drag divergence Mach number, is traditionally divided into lift-induced drag and minimum drag. Minimum drag is further divided into friction drag, profile drag, and interference drag.

The accuracy of the drag divergence Mach number, $M_{DD}$, prediction varies widely with configuration. Rooney and Craig's[119] detailed assessment of the Tomahawk flight-test data indicates that $M_{DD}$ cannot be accurately predicted, while Henne, Dahlin, and Peavey[57] conclude that a substantial capability exists for the determination of drag divergence Mach number for transport-type geometries. The determination of drag levels and increments is rated marginal. McGeer and Shevell[94] were able to correlate transonic drag rise data for peaky-type airfoils on older Douglas transport aircraft wings. Their correlation was not accurate, however, for the Boeing 747 or the F-111A/TACT aircraft.

### IV.2.1   Empirical Correlations (C. R. James)

In the preliminary design phase, gross or overall correlations are used to estimate the size and geometric features of the aircraft. These correlations are traditionally based on wind-tunnel and flight-test data from older, successful aircraft. The key to this method is to obtain sufficient relevant data from aircraft of the same general type (bomber, transport, or fighter, etc.) being designed in order to form trends and extrapolations with confidence. Data collections of this type are continuously sought and maintained within the preliminary design groups of the aircraft industry. The data from a single airframe manufacturer for a particular series of aircraft similar to the new design usually produce the most accurate drag prediction.

Correlations of flight-test data based on aircraft geometry provide insight as to the relative importance of factors influencing subsonic minimum drag. This section treats two types of aircraft: bomber/transport and fighter/attack. Minimum drag is related to wetted area through the use of an equivalent

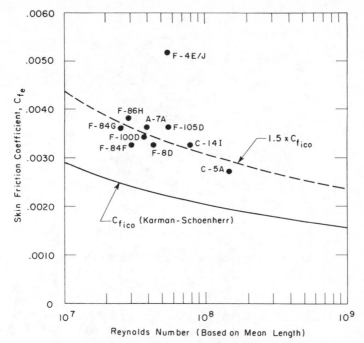

**Fig. IV.7   Equivalent skin friction.**

skin-friction coefficient, $C_{fe}$. Figure IV.7 compares flight-derived values of $C_{fe}$ with incompressible flat-plate turbulent flow skin-friction coefficients for jet aircraft of both types. This comparison shows that, in general, about two-thirds of subsonic minimum drag may be attributed to skin friction. The balance is due to form drag and interference. Form drag results from flow over curved surfaces and flow separation induced by viscous effects. Interference drag is the result of mutual interaction of the flowfields developed by the major configuration components. The equivalent skin-friction coefficient, $C_{fe}$, is a convenient method to relate total drag (including form and interference drag) to wetted area:

$$C_{fe} = D/qS_{wet}$$

where $D$ = minimum drag, lb; $q$ = dynamic pressure, lb/ft$^2$; and $S_{wet}$ = total wetted area, ft$^2$. The quantity $D/q$ is defined as equivalent parasite area $f$.

*IV.2.1.1   Bomber/Transport Aircraft*

The relationship of $C_{fe}$ and $f$ to wetted area is illustrated by Fig. IV.8. Experimental points derived from flight tests of several transports and bombers are indicated. Wetted areas vary from approximately 5800 ft$^2$ for the Breguet 941 to 33,000 ft$^2$ for the C-5A. Equivalent skin-friction coefficients

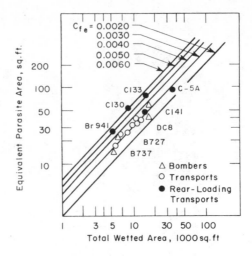

**Fig. IV.8   Subsonic parasite drag—bomber/transport.**

vary from 0.0027 for the C-5A to 0.0060 for the C-130. The variation in $C_{fe}$ reflects factors that influence form and interference drag as well as Reynolds number effects resulting from variations in aircraft size. For example, the three turboprop military transports (Br-941, C-130, and C-133) exhibit values of $C_{fe}$ in the 0.0050–0.0060 range. These aircraft have upswept afterbodies characteristic of rear loading arrangements and consequently incur form drag penalties which increase $C_{fe}$. Interference effects of nacelles and propeller slipstreams also contribute to the higher drag levels. Jet transports (737, 727, and DC-8), with a higher degree of aerodynamic cleanness, fall in the 0.0030 range. The low $C_{fe}$ of the C-5A is due to its aerodynamic cleanness and very low aftbody drag increment, despite its military upsweep.

### IV.2.1.2   Fighter/Attack Aircraft

Correlation of flight-test data of fighter/attack aircraft provides additional insight as to the relative importance of factors influencing subsonic minimum drag. This section illustrates effects of four parameters: wetted area, relative wing size, fuselage shape, and aerodynamic cleanness. Correlations were developed from flight tests of ten jet fighters. Table IV.3 lists the aircraft and some key parameters. Wetted areas vary from slightly over 1100 ft$^2$ for the F-84G to approximately 3100 ft$^2$ for the F-14. Equivalent fuselage fineness ratios vary from about 4.5 for the F-84H to 9.3 for the F-18. These aircraft have single engines with the exception of the F-4E/J, F-14, and F-18. The relationship among $f$, $C_{fe}$, and wetted area is illustrated in Fig. IV.9. Experimental points derived from flight tests of the aircraft listed in Table IV.3 are indicated in this figure. Equivalent skin-friction coefficients vary from 0.0032 for the F-8D to 0.0052 for the F-4E/J. The variation in $C_{fe}$ results primarily from factors that influence form and interference drag. Significant factors are: 1) relative wing

Table IV.3   Aircraft characteristics, fighter/attack geometry

| Model designation | Areas, ft$^2$ | | Equivalent fuselage fineness ratio |
| | Wing reference | Total wetted | |
|---|---|---|---|
| A-7A | 375 | 1691 | 6.43 |
| F-4E/J | 530 | 2092 | 5.88 |
| F-8D | 375 | 1821 | 8.16 |
| F-84F | 325 | 1257 | 5.92 |
| F-84G | 260 | 1104 | 5.08 |
| F-86H | 313 | 1186 | 4.57 |
| F-100D | 400 | 1571 | 6.61 |
| F-105D | 385 | 1907 | 8.86 |
| F-14 | 565 | 3097 | 8.22 |
| F-18 | 400 | 2046 | 9.28 |

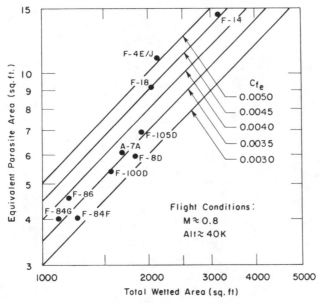

Fig. IV.9   Subsonic parasite drag—fighter/attack.

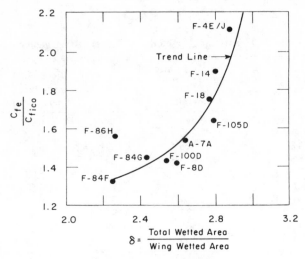

**Fig. IV.10   Relative wing size.**

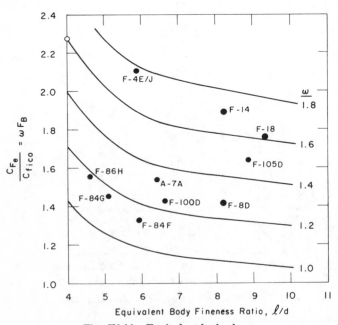

**Fig. IV.11   Equivalent body shape.**

size, 2) fuselage shape, and 3) aerodynamic cleanness. Figures 10–13 illustrate the effects of these factors.

The effect of relative wing size is illustrated in Fig. IV.10. The ratio of the equivalent skin-friction coefficient, $C_{fe}$, to the incompressible mean flat-plate value, $C_{f_{ico}}$, is shown as a function of $\delta$, the ratio of the total wetted area to wing wetted area. The increase in this ratio with increasing $\delta$ reflects the tendency for wing surface tolerances normally to be held much tighter than other components, thus protuberance, roughness, and leakage penalties are smaller for the wing than for other components.

An important factor affecting fuselage drag is the degree of pressure or form drag developed due to flow separation induced by viscous effects. In general, flow separation is more pronounced with bodies of low fineness ratio. This effect is illustrated in Fig. IV.11 by defining an equivalent body which includes

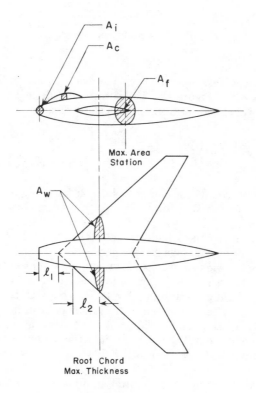

$$\left.\frac{\ell}{d}\right)_{equiv} = 2 \times \ell_f \Big/ \sqrt{\frac{4 \cdot A_{eq}}{\pi}}$$

$$\ell_f = \ell_1 + \ell_2$$

$$A_{eq} = A_f + A_c + A_w - A_i$$

Fig. IV.12   Equivalent body concept.

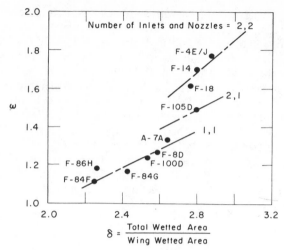

**Fig. IV.13   Aerodynamic cleanness.**

frontal areas of the fuselage, canopy, and wing to establish an equivalent fuselage fineness ratio. Figure IV.12 illustrates and defines this parameter. A form factor, $F_B$, is defined from the empirical expression

$$F_B = 1.02\left[1 + \frac{1.5}{(\ell/d)^{1.5}} + \frac{7}{(\ell/d)^3(1 - M^3)^{0.6}}\right]$$

Mach number $M$ accounts for compressibility effects; Fig. IV.11 also includes an aerodynamic cleanness factor, $\omega$, which is related empirically to relative wing size (Fig. IV.13).

The aerodynamic cleanness factor accounts for form drag and interference effects not treated by fuselage shape parameters. Examples are form and interference drag for wings and tails, protuberance and leakage drag, manufacturing tolerance effects, and surface roughness. Figure IV.13 illustrates the magnitude of $\omega$ and the variation with relative wing size. Separate trend lines are identified as a function of the number of inlet and exit flows. The lower line is determined by the single-inlet/single-nozzle configurations; the middle line by the data from the F-105D with two inlets and one exit nozzle; and the upper line by the data from the F-4E/J, F-14, and F-18 with two inlets and two nozzles.

These correlations suggest an initial approach for estimating subsonic minimum drag for similar fighter/attack configurations:

1) Calculate the wetted area and a characteristic length for each component.

2) Calculate an area-weighted mean length $l_m$ for the complete configuration, or estimate $l_m$ from experience.

3) Use this mean length to calculate Reynolds number at the flight condition to be evaluated.

4) Determine an incompressible skin-friction coefficient, $C_{f_{ico}}$, using this Reynolds number. The von Kármán-Schoenherr equation for turbulent flow or a similar expression may be used.

5) Calculate the wetted area ratio $\delta$, and use Fig. IV.13 to establish the value of $\omega$ for the appropriate inlet/nozzle arrangement.

6) Calculate the equivalent fuselage fineness ratio $l/d$, and determine the form factor $F_B$.

7) Determine equivalent parasite area $f$ from the expression:

$$f = C_{f_{ico}} \times \omega \times F_B \times S_{wet}$$

A similar approach developed for bomber/transports would highlight these factors influencing subsonic minimum drag for this type of aircraft.

### IV.2.2  Detailed Drag Estimates—Component Buildup

As the design project progresses, the geometry of the aircraft is better defined, and more detailed drag prediction methods are used to assess drag subtargets or design goals for each aircraft component. These goals must be consistent with the overall drag target previously determined since the total configuration drag is the sum of the component drags including interference effects.

Each component will have an initial run of laminar flow, followed by a transition zone and fully developed turbulent flow thereafter. The extent of each flow region is difficult to determine with precision. Fully turbulent flow was usually assumed from the leading edge aft on full scale flight vehicles at altitudes below about 20 km. The smoothness and rigidity of newer composite materials may provide longer runs of laminar flow on future flight vehicles. Transition is usually fixed by trip strips or grit on subscale wind-tunnel models that are tested at lower Reynolds numbers. The positioning and selection of boundary-layer trip strips that accurately simulate full scale conditions is an art itself.[8,14,15]

#### IV.2.2.1  Friction Drag

There is little controversy concerning calculation of the average laminar skin friction on one side of a doubly infinite flat plate. It is given by Blasius'[9] formula from his exact solution to the laminar boundary-layer equations for zero pressure gradient

$$C_f = 1.32824/R_e^{\frac{1}{2}}$$

where $R_e = U_\infty \ell / \nu$ and $\ell$ is the distance from the stagnation point to the transition zone.

Transition takes place when the length Reynolds number nominally exceeds $3.5 \times 10^5$ to $10^6$ for a flat plate.[123] This is the critical Reynolds number, often assumed to be 0.5 million for many flows, although pressure gradients strongly

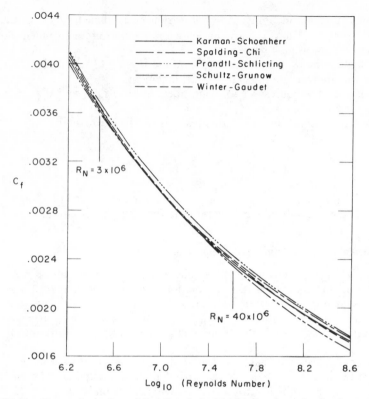

**Fig. IV.14  Comparison of empirical flat-plate skin friction formulae for incompressible turbulent flow.**

influence the location of transition in the boundary layer. It is the objective of boundary-layer stability theory to predict the value of the critical Reynolds number for a prescribed main flow. Success in the calculation of the critical Reynolds number for flows of general aeronautical interest has eluded workers in this field for many decades despite many dedicated efforts.

Many correlations of turbulent boundary-layer data exist. Figure IV.14 displays the variation in $C_f$ with Reynolds number predicted by five different methods. Significant changes in $C_f$ can occur when scaling from wind tunnel ($R_n = 3 \times 10^6$) to flight ($R_n = 40 \times 10^6$) if different correlations are used. Two well-known and widely used formulas are Prandtl-Schlichting,

$$C_f = \left[ 0.455/(\log R_e)^{2.58} \right] - (A/R_e)$$

where $A$ depends on the position of transition, and von Kármán-Schoenherr,

$$C_f^{-\frac{1}{2}} = 4.13 \log(R_e C_f)$$

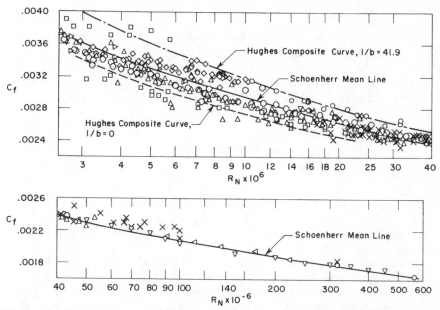

**Fig. IV.15  Summary of experimental research on flat-plate skin friction, incompressible speeds.**

The Prandtl-Schlichting formula, being explicit in $C_f$, has been more widely used in the past, although Paterson[108] has shown that the von Kármán-Schoenherr formula is a good representation of existing test data (Fig. IV.15). Both formulas have been used as the basis for aircraft drag prediction methods.[63,107,135] A new method based on explicit Prandtl-Schlichting-type relations has been developed by White and Christoph[153,154] for compressible turbulent skin friction. Their explicit approximation

$$C_f = 0.42 / \ell n^2 ( 0.056 R_e )$$

is accurate to $\pm 4\%$ in the Reynolds number range $10^5$–$10^9$. It is the formula selected by Schemensky[122] for his complete drag prediction method.

The reference length used in the Reynolds number computation is the body length for near bodies of revolution, and MacWilkinson, Blackerby and Paterson[86] recommend use of the mean aerodynamic chord for lifting surfaces. This choice of reference length will be less accurate (about 2% low) for highly swept delta wings. A strip method, where a reference length is calculated for each streamwise strip of the wing, should be used.[63]

Compressibility corrections to the skin friction are less than 10% at Mach numbers less than 1 and are often ignored at subsonic speeds unless continuity with supersonic predictions is desired.

*IV.2.2.2   Form Drag*

Subsonic minimum drag is the sum of the friction, form, and interference drag when the component buildup method is used. The form drag, or pressure drag, resulting from the effects of nonzero pressure gradient (component thickness) is usually accounted for by a multiplicative factor applied to the skin-friction drag. Following Schemensky's drag accounting system, for example,[122]

$$C_D = \sum_{\text{components}} \left( C_f * \frac{A_{\text{wet}}}{S_{\text{ref}}} * \text{FF} * \text{IF} \right) + C_{D_{\text{camber}}} + C_{D_{\text{base}}} + C_{D_{\text{misc}}} \qquad \text{(IV.1)}$$

where $A_{\text{wet}}$ is the component wetted area; FF is the component form factor, also called the shape factor, SF; and IF is the component interference factor. The interference, camber, base, and miscellaneous drags are treated in later sections.

The prediction of the form drag, or FF, is dependent on empirical correlations as given by Hoerner,[64] Schemensky,[122] DATCOM,[63] O'Conner,[107] Snodgrass,[135] MacWilkinson, Blackerby, and Paterson,[86] and others. Example formulas for the computation of FF for bodies and for nacelles, respectively, are:

$$\text{FF} = 1 + 60/\text{FR}^3 + 0.0025 * \text{FR}$$

$$\text{FF} = 1 + 0.35/\text{FR}$$

where FR = component length$/\sqrt{\text{width} * \text{height}}$, and for airfoils:

$$\text{FF} = 1 + K_1(t/c) + K_2(t/c)^2 + K_3(t/c)^4$$

where $t/c$ is the streamwise thickness-to-chord ratio and the constants $K_1$, $K_2$, and $K_3$ depend on airfoil series. An additional term, dependent on the section design camber, is added for supercritical airfoils.

MacWilkinson, Blackerby, and Paterson[86] have shown that fuselage profile drag (the sum of skin-friction drag and form drag) can be correlated for bodies of revolution by the method of Young[169] (Fig. IV.16). This theory agreed with experimental data from four transport fuselages that are radical departures from the ideal body of revolution, if the proper fineness ratio is used as shown in Fig. IV.17a. However, excess drag due to upsweep,[93] which may amount to 10% of the total cruise drag of military transport aircraft, is not apparent from these correlations. A new prediction method for reasonable estimates of the drag of afterbodies for military airlifters has recently been published by Kolesar and May[72] to replace the method of Hahn et al.[50]

Methods of estimating airfoil shape factors (or form factors) were also compared by MacWilkinson, Blackerby, and Paterson.[86] There is wide scatter

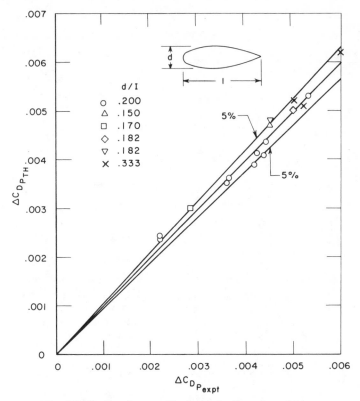

**Fig. IV.16   Fuselage profile drag—bodies of revolution.**

in both the theoretical and experimental values of the shape factor, as shown in Fig. IV.17b, particularly at 12% thickness-to-chord ratio, a nominal value for modern transport wings. The data on the earlier NACA airfoils were not suited to accurate assessment of airfoil form drag because of the testing techniques used. These earlier data, however, form the basis of many older shape factor correlations.

Formulas for computing the form factors for other aircraft components, canopies, stores, struts, etc., are available from Hoerner,[64] Schemensky,[122] DATCOM,[63] O'Conner,[107] and Snodgrass,[135] along with examples illustrating their use for drag estimation.

Formally, the form drag follows the same dependence on Reynolds number and Mach number as the skin friction since the form factors are constant for a given geometry in most drag prediction methods. Hoerner[64] has developed compressible form factors that explicitly account for Mach number effects by transforming the appropriate thickness or slenderness ratio using the Prandtl-Glauert transformation factor $(1 - M^2)^{\frac{1}{2}}$. The magnitude of these corrections, however, precludes verification of their accuracy due to lack of precision in experimental data.

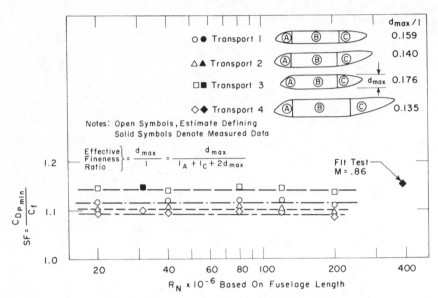

Fig. IV.17a    Fuselage shape factors $M \leq 0.700$.

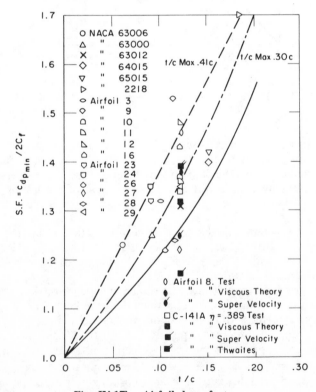

Fig. IV.17b    Airfoil shape factors.

*IV.2.2.3   Interference Drag*

Aerodynamic interference in aircraft is the change in flow over given components of the configuration due to the presence of one or more of the other components. Interference can be unfavorable with an attendant drag increase or, in the case of skillful design, favorable where the sum of component drags is greater than the total drag of the configuration. The importance of interference to the optimization of aircraft to meet performance requirements was delineated at a recent AGARD conference.[167] Generally, important interference effects involve fluid-dynamic phenomena far too complex to be analyzed by existing computational methods.

Interference factors, to account for the mutual interference between aircraft components, are given constants based on experimental evidence for the smaller components and are usually presented as plotted data for the major components. Representative examples can be found in Schemensky[122] and DATCOM.[63]

Interference factors, unlike the form factors, are correlated in terms of Mach and Reynolds numbers for the major aircraft components. These factors are usually greater than unity at flight Reynolds numbers since the resultant total drag is usually larger than the sum of the component drags when tested singly. The factors also increase with Mach number to account for compressibility effects. The interference factors for the smaller components are generally considered constants.

*IV.2.2.4   Camber Drag*

This increment to the minimum drag accounts for camber and twist in the wing and, in some drag accounting systems, for the fact that all aircraft components are not mounted relative to each other to attain zero lift simultaneously. This increment is also related to the drag-due-to-lift methodology through the drag accounting. A correlation consistent with Eq. (IV.1) is

$$C_{D_{\text{camber}}} = \frac{1}{1-e} K(C_L)^2$$

where

$e = 1/(\pi * \!\!A\!\!R * K)$ span efficiency factor
$K$ = induced drag factor
$C_L$ = a constant (the lift coefficient for a minimum drag, often the design lift coefficient)

The factors $e$ and $K$ are determined from the variation of drag with lift, to be considered later.

A different drag predicting method has been shown in Fig. IV.2. In that system all changes in profile drag with lift are derived from experimental data correlations, and the basic profile drag is not incremented for camber directly. The method of Fig. IV.2 is most accurate when the aircraft designs are closely related members of the same series, for example, transports from a single manufacturer.

Thus, drag data reduced using one bookkeeping system may not include all the interference, camber, etc., effects in the resulting correlations that the drag buildup method assumes are included, unless extreme care is taken to achieve consistency.

*IV.2.2.5   Base Drag*

This term is a weak function of Mach number at low speed and is given by Hoerner[64] as

$$C_{D_{base}} = ( 0.1 + 0.1222M^8 ) ( S_{base} / S_{ref} )$$

It obviously gains significance with increasing Mach number; however, it is independent of lift and is often not correlated separately from the body drag. Power effects on base drag are discussed in Chap. V.

*IV.2.2.6   Miscellaneous Drag*

This term accounts for the differences between the corrected aerodynamic reference model drag and the full scale airplane drag due to surface irregularities, such as gaps, mismatches, fasteners and small protuberances,[23] and leakage due to pressurization. It is estimated, from experience and aircraft type, as some percentage of the total friction, form, and interference drags in the preliminary design stages. In the later design stages, when configuration details are known, each of these small drag terms may be accounted for separately. An AGARD conference,[170] devoted to this subject, was held in 1981. The transfer of wind-tunnel data to full scale flight conditions is discussed subsequently in this chapter under wind-tunnel/flight correlation of lift, drag, and pitching moment.

It would be highly desirable to assess the credibility of each term in the drag buildup equation. However, for complete aircraft configurations this would require the ability to determine accurately the variation, with Reynolds and Mach numbers, of the drag forces due to viscous and pressure effects of each aircraft component—an intractable task.

*IV.2.2.7   Drag Due to Lift*

Various methods have been used to correlate subsonic drag due to lift, or induced drag, depending upon the design project phase, the configuration, the anticipated range of angle of attack, and the sophistication of the prediction method.

The complexity of drag-due-to-lift prediction methodology is illustrated in Fig. IV.18, from Schemensky,[122] based on the extensive systematic correlations of Simon et al.[131] The drag is predicted in each of the regions, 1 through 7, by modified or distinct equations and, perhaps, additional correlations (terms) to account for increasing complexity in the flowfield. The following section describes drag calculation in regions 1, 5, and 6 below the critical Mach number $M_{CR}$. The detailed drag methodology is contained in Schemensky's report.[122]

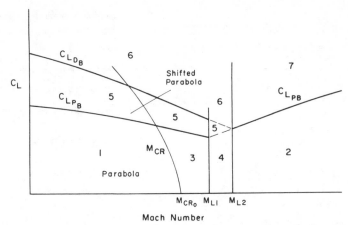

Fig. IV.18   Lift and speed regions for calculations of drag due to lift.

Classical aerodynamic theory predicts that the drag due to lift is a function of the vortex distribution shed behind the wing (downwash). It is parabolic with lift and given by

$$C_{D_i} = C_L^2/\pi R$$

for an elliptic lift distribution spanwise. Wing taper and sweep produce nonelliptic loadings; thus

$$C_{D_i} = C_L^2/\pi R e'$$

where $e' < 1.0$ is the wing alone efficiency factor. An empirical modification to $e'$ to account for the presence of the fuselage in the spanwise lift distribution results in

$$C_{D_i} = C_L^2/\pi R e = N C_L^2$$

where

$$e = e'\left[1 - (d/b)^2\right]$$

and $d/b$ is the body diameter-to-wingspan ratio. The effects of camber are not included in these equations. Camber causes a shift in the polar so that minimum drag occurs at a finite lift coefficient, $C_{L_0}$ or $\Delta C_L$, depending on the author of the method. Simon et al.[131] illustrated this camber effect and the nonparabolic nature of the "drag polar" at higher lift coefficients where flow separation effects become important, as shown in Fig. IV.19.

Simon et al.[131] found that a further empirical modification to this equation was necessary to account for local supervelocities at the wing leading edge,

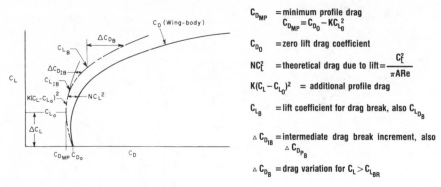

Fig. IV.19   Subsonic drag polar.

while Schemensky,[122] using the camber drag term, found that

$$K = \frac{1-R}{C_L} + \frac{R}{\pi \! \mathcal{R} e}$$

Here $R$ is the leading-edge suction parameter and properly bounds $K$ between $1/\pi \! \mathcal{R} e_0$ for full leading-edge suction, or fully attached flow ($R=1$) and the upper bound of drag, $1/C_{L_\alpha}$, for zero leading-edge suction, fully separated flow, where $C_D = C_L \tan \alpha$. See also Sec. IV.4.3. The suction parameter $R$ is significantly affected by airfoil camber, conical camber, leading-edge radius, Reynolds number, and sweep. The proper determination of $R$ is an involved procedure.[122]

It is within region 1, bounded by the onset of leading-edge separation with reattachment, i.e., the $C_L$ for initial polar break boundary, and the critical Mach number boundary, that surface paneling methods are accurate and can be applied. These methods are all based on the Prandtl-Glauert equation, a linearized form of the complete fluid-dynamics equations of motion, and can predict the lift coefficient for minimum drag and the variation of drag with lift for complex, complete aircraft configurations quite accurately.

Four panel methods were systematically evaluated in 1976 at an AGARD meeting.[140] However, many new methods are currently in use; for example, DOUGLAS-HESS,[58,59] MCAERO,[16] PAN AIR,[26,39,130,146] QUADPAN,[171] USSAERO,[41,165] and VSAERO.[38,88] Each has its particular advantages and limitations[143] in addition to those due to the Prandtl-Glauert equation because of the particular numerical method used. An evaluation of these six newer methods against the datum cases of Ref. 140 would be enlightening. Drag prediction, not specifically considered during that meeting, should be added as an evaluation criterion.

Basically all newer panel methods use Green's theorem and assume an algebraic form for the source and doublet singularity strengths on each of the panels used to represent the aircraft's surface. Higher-order methods[16,26,59] use polynomials to represent the singularities while lower-order methods[38,41,171] assume that the singularities are constant (or vary linearly in one direction) over each panel. The accuracy of each type of representation increases as the paneling density is increased; however, so does the computer time and cost, thus forcing the usual tradeoff between accuracy, schedule, and cost.

Panel methods have primarily been used as flow diagnostic tools,[31,146] by performing systematic comparative studies of alternative geometries and noting regions of high velocity or velocity gradient for further detailed analysis.

The proper application of any panel method remains an art due to the myriad of paneling layouts that can be used to represent the configurations,[26,149] the impact of various types of boundary conditions,[28,142] and the many numerical ways of enforcing the Kutta condition.[59,143,146,171] The calculation of the induced drag from the velocity potential also presents problems. The near-field summation of pressure multiplied by the area projected in the drag direction results in differences of large numbers and a loss of significant figures. Control volume[26] and the associated far-field or Trefftz plane wake methods[68,74] for drag calculations also present conceptual and numerical problems when applied to complete wing, fuselage, and tail-plane configurations. The optimum panel method primarily depends upon the skill of the user, the computational resources available, the geometric configuration to be analyzed, and the type and accuracy of the results desired.

Typically drag prediction results are not published in the open literature; however, two examples have been found to demonstrate the induced drag prediction capabilities of panel methods.

The first, from Miller and Youngblood,[98] is shown in Fig. IV.20. The intricate paneling required for the advanced fighter configuration is shown, along with the drag polars predicted using PAN AIR[130] and a lower-order method, USSAERO-B.[165] The minimum drag was predicted using the empirical methods of the previous sections. The comparison with experimental data is good, with the results favoring use of the higher-order PAN AIR method; however, the computer resources necessary to achieve the increased accuracy were considerably higher. Modeling studies, particularly with respect to the placement of the canard wake, were also necessary to achieve these results.[98]

Figure IV.21, the second example, from Chen and Tinoco,[31] demonstrates the versatility of the PAN AIR method in predicting the effects of engine power on the lift and drag of a transport aircraft wing, body, strut, and nacelle wind-tunnel model. The computed lift-and-drag increments on various components of the aircraft due to increasing the fan nozzle pressure ratio from "ram" to "cruise" shows that the wing, the fuselage, and the strut all had favorable contributions, while the nacelle was the only component that contributed to a drag penalty. When the lost lift was restored and the associated induced drag was included, the computed blowing drag was 1.3 drag counts. About two drag counts were measured in the wind tunnel.

Additional applications of panel methods include: fighter aircraft with externally carried stores[28] (mutual interference effects), transport aircraft with flaps deployed,[102] and separated flow modeling using free vortex sheets.[88] Panel methods have also been used with boundary-layer calculation

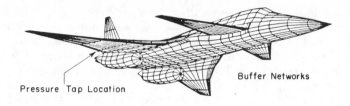

Pressure Tap Location

Buffer Networks

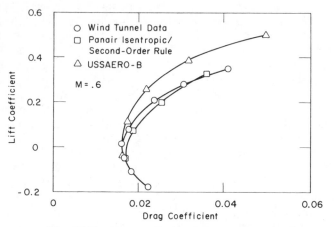

Fig. IV.20   Drag predictions using panel methods.

methods[31,70] to simulate viscous displacement effects and the parabolized Navier-Stokes equations.[111] These hybrid methods, designed to conserve computer resources by using only the equations necessary to analyze the flow in each appropriate zone, have had limited success in predicting the fluid velocities and pressure[111] and lift loss due to viscous effects.[70] Drag prediction is presently well beyond their grasp.

Drag-due-to-lift prediction in regions 5 and 6 of Fig. IV.18 is an empirical art, and it is likely to be many years before satisfactory physical and mathematical models are developed for the complex separated flows that occur in these regions.[61] For example, the lift coefficient for polar break $C_{L_{PB}}$ and the initial stall lift coefficient $C_{L_{DB}}$ coincide for thin wings, low Reynolds number, or highly swept wings, and region 5 does not exist.[122]

The lower bound for region 5 is the lift coefficient at which leading-edge separation and subsequent reattachment occur—the bubble-type separations. The upper bound is reached when stall occurs and the entire upper surface flow is separated. Blunt, thick airfoils generally exhibit trailing-edge separation, while very thin airfoils exhibit leading-edge separation. Airfoils of moderate thickness are likely to separate and reattach at the leading edge, followed by trailing-edge separation and stall at higher lift coefficients. The leading-edge

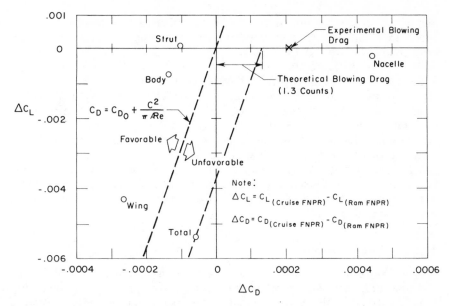

**Fig. IV.21   Effects of engine power on lift and drag of various components of an airplane.**

bubble separation produces an increase in drag due to lift because of decreased leading-edge suction. Above drag break the flow separates completely from the wing, and the drag rapidly increases.

The drag correlations developed by Schemensky,[122] Simon et al.,[131] and Axelson[4] can be used for rough estimates during the preliminary design phase but lack the logical and consistent framework necessary for confident, detailed-design tradeoff studies. As shown in Fig. IV.19, the drag due to lift in region 5 varies with lift at a greater rate than in region 1. The additional drag in region 6 is a separation drag increment that has been correlated with the lift coefficient at drag break.[131]

Numerous symposia have been devoted to the prediction of viscous/inviscid interactions, separation, and stall. Gilmer and Bristow[45] accurately predicted the lift and drag of three airfoils through stall by combining a two-dimensional panel method, an integral boundary-layer method, and an empirical relationship for the pressure in the separated zone and wake recovery region. Maskew, Rao, and Dvorak,[89] also using a panel method with a simple wake model with prescribed separation line and wake geometry, were able to predict the pressure distribution over an $R = 6$, 10-deg swept wing at 21-deg angle of attack.

The subject of leading-edge separation from slender wings has recently been summarized by Hitzel and Schmidt.[62] Potential flow methods that model the leading-edge vortex and recently developed Euler methods can predict the flowfield and lift, but drag comparisons are not included. The Pohlhamus suction analogy in conjunction with linear lifting surface theory[24] has produced excellent comparisons with experimental drag data for thin slender wings; however, the method fails for higher aspect ratio wings typical of subsonic transport aircraft.

The use of strakes and other leading-edge devices to reduce the drag at high lift and transonic speeds typical of maneuvering fighter aircraft was the subject of an extensive AGARD conference.[61] Design guidelines and empirical rules for strakes were determined from extensive wind-tunnel data bases. A fundamental understanding of the effects of strakes on the wing flowfield, lift, and drag was not evident in the proceedings.

High angle-of-attack aerodynamics[61] is certainly one area of aerodynamics that is not a mature science. Considerable theoretical and experimental effort will be required to develop accurate force and moment prediction methods for general configurations.

### IV.2.3 Cruise Efficiency

The Cruise Figure of Merit is a correlation parameter which includes airframe drag and engine installation factors. This parameter is determined from flight-test results of specific range (miles per pound of fuel). By establishing equilibrium flight for a range of weights, speeds, and altitudes, the conditions that maximize specific range can be determined. The Figure of Merit is then calculated from the expression:

$$FM = (SR \times W \times TSFC)/V$$

where

$SR$ = optimum specific range, nautical miles/pound of fuel
$W$ = gross weight, pounds
$TSFC$ = uninstalled thrust specific fuel consumption, pounds per hour fuel flow/pound of thrust
$V$ = cruise velocity, nautical miles per hour

In terms of aerodynamic parameters, the Figure of Merit is an "effective" lift-to-drag ratio, reflecting the match of engine-airframe characteristics for cruise efficiency. Included in this effective ratio are the thrust losses associated with engine installation. The maximum value of the ratio of lift to drag may be expressed as follows:

$$(L/D)_{max} = \sqrt{\frac{\pi}{4}\left(\frac{b^2}{S_{wet}}\right)\frac{e}{C_{fe}}}$$

where

$b$ = wingspan, ft
$S_{wet}$ = total wetted area, ft$^2$
$e$ = Oswald's induced drag efficiency factor
$C_{fe}$ = equivalent skin-friction coefficient

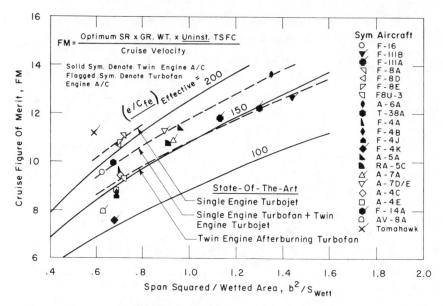

**Fig. IV.22   Cruise efficiency comparison.**

Figure IV.22 correlates Figure of Merit for several turbojet-powered aircraft as a function of these parameters.

Several interesting trends are apparent. The highest Figure of Merit is achieved by the A-6A, primarily because of a large span combined with a good effective aerodynamic efficiency ($e/C_{fe}$ about 180). The F-111B has the highest span/wetted-area ratio, but lower aerodynamic efficiency than the Tomahawk cruise missile, a point design extremely clean aerodynamically, with an $e/C_{fe}$ of about 270. The single-engine turbojet F-8 has the highest aerodynamic efficiency of the fighters shown with an $e/C_{fe}$ of about 220; the lowest is the twin-engine turbofan F-4K ($e/C_{fe} \approx 105$). The dashed lines indicate the state of the art for single-engine turbojets, single-engine turbofans, twin-engine turbojets, and twin-engine afterburning turbofans. The penalties in aerodynamic efficiency arising from twin-engine and turbofan installations are readily apparent.

## IV.3   Transonic Drag

The format of this section parallels that of Sec. IV.2. A rapid, strictly empirical method for determining the overall drag rise for preliminary design is presented, followed by a drag component buildup method to assess drag subtargets for each aircraft component. This section concludes with a summary of the status of numerical aerodynamics methods for detailed drag cleanup during the later design phases.

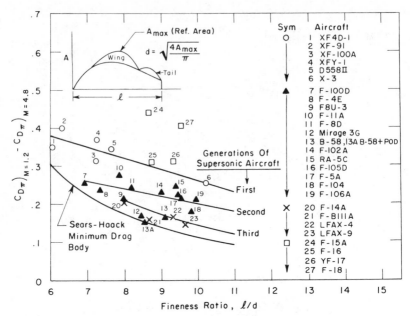

**Fig. IV.23   Historic correlation of transonic drag rise.**

### IV.3.1   The Drag Rise (C. R. James)

A correlation of the drag difference between Mach numbers 1.2 and 0.8 for many fighter aircraft is shown in Fig. IV.23. The shape of the drag curve between these Mach numbers, where the drag may attain a maximum, cannot be determined by this method.

The increase in minimum drag coefficient at transonic speeds is determined primarily by configuration slenderness. Body fineness ratio (length/diameter), lifting surface sweep, and thickness ratio (thickness/chord length) are primary parameters for expressing slenderness of major configuration elements. The area rule concept[97,167] provides a convenient approach to combining body and wing cross-sectional areas to define an equivalent body of revolution. Fineness ratio of this equivalent body is used as an index of configuration slenderness to derive a correlation of transonic drag rise.

Figure IV.23 compares historic data from 28 aircraft where the cross-sectional area distribution normal to the body centerline (Mach 1.0 cut) is used to define an equivalent body fineness ratio (length/diameter). In all air-breathing configurations, the inlet capture area is subtracted from the body total area aft of the inlet station to define the equivalent body shape. Minimum drag coefficient, $C_{D_\pi}$, is based on the maximum cross-sectional area of the equivalent body, and the difference between drag values at $M = 1.20$ and $M = 0.80$ is defined here as the drag rise.

Three generations of aircraft are shown: 1) early experimental aircraft, 2) the Century series aircraft, and 3) subsequent developments. Most of the data are

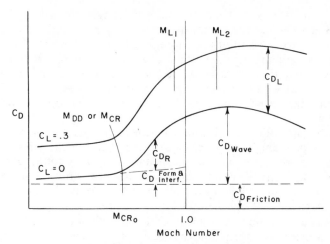

Fig. IV.24    Transonic drag buildup.

taken from flight-test results, although data for some aircraft (the NASA LFAX configurations and the XFY-1, for example) are taken from wind-tunnel tests. The theoretical slender-body value determined for Sears-Haack[126] minimum drag body shapes, pointed on both ends, is indicated as a lower limit. An overall progression toward this lower limit is discernable. The first-generation supersonic aircraft with fineness ratios of the order of 7.0 had drag rise values about 60% higher than the Sears-Haack limit, the second generation about 50%, and the third about 25%. The B-58 with an external store pod is the configuration nearest the lower limit.

Some of the newer, currently operational aircraft, notably the F-15A and the F-18 attain drag rise values greater than the trend established by first-generation aircraft. This is due to the increased emphasis placed on transonic maneuverability during the design tradeoff studies that resulted in these aircraft.

### IV.3.2    Detailed Drag Estimates—Component Buildup

*IV.3.2.1    Zero Lift Drag*

A schematic diagram of the component buildup method is shown as Fig. IV.24. The skin-friction drag may be computed by the methods of Sec. IV.2.2 at the appropriate Reynolds and Mach numbers for the flow. The friction drag is nearly invariant over the small range of Mach numbers that are considered transonic, unless the localized details of shock/boundary-layer interactions are considered.

The form and interference drags are also computed by the methods of Sec. IV.2.2; however, both are assumed to terminate abruptly at sonic speed. Recall that some form factors are constants over the entire subsonic range while others are functions of both Reynolds and Mach numbers. Similarly the wave drag, to be treated in Sec. IV.4.2, is assumed to begin abruptly as Mach 1 is exceeded. Since neither assumption is physically realistic, the zero lift drag curve is forced to be continuous through the first derivative by fitting a polynomial through the zero lift drag value at critical, or drag divergence, Mach number. Drag creep is accounted for by the variation of form and interference factors with subsonic Mach number given in Sec. IV.2.2. Thus, the drag rise and the interference plus form drag are replaced by wave drag beyond Mach 1.

The decrease in critical Mach number with increasing lift for most airfoils causes the subsonic drag polar to begin its rise earlier at higher lift coefficients. The drag rise is usually separated into a minimum drag contribution and a contribution due to lift to account for this change. The drag rise due to lifting surfaces begins at $M_{CR}$, while the drag rise due to all other components begins at $M_{CR_0}$ as correlated by Schemensky.[122]

Certainly many unresolved problems and contradictions are evident in this method. It is merely an acceptable artifice to provide continuous drag curves through the transonic regime. For example, the component critical Mach numbers depend on airfoil section, thickness ratio, or body slenderness ratio.[122] It may be different for each aircraft component, causing each to enter the transonic regime at a different freestream Mach number, or interference between closely spaced components; for example, aft pylon-mounted engines may dominate the entire flowfield and drag.[167] The drag curves generated by this curve fitting also do not account for the pressure drag slope relationships derived from transonic similarity theory,[105] and so on. Transonic drag prediction is not yet a reliable, consistent art.

*IV.3.2.2   Drag Due to Lift*

In transonic region 3 of Fig. IV.18, bounded by the critical Mach number and the limit Mach number $M_{LI}$, where $0.95 \leq M_{LI} \leq 1.00$, the induced drag is

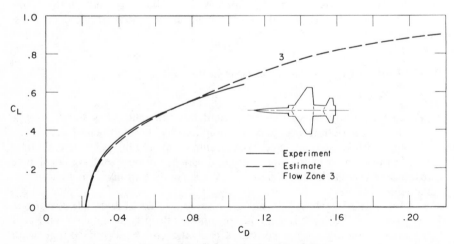

**Fig. IV.25   F-5 $C_L$ vs $C_D$; $M = 0.9$.**

computed in the same manner as described for region 1, Sec. IV.2.2. The drag rise for lifting surfaces from the previous section shifts the basic polar, depending upon $M_{CR}$. In transonic region 4 above $M_{LI}$, the drag polar is calculated by interpolation of the polar shape factors $K$ in regions 2 and 3.

A newer method for transonic drag-due-to-lift prediction to extremely high angles of attack has been programmed by Axelson.[4] This method is based on new forms of compressible wing theory covering potential and nonpotential flows, and it requires, in addition to the usual geometric inputs, the limiting forward and aft chordwise locations of the shock wave. The method predicts the drag due to lift reasonably well for quite general, assumed limit shock positions, as shown in Fig. IV.25. The drag at zero lift is not calculated by this program and must also be supplied by the user.

### IV.3.3    Numerical Transonic Aerodynamics

The progress in computational transonic aerodynamics has been reviewed in several recent publications.[105,145,167,168] *Transonic Aerodynamics*, edited by D. Nixon,[105] describes steady transonic flow research from both industry and research viewpoints, including recent advances in experimental techniques and prediction methods. Yoshihara's AGARD reports[167,168] review transonic aerodynamics for the applied aerodynamicist engaged in the design of combat and airlift aircraft with proper emphasis on aircraft component interference.[167]

The numerous authors in the field of numerical transonic aerodynamics have reached a consensus: transonic drag predictions are currently unreliable by any method. Neither the polar shape nor drag levels are acceptably predicted. The uncertainty in the magnitude and often even the sign of incremental drag due to small geometric changes precludes the use of drag as an object function in minimization procedures. The surrogate transonic design criterion is the pressure distribution, which places a heavy burden on the aerodynamicist. The target pressure distribution must be chosen such that boundary-layer separation is avoided, wave and vortex drag are minimized, and the resulting geometry is acceptable.

Transonic flow is highly three-dimensional and inherently nonlinear. A hierarchy of approximate equations, in addition to the full Navier-Stokes equations, have been used to analyze these flows with varying degrees of success.[168] Unfortunately the geometric fidelity of the numerical representation necessary to distinguish drag differences due to design changes diminishes as the higher approximations are used. This is a result of increased grid generation and computational storage requirements. Grid resolution has been a primary cause for poor drag predictions using even the full Navier-Stokes equations with various turbulence models.[32]

The theoretical and numerical bases for drag calculations based on the approximate equations (transonic small disturbance, full potential, and Euler) have been questioned[103,167,172] and remain controversial. The addition of a coupled boundary-layer solver to the inviscid transonic flow calculation further complicates the problem of accurate drag determination.[168]

Transonic flow calculations are presently limited in geometric capability to various collections of aircraft components and lack the full-configuration

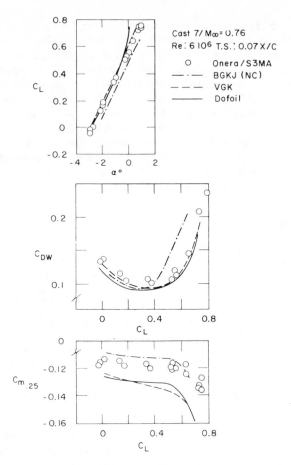

**Fig. IV.26 Total forces and moments comparison between different computer programs.**

geometric capability of panel methods (Sec. IV.2.2.7). The following examples indicate the current limitations and progress that is being made in this area.

Figure IV.26, from Longo, Schmidt, and Jameson,[84] illustrates the improvement in transonic airfoil drag prediction produced by modifying the original Bauer-Garabedian-Korn-Jameson (BGKJ) nonconservative full-potential method with weak viscous interaction to account for strong viscous interactions. The Dofoil method consists of a conservative, full-potential flow solver coupled to a set of integral boundary-layer methods with special models for separation bubbles and the trailing edge. Further improvements (unpublished) have been obtained by substituting a finite-volume Euler flow solver for the inviscid method.

Henne, Dahlin, and Peavey,[57] also using the BGKJ method, obtained an excellent correlation of drag divergence Mach number but found significant scatter in the drag level and compressibility drag increment results for eight airfoils.

Hicks'[60] extensive review of transonic wing design using both conservative and nonconservative potential flow codes did not use drag as an evaluation criterion since it was considered to be the least accurate quantity calculated by the aerodynamic codes. Poor experiment-to-theory correlations of the wing pressure distributions were found for configurations with low fineness ratio bodies. Henne, Dahlin, and Peavey[57] again found drag divergence Mach number could be correlated while drag level and compressibility drag increment could not for three high aspect ratio transport-type wings, as shown in Fig. IV.27. Waggoner,[150] using a small-disturbance transonic analysis code

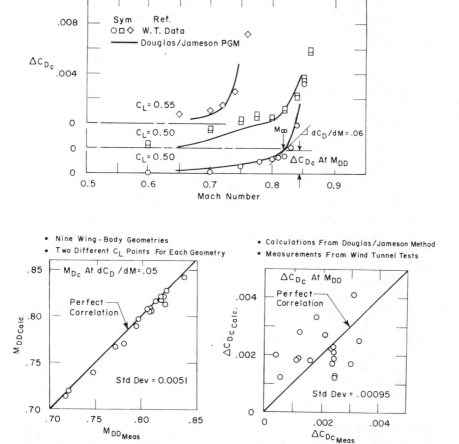

Fig. IV.27  Correlation of calculated and measured wing-body drag divergence Mach number, compressibility drag at drag divergence, and drag rise for three supercritical wings.

coupled with a two-dimensional boundary-layer code for the analysis of a supercritical transport wing-body configuration, found the method sensitive to design changes; but an increment of 41 drag counts was added to the predicted drag level to obtain agreement with test data. The computational grid resolution was coarse, particularly on the body nose and boattail region, which may account for this drag increment.

The preceding results show that transonic aerodynamics has enjoyed limited success when applied to high aspect ratio transport configurations. However, when applied to a high-performance fighter configuration with a wide complex fuselage, or canards, none of the full-potential methods is capable of straightforward prediction of the wing pressures, even at the outboard span stations.

With the possible exception of the development of a new transonic method based on paneling technology,[39] geometric generality has been limited to the small-disturbance equations. Shankar and Goebel[127] have developed a numerical transformation capable of modeling the closely coupled canard/wing interactions typical of proposed highly maneuverable fighter configurations. Drag minimization for these complex configurations has been addressed by Mason,[90] using a modified vortex lattice method.

Transonic aerodynamic theory and the associated numerical methods are currently employed for design guidance. It will be some time before transonic drag predictions are accurate and reliable enough for detailed design data.

### IV.4  Supersonic Drag

Somewhat contrary to the state of transonic drag prediction, supersonic drag is firmly based on theoretical and numerical methods. In the 1960s and early 1970s, the supersonic transport and the B-70 programs provided the impetus for the extensive development and numerical implementation of linearized supersonic theory that had begun in the early 1960s. The dedicated pioneering efforts of Bonner et al.,[10-12] Carlson and Mack,[21] Carmichael and Woodward,[25] Harris,[54] Harris and Landram,[55] Middleton and Lundry,[97] Woodward,[165] and others have produced automated numerical design procedures comprised of many individual computer codes under the control of an executive coupling routine. These numerical procedures are continually evolving to reduce the shortcomings uncovered during the succeeding years of intensive application to fighter, bomber, and transport configurations. The drag predictions produced by the current versions of these programs are excellent for many slender supersonic configurations at low lift coefficients and are widely used throughout the aircraft industry for preliminary design.

As shown in Fig. IV.5a, the supersonic cruise drag components are dramatically different from the subsonic cruise drag components. The large, obvious difference is the wave drag, a component that does not have a counterpart in subsonic flow. The lift-dependent and skin-friction drag components are analogous to their subsonic flow counterparts and are calculated by the following modifications in the previous subsonic methods. The supersonic drag buildup, using linearized theory, is illustrated in Fig. IV.28.

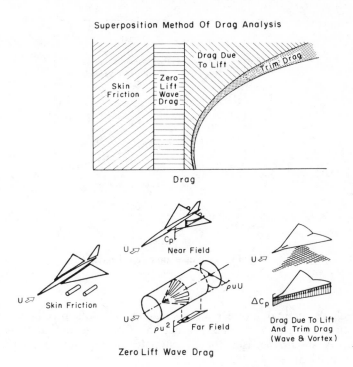

Fig. IV.28   Supersonic drag buildup.

## IV.4.1   Friction Drag

The effects of compressibility and heat transfer must be accounted for at supersonic speeds. The mean adiabatic skin-friction coefficient decreases by about 30%, regardless of Reynolds number, in changing from incompressible speeds to Mach 2, as shown in Fig. IV.29. The relative predictions from the four theories vary with both Reynolds and Mach numbers, illustrating several drag prediction problems. Certainly, the basic full scale skin-friction drag prediction during initial design depends on the theory selected. The correlation from wind-tunnel model scale to full scale flight, as illustrated by the Concorde data,[81] depends on the theory selected. Finally the reduction of flight-test results, in order to define actual drag as a function of Mach number, altitude, ambient temperature, and lift coefficient, is also theory-dependent. The magnitude of the drag change produced by selection of alternate prediction methods is, however, small. Leyman and Markham[81] have shown that, compared with the later method of Winter and Gaudet, Michel's method predicted lower aircraft drag directly by 2.5% $C_F$ (0.8% total drag) and lower aircraft drag from the 1/45th scale-model data by 5% $C_F$ (1.5% total drag) for the Concorde. The Winter and Gaudet method is similar to that of Sommer and Short,[136] and each reproduced available, acceptable, experimental results more accurately

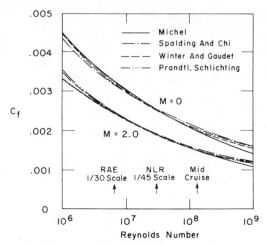

**Fig. IV.29   Supersonic skin friction.**

than Michel's method; however, the scatter in all data precluded changing skin-friction prediction methods during the Concorde design process. The same theory should be used for all estimates during the design process to achieve consistent results.

The Sommer and Short $T'$ method[136] calculates the compressible skin-friction coefficient $C_F$ from a reference skin-friction coefficient $C_F'$ for a selected freestream Mach number $M_\infty$, Reynolds number $R_\infty$, and adiabatic wall temperature $T_w$, using the incompressible von Kármán-Schoenherr formula (Sec. IV.2.2). The details of the theory and original experimental verification are given in Ref. 136. This method has subsequently been extensively tested against newer experimental data and found to be accurate by Peterson,[112] Sorrells, Jackson, and Czarnecki,[137] Stallings and Lamb,[138] and Tendeland[141] for a wide variety of geometries, flow conditions, and heat-transfer rates. The method has been programmed for many computers and forms the skin-friction module of the Integrated Supersonic Design and Analysis System.[97]

Consistent use of the Sommer and Short $T'$ method from the preliminary design phase through flight testing will improve drag correlations. Note that drag component correlations used during the preliminary design phase are ultimately based on older wind-tunnel and flight-test data. A considerable amount of these data was reduced using the Sommer and Short $T'$ method as a basis. It is therefore consistent to continue use of this method for subsequent drag buildups on new configurations.

### IV.4.2   Wave Drag

The wave drag produced by standing pressure waves that are not possible in subsonic flow has been traditionally decomposed into two components[5]: a zero lift wave drag and a wave drag due to lift. The decomposition is a result of the supersonic area rule method used for wave-drag prediction and is not permitted if nonlinear methods are used. These drag components vary as:

$$D_{0\text{ wave}} \sim (\text{volume})^2/(\text{length})^4$$

$$D_{i\text{ wave}} \sim (M^2 - 1)(\text{lift})^2/(\text{lifting length})^2$$

and must be added to the supersonic counterparts of the subsonic skin-friction drag and induced or vortex drag. Thus, the supersonic design problem of obtaining maximum lift-to-drag ratio is more complicated due to constraints arising from these additional drag terms.

Linear supersonic aerodynamic methods are the mainstay of the aircraft industry and are routinely used for preliminary design because of their simplicity and versatility despite their limitations to slender configurations at low lift coefficients. Not surprisingly most successful supersonic designs to date have adhered to the theoretical and geometrical limitations of these analysis methods. Linearized theory has not been supplanted since second-order theories have not been as successful at predicting experimental data, and practical application of nonlinear methods has been precluded by severe geometrical restrictions, i.e., axisymmetric bodies, conical flow, etc.

The normal pressure-drag components, consisting of form drag, vortex drag, lift-dependent wave drag, and volume-dependent wave drag are shown in Fig. IV.6.

The supersonic form drag, or drag due to adding the boundary-layer displacement thickness to the physical areas to obtain the effect of the boundary layer on the wave drag, is conventionally ignored.[19,81] Calculations have shown this contribution to be small ($\pm 1\%$) and dependent upon longitudinal area distribution for bodies of revolution.[19] This assumption is well within the accuracy of the prediction capability for the body alone,[55,67] particularly for low fineness ratio bodies.[80] The accuracy of this assumption is unknown for wings.

As shown in Fig. IV.28, both near- and far-field methods are used to calculate the components of inviscid normal pressure drag (Fig. IV.6).

A surface panel method, such as PAN AIR,[39] that represents the configuration's thickness, camber, and angle of attack with surface distributions of sources and doublets and that sums the resulting pressure times the streamwise component of area (near-field method), obtains the sum of the vortex drag and the lift-dependent and volume-dependent wave drags. These methods offer improved geometric representation of the configuration but do not permit independent consideration of each drag component for optimization purposes.

Far-field methods relate the wave drag to the lateral convection of streamwise momentum through a cylindrical control surface, Fig. IV.28, placed a large distance (several wingspans) from the configuration. The vortex drag is similarly related to the transverse components of momentum through a cross-flow plane far downstream in the wake—the Trefftz plane. The supersonic vortex drag is identical to the induced drag for subsonic flow since the trailing vorticity remains essentially stationary with the fluid, regardless of flight speed.

Due to the inherent theoretical limitations of each approach, a combination of near- and far-field methods, along with semiempirical modifications, are currently used for drag prediction and design optimization.[5,97]

The zero lift, or volume-dependent, wave drag is commonly computed using a numerical implementation of the supersonic area rule concept. This far-field method relates the zero lift wave drag of a wing, fuselage, empennage, nacelle, etc., to the number of developments of the normal components of cross-sectional areas as intersected by inclined Mach planes through the von Kármán slender-body formula (Fig. IV.28). The use of linear source-sink distributions to represent the configuration precludes representation of both local lift and total integrated lift. The wave drag due to lift and the vortex drag are, perforce, calculated by another method, conventionally the near-field Mach box method for thin (zero thickness) wings.

With an additional restriction on the slenderness of the configuration, the Transfer Area Rule can be obtained, which permits optimization of the wing/body for minimum wave drag due to volume independent of the wing. Since the fuselages of military aircraft are usually not slender, the use of the Transfer Area Rule in this case would represent a greater violation of theory than use of the Supersonic Area Rule.

Numerous comparisons of this theory with experimental data from the slender XB-70 and supersonic transport models[5,10] have demonstrated that it is remarkably accurate (less than $\pm 10\%$ error in total drag) over a range of Mach numbers from 1.2 to 3.0.

Bonner,[10] using an integral similar to von Kármán's slender-body formula derived by Hayes, obtains both the zero lift and lift-dependent wave drag. This procedure, however, requires a priori knowledge of the longitudinal distribution of lift. The use of this procedure is relatively limited since a near-field method capable of obtaining both the vortex and lift-dependent wave drags must be used to supply the required lift distribution, and lifting surfaces are much less likely to satisfy the slenderness requirements. The distant point of view does, however, incorporate lift-volume interference not reflected in many of the theoretical techniques used to estimate the surface pressure.

### IV.4.3   Lift-Induced Drag

The drag due to lift, both wave and vortex, has conventionally been calculated using the Mach box method of Middleton and Carlson,[96] currently used with many semiempirical modifications to alleviate known shortcomings.[97] A comprehensive review of this method, as of 1974, is contained in Ref. 23, and later modifications are contained in Refs. 21 and 97. The method subdivides the zero thickness planform, often with the fuselage outline (Fig. IV.28), into rectangular panels whose diagonals are aligned with the freestream Mach lines of the flow. Alternatively the sides of the element could be parallel to the Mach lines and the Mach diamond or the characteristic box method would result. Generally the Mach box method is more accurate for supersonic edges, while the characteristic box gives better results for wings with subsonic edges.[81]

The linear theory integral equation relating camber surface slope to lifting pressure difference $\Delta C_p$ across the planform is numerically evaluated by considering $\Delta C_p$ a constant within each box or element and reducing the integral equation to an algebraic summation. The summation can be solved for either the camber surface slope for a known pressure distribution (wing design case) or the lifting pressure coefficient in terms of known camber surface slopes (lift analysis case). A number of component loadings (up to 17) can be combined in the design mode to determine optimum camber shapes for minimum drag due to lift at a given lift coefficient, subject to optional pitching-moment constraints. The effects of the fuselage and nacelles may be included in the component loadings and additional constraints may be applied to the design pressure distribution and local camber surface to provide physical realism. Linear theory consistently overestimates expansion pressures (to less than vacuum pressure!) and underestimates the compression pressure.[114] Proper selection of the component loadings and constraints requires complete understanding of the computer programs, backed by critical operational use, otherwise occasional endless loops may result.[97] There may also be small

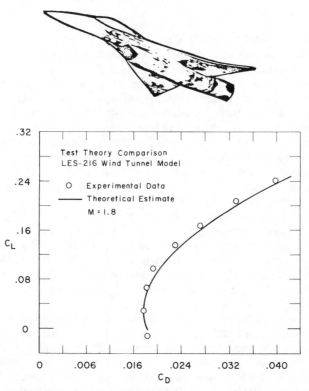

Fig. IV.30  Supercruiser type configuration—test-theory comparison.

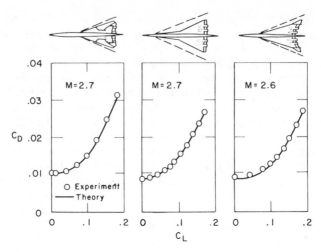

Fig. IV.31   Experimental and theoretical drag polars of models of supersonic-cruise airplanes; $R_c \approx 4.8 \times 10^6$.

discrepancies between wing loadings and forces determined for an optimized wing and the loadings and forces for the same shape upon submittal to the evaluation program.[23]

The preceding supersonic aerodynamic design and analysis methods have been substantiated for numerous wings, wing/body, and wing/body/nacelle configurations at the lower lift coefficients typical of supersonic cruise conditions. A comparison with wind-tunnel data from the Lightweight Experimental Supercruise (LES) model at the design Mach number and lower lift coefficients is shown in Fig. IV.30. Test-theory comparisons for supersonic transport and bomber models at typically higher cruise Mach numbers and lower design lift coefficients are equally as accurate,[10,116] if performed by experienced designers familiar with the methods (Fig. IV.31).

The major unresolved drag prediction problem of thin, highly swept wings typical of efficient supersonic flight is the evaluation of leading-edge thrust or leading-edge suction. The large influence on the drag polar of the various assumptions used to determine the magnitude of this force is shown in Fig. IV.32 for the same configuration as shown in Fig. IV.30. The polar shown in Fig. IV.32 is for $M = 1.35$, an off-design condition for the $M = 1.8$ supercruiser configuration.

The leading-edge thrust results from the low pressure induced by the high flow velocities around the leading edge from the stagnation point on the undersurface of the wing to the upper surface. For the high aspect ratio wings typical of low subsonic flight speeds, this force largely counteracts the drag from the pressure forces acting on the remainder of the airfoil.[22,116] The thrust force diminishes with increasing speeds but exists as long as the leading edges are subsonic.[163] It was found to be negligible at $M = 3.0$ for the configuration analyzed by Kulfan.[76] Thus, thrust effects have conventionally been ignored

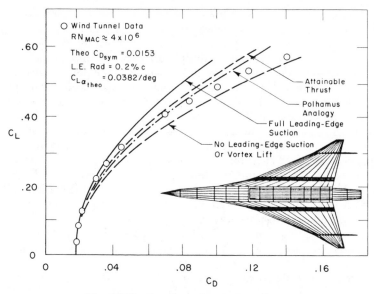

Fig. IV.32   Leading-edge suction effects.

for supersonic wing design and analysis at Mach numbers approaching 3,[116] with excellent results, as shown in Fig. IV.31.

The renewed interest in this phenomenon resulted from the increased sophistication in supersonic wing design required to achieve higher design lift coefficients for the lower Mach numbers typical of proposed fighter designs with sustained supersonic cruise capability.[164] The maneuver requirements also demand higher available lift coefficients and the attendant thickness and complex camber shapes to minimize drag.

The current complex of NASA computer programs[97] contains four user-selected options to estimate leading-edge thrust (Fig. IV.32). The basic Mach box zero thickness wing design and analysis method provides the no leading-edge suction drag estimate since the low pressures do not have a forward-facing area to act upon. The Polhamus analogy assumes the flow is separated from the entire leading edge and has formed the spiral vortex sheets characteristic of slender wings at higher angles of attack.[76] The Polhamus analogy has been extended by Kulfan[76] to account for airfoil shape, wing warp, and planform effects in order to determine the influence of wing geometry on leading-edge vortices. Kulfan demonstrates that, except for very slender wings, lowest drag due to lift is achieved with attached flow.

The attainable thrust option (Fig. IV.32), developed by Carlson, Mack, and Barger,[22] has been found to work well for wings with standard NASA airfoil sections, from which the correlations were developed.[163] However, the attainable thrust forces predicted did not agree with those experimentally measured on wings with sharp and varying leading-edge radii; large differences in both levels and trends were evident.[163]

The magnitude of the full leading-edge thrust is dependent upon the upwash in the vicinity of the leading edge and its effect on the pressure distribution. It is determined by calculating the limit of the pressure coefficient (which theoretically approaches infinity) multiplied by the distance from the leading edge. The evaluation of this limit is difficult because the pressure distributions, as calculated by panel methods, vary from the exact theoretical values required. This limit problem and its impact on wing design and drag prediction has been extensively studied by Carlson and Mack.[21]

Calculations using near-field surface-panel methods should predict the full-suction drag polar; however, the pressure-area integral is sensitive to leading-edge paneling density. Near thin wing tips, the paneling may be insufficient to accurately evaluate this integral, and the resulting force will be bounded between the full and zero leading-edge suction values. The force will thus depend on the numerical representation of the wing rather than aerodynamic assumptions, a disconcerting situation for the designer.

As observed by several investigators, a reasonable estimate for the drag polar could be obtained by numerically averaging the zero and full leading-edge suction drag polars.

Drag polars calculated by two near-field methods for Mach numbers of 1.196, 1.6, and 2.2 are shown in Fig. IV.33.[98] The complex canard/wing/fuselage configuration and paneling arrangement are shown in Fig. IV.20. The skin-friction drag was estimated using the methods of Sec. IV.4.1. Considerable scatter is evident in the zero lift drag, which may be attributed to either the skin-friction estimate or, as is more likely, the zero lift wave drag predictions of the near-field methods. The variation in vortex and lift-dependent wave drag with angle of attack was correctly predicted by the higher-order panel method, PAN AIR.[26] The lower-order method, USSAERO-B, has recently been updated to improve supersonic prediction capability by inclusion of the Triplet Singularity.[41,165] The wing camber design Mach number for this configuration was 1.8. Many additional test-theory comparisons, exhibiting much closer agreement, are contained in Tinoco, Johnson, and Freeman[144] and Tinoco and Rubbert.[146] The prediction methods included a pilot version of PAN AIR, the Mach box method, and the Woodward constant-pressure panel method (USSAERO) for a Mach 3 Recce-Strike design and the supercruise configuration shown in Figs. IV.30 and IV.32.

Correction procedures to remedy the observed shortcomings of linear theory are under continuous development. Stancil[139] has modified near-field linear theory to improve the prediction of compression and expansion pressures and to eliminate local singularities. His modifications—the use of exact tangency surface boundary conditions and the corrected local value of Mach number—cause the solution technique to become iterative. The modifications improve the zero lift wing wave-drag prediction capability of linear theory for the F-8C.[139] Shrout and Covell[129] compared both Stancil's modified linear theory and the far-field slender-body theory zero lift wave-drag predictions with their experimental data for a series of forebody models having various degrees of nose droop. The far-field method correctly predicted trends for Mach numbers of 1.2 and 1.47 and was erroneous at 1.8 for the cambered

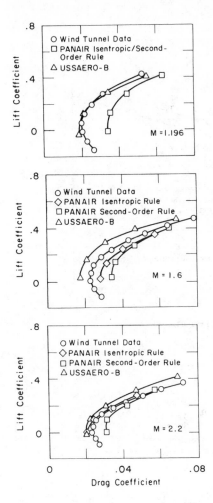

Fig. IV.33   Drag polars, CDAF configuration.

models. Stancil's method[139] correctly predicted wave-drag trends for Mach numbers of 1.47, 1.80, and 2.16.

Another breakdown of linear theory occurs when the cross-flow velocities exceed sonic value, a situation encountered on fighter aircraft at maneuver lift coefficients. The formation of cross-flow shock waves causes the experimental drag-due-to-lift performance of wings designed by linear theory to fail to meet predicted values. The supercritical conical camber concept $(SC^3)$[91] has been proposed to regain wing performance for wings swept such that the Mach number normal to the leading edge is transonic. The design method employs repetitive application of a transonic full-potential flow solver, COREL, and a specially adapted version of the Woodward USSAERO linear theory paneling code.[92] Comparison of this design procedure with data from the wind-tunnel

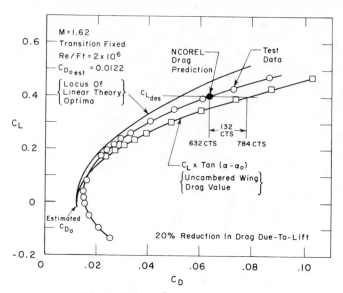

**Fig. IV.34   SC³ drag performance.**

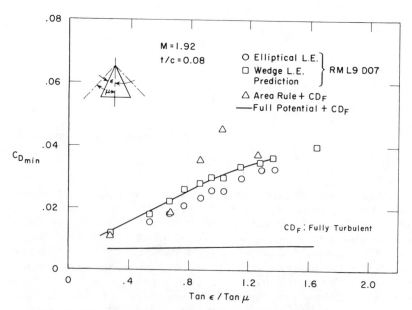

**Fig. IV.35   Supersonic volumetric efficiency.**

tests of a demonstration wing show that substantial reductions in drag due to lift are possible (Fig. IV.34) at the maneuver design condition.[91]

Bonner and Gingrich[12] approach fighter wing design through the use of variable camber and a design compromise between the transonic maneuver configuration and an optimum supersonic cruise design. The unconstrained supersonic design was initially accomplished by conventional linear theory while both small-disturbance and full-potential theories were used for the transonic design. These designs established the boundaries for the design cycle. Experience then guided the selection of the compromise camber shape, followed by alternating constrained optimizations at the supersonic and transonic design points. This multipoint design process illustrates the need for accurate drag prediction. Spurious high-drag regions entirely due to the use of linear theory, shown in Fig. IV.35 at the sonic leading-edge condition, could eliminate the optimum design from consideration on the first cycle. Significant potential for further improvement in wing design for minimum drag due to lift exists through the use of nonlinear computational methods early in the design cycle.[46] A comprehensive set of design constraints for efficient, highly swept wings has been determined by Kulfan and Sigalla.[77]

The problems of fighter design and analysis are further compounded by the inapplicability of linear theory to relatively large fuselages with nonsmooth area distributions, as shown by Wood et al.[164] Four wings were designed and

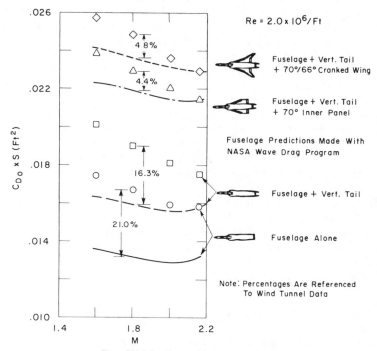

Fig.  IV.36   Zero lift drag buildup.

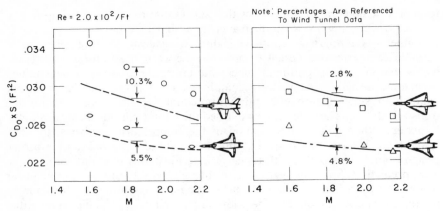

Fig. IV.37   Test-theory zero lift drag comparison.

tested with a common fuselage in an effort to adapt supersonic technologies to a $M = 1.8$ fighter design. Their test-theory comparisons (Fig. IV.36) clearly show the inaccuracy in zero lift drag prediction using the Middleton and Lundry[97] complex of computer programs. This fuselage severely violates the assumptions of linear theory concerning smooth, slender area distribution bodies; however, further inaccuracies as the configuration is built up are opposite in sign and cancel most of the fuselage error—illustrating the somewhat forgiving nature of linear aerodynamics. The zero lift drag comparison for each of the four wings is shown in Fig. IV.37. The prediction capabilities of linear theory are excellent for the complete configurations and vary with slenderness and smoothness of the area distribution. The drag-due-to-lift comparisons were equally as accurate; however, they were performed using in-house empirical corrections.

Linear theory, as developed by Middleton and Lundry,[97] has become the standard preliminary design tool for supersonic aircraft. Correction procedures to alleviate the well-known restrictions of the theory are under constant development. Higher-order panel methods, necessary for supersonic flow analyses, along with nonlinear numerical methods, are also used during the latter design stages, particularly for lower design Mach number, blunt fighter configurations.

An innovative, statistical method for the prediction of fighter aircraft drag due to lift was developed by Simon et al.[131] for Mach numbers of 0.4 through 2.5. A large experimental data base was fit with regression equations based on easily calculated nondimensional geometric parameters. The accuracy of the resulting regression equations was only limited by the expanse of the data base from which they were obtained, as shown by Johnson[69] and Tomasetti.[148]

## IV.5   Numerical Aerodynamics

Here the term numerical aerodynamics is used to delineate the full spectrum of aerodynamic methods requiring large computers—from the linearized theory panel methods to the full Navier-Stokes equations. The term computational fluid dynamics (CFD) is often used in this context, and just as often limited to mean only Navier-Stokes computations, particularly with respect to future goals. Numerical aerodynamics now permeates nearly every stage of the design process, as well as each speed regime. It is assuming an ever-increasing role; however, it is not limited to CFD.

Many prognostications concerning the possible roles of, primarily, CFD and the wind tunnel have been published (Refs. 27, 29, 30, 42, 48, 53, 73, 79, 83, 99, 110, 115, 121, 124, 125, 132, 167) since Chapman, Mork, and Pirtle's[29] 1975 article and the prediction that CFD would supplant traditional wind-tunnel tests within a decade. The pacing item in CFD technology at that time was computer size and speed. A later article[30] recognized turbulence models to be a pacing item. Computer hardware has advanced tremendously in the last decade,[79] as have flow simulations using the Reynolds-averaged Navier-Stokes equations; however, the wind tunnel remains the primary tool for design verification and a large portion of design development work. CFD analysis of full three-dimensional configurations capable of flight may still be a decade away.[48,73] The fundamental limitation of CFD during the last decade—turbulence modeling—remains the pacing item. Progress in this area in the next 15 years is not projected to remove this limitation,[73] and recent reviews of turbulence models[52,82,87,117,133] reinforce this contention. The fundamental physics of turbulence is not known well enough to permit accurate mathematical modeling and flow simulation. If the present is any indication of the future, drag prediction and design optimization with drag as an objective function using CFD are even further in the future.

Recent publications recognize the complementary roles of wind tunnels and CFD[73]; however, the role of the lower-order approximations (linear, potential, Euler) is often understated. Each of these methods will assume a complementary place in the design process, depending on configuration, schedule, cost, availability of hardware and software, and the persuasion of the designer and his management. Confidence in the accuracy of the predictions is paramount.[79,99] Aircraft design departments should evolve a full spectrum of numerical aerodynamics methods, coupled through a data management system, to provide accurate, timely, cost-effective design data.[110,121,124,168]

### Acknowledgments

I would like to acknowledge the following for their assistance: E. "Woody" Bonner, Rockwell International/LA; Tom Gregory, NASA ARC; Roy Harris, NASA LaRC; Brian Hunt, Northrop; Don Kotansky, McDonnell Aircraft Co.; A. "Del" Nagel, Boeing; Charles W. Smith, General Dynamics/Fort Worth; and the Applied Aerodynamics Technical Committee of the AIAA.

# Prediction and Verification of Aerodynamic Drag
# Part II: Wind-Tunnel/Flight Correlation
# of Lift, Drag, and Pitching Moment

A. Wayne Baldwin* and Donald W. Kinsey*
*Flight Dynamics Laboratory, Wright Patterson Air Force Base, Ohio*

## Background

### Definition/Objective

THE definition of correlation as stated in Webster's dictionary is "to establish a mutual or reciprocal relation; to relate so that to each member of one set or series a corresponding member of another is assigned." Wind-tunnel/flight-test correlation attempts to establish a mutual relation between the two sets of data; i.e., each set of data is adjusted for the differences known to exist. The flight-test data represents the final desired information of the aircraft system; however, accurately determining and recording this information requires specialized expertise. This is obviously true for wind-tunnel testing as well. An accurate understanding of how each set of data is obtained and the limitation of each is required if one is to plan and achieve an adequate correlation.

The need or objective of correlation is to check the procedures used to obtain the estimated flight data and to obtain a better understanding of the aerodynamics and propulsion interactions. Correlation either will verify the procedures used and add confidence to their application in the early design stages of new aircraft or should identify areas where improved or modified procedures are needed.

The complexity of the correlation will be in direct relation to the complexity of the aircraft configuration. The correlation of nonpowered flight results with

---

*Aerospace Engineer.

wind-tunnel data is relatively straightforward; however, when propulsive forces are introduced, correlation becomes much more difficult. Both the in-flight net thrust and the interactions of the propulsive force with the aircraft external aerodynamics must be accurately determined. The type and complexity of the force accounting system must be matched to each specific airplane configuration. Correlation procedures could be established to include forces in one, two, or all three axes (longitudinal, lateral, directional) as needed or could be established for specific local aerodynamic phenomena, such as static surface pressure or boundary-layer conditions.

Successful correlation of total drag for a particular aircraft requires a complex force accounting and correlation procedure. Considerable attention given to detail in both wind-tunnel test and flight test is required. The rewards of this effort are great. Flight-test drag polars so obtained and combined with flight-derived power available can be used to calculate any flight-test performance desired. The majority of the performance evaluations in the past were for point performance ($P_s$, range, $V_{max}$, etc.), which were corrected for test atmospheric conditions to standard day conditions.

### Correlation—Theoretical Procedures to Flight Test

Accurately defining the complex, three-dimensional geometry of an aircraft in a form most theoretical programs require is very time-intensive for the user and consumes large amounts of computer resources. Even the most comprehensive programs and most diligent engineer must exclude such subtle items as antennas, steps and gaps, and nonflush fasteners. For flexible aircraft (and they are all flexible to some degree) the engineer must define the baseline or reference shape. The alternatives include the jig shape, cruise shape, or a maneuver shape. Some theoretical procedures (FLEXSTAB,[7] for example) attempt to account for flexibility, but they typically rely on simplified aerodynamic prediction routines that cannot accurately predict transonic aerodynamic drag. The more comprehensive theoretical procedures, particularly those with transonic prediction capability, ignore flexibility, leaving the engineer to apply some empirical corrections after the fact.

It is equally difficult to establish proper viscous corrections to these theoretical procedures. While, strictly speaking, this is not a geometric correction, many codes allow for boundary layers and local separated regions, etc., by adjusting the geometry (if any correction at all is made).

There is very little published information on the correlation of theoretical procedures with flight data for total configuration drag. Indeed, Firmin and Potter[40] conclude that CFD codes are, in general, not yet accurate enough to estimate full scale aerodynamic drag on complex airplane configurations. This is particularly true for transonic flows with strong viscous coupling. There appears, therefore, to be little point in pursuing a theoretical/flight-correlation analysis beyond that provided in the first part of this chapter. The remainder of this section will deal with analysis of wind-tunnel/flight-correlation efforts.

## Correlation Analysis—Data Arrangement

### Force Accounting System

A force accounting system such as that described in Chap. II is needed to correlate wind-tunnel and flight-test data or to establish flight-test aerodynamic performance. Such a system identifies and accounts for all the interacting forces that exist between airframe aerodynamics and the propulsion systems. These interacting forces occur mainly at the engine inlet and exhaust regions and are most significant for configurations where the inlet and exhaust are closely coupled to the airframe. The majority of useful force accounting systems separate the forces that are invariant with power setting from those forces that vary with power setting, and assign the invariant forces to the airframe aerodynamics and the varying forces to the propulsion system. This permits the establishment of a reference aerodynamic drag level at a selected set of engine operating conditions. As engine operating conditions vary from this selected baseline condition, either through throttle adjustment or pressure, density, or temperature variation, the external drag of airframe components in the interface region may change. These drag variations are introduced into the engine thrust package. This approach precludes the necessity for thrust-drag iteration in the calculation of aircraft performance. A schematic presentation of the TACT[6] force accounting system is shown in Fig. IV.38. It embodies all the important characteristics described in Chap. II but is tailored to the F-111 aircraft used in the TACT program.

### Reference Conditions

*Inlet Spillage Drag*

One of the more common aerodynamic/propulsive system interface forces occurs at the engine inlets and is commonly referred to as "spillage drag." This item consists of two partially canceling components, additive drag, and lip suction. For convenience, conventional thrust calculations assume that freestream pressure and velocity exist at the inlet face. Except near takeoff, engine/inlet combinations generally operate at capture area ratios less than unity. The stream tube entering the inlet is expanding and generates a preinlet momentum loss defined as additive drag that must be accounted for. Theoretical additive drag can be evaluated by the method of Ref. 35. Lip suction is the reduction in external drag due to the change in cowl external pressure resulting from operating at capture area ratios less than unity. The lip suction force partially cancels the additive drag. Airframe external drag should be determined at some baseline, engine-airflow condition, and the drag variation from this baseline should be incorporated into the propulsion package as described in Chap. II.

The inlet spillage drag is usually determined by varying the mass flow in the flow-through aerodynamic force model. Duct exit plugs can be used to adjust the mass flow, and total head pressure measurements across the exit are

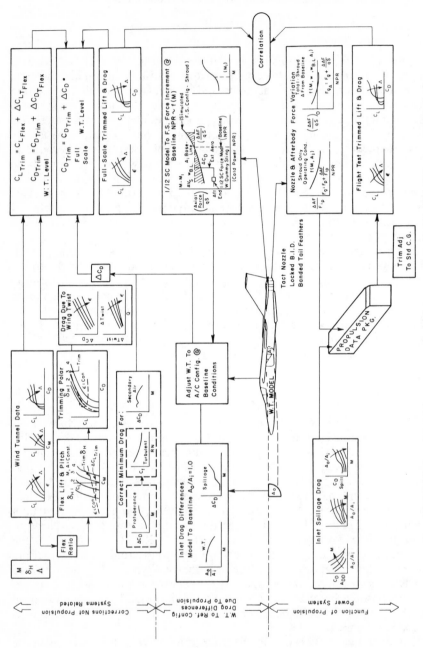

Fig. IV.38  Schematic of force accounting system used on F-111/TACT program (Ref. 6).

recorded to determine the magnitude of the mass flow. The change in stream-tube momentum between freestream and nozzle exit is called internal duct drag ($C_{D_{INT}}$). When the pressure area forces ($[P_B - P_0]A_B$) on the base of the model are removed from the drag data ($\Delta C_D = [P_B - P_0]A_B/qS$), along with $C_{D_{INT}}$, the remaining items are external model drag and additive drag. The variation of model and additive drag with duct mass flow is the spillage drag ($C_{D_{SPILL}}$). The magnitude of $C_{D_{INT}}$ can be 20–30% of the external model drag; therefore, care should be taken in its evaluation. The determination of exit momentum is directly related to the measurement of the exit total and static pressures (including wall static), and care should be exercised in the area weighting of these pressures and their integration across the exit. Duct mass flow calibrations vs exit pressure will improve the overall accuracy of exit momentum and thus $C_{D_{INT}}$.

Spillage drag can be used to adjust the wind-tunnel model and flight-test data to a common baseline level, normally a capture area ratio of unity. Data indicate no measurable variation in spillage drag for small changes in Reynolds number (two to seven million per foot). These data are normally assumed reliable at higher flight Reynolds numbers. This procedure allows the wind-tunnel drag to be established at a baseline or reference level with the spillage drag being that associated with a capture area ratio of unity. The flight-test drag, if recorded at a mass flow ratio other than unity, can be adjusted to unity for commonality or correlatable results.

*Jet-Effects Drag*

Another common interface between aerodynamic and propulsion system forces is the nozzle/aftbody interaction, which varies with test conditions, nozzle geometry, nozzle pressure ratio, etc. These effects are usually determined from a propulsion model on which the overall axial forces on the aftbody (separate from the thrust) are measured as a function of engine operating conditions. These powered models present unique problems in bridging the balance and accounting for all the pressures and flow momentums of the high-pressure air or chemical fuels brought onboard the model to simulate the high-pressure/mass-exhaust flows. The models are normally blade-mounted in the wind tunnel so that the nozzle exhaust area is free of distortion. They are normally capable of duplicating both real airplane aft-end lines and the force and moment flow-through model aft-end lines. The difference in measured drag between these two geometries is related to the force model distorted aft-end drag and the real airplane boattail drag. The simulated thrust associated with the force model lines duplicates the mass flow of the flow-through ducts, while the thrust associated with the real airplane lines should vary across the expected flight nozzle pressure ratio (NPR). This aftbody drag, or jet-effects drag, as it varies with thrust, can be used to establish the baseline or reference jet-effects drag, which is applied to the force model data. The variation from the baseline can be applied directly to the engine thrust as it varies with NPR. The reference thrust condition is usually chosen to be the estimated thrust required for cruise, and it varies with Mach number. The flight-test drag, if recorded at a thrust level other than the reference condition, can be adjusted with the same jet-effects data to the reference thrust condition for commonality or correlatable results. The increments in lift and pitching moment should also be evaluated, in much the same way as the drag, as a function of inlet and nozzle variations.

### Trim Drag

A third area where a reference condition for wind-tunnel and flight-test data correlation is normally chosen is the trim center-of-gravity position (CG). Models can be supported in a wind tunnel at a wide range of angles of attack and many control surface deflections. An airplane in free air, on the other hand, can maintain unaccelerated flight only when all the forces are balanced. Traditionally the balancing device for pitch is a horizontal tail located at the rear of the fuselage. For a statically stable configuration, with a negative pitching moment coefficient $C_{M_0}$, down loads are required on the tail to balance the airplane. The drag associated with the load on the tail plus the induced drag resulting from the increase in wing lift required to overcome the tail force is called trim drag. Both the wind-tunnel and flight lift and drag must be trimmed to the same CG position for correlation.

The amount of tail load required to balance the aircraft in flight changes as the center of gravity of the configuration changes (from burning of fuel, for example). Therefore, it is convenient to select a reference CG position close to that which is expected in flight and trim the wind-tunnel data to that position.

If the flight lift and drag are obtained at a CG position different from the anticipated reference position, they must be adjusted for the changes that exist between trim at the "as-flown" position and the reference position. Paterson, MacWilkinson, and Blackerby[109] chose 30% mean aerodynamic chord (MAC) for the C-5A, whereas Baldwin[6] chose a different location for each of six wing sweeps on the TACT airplane in order to closely match the expected flight CG and minimize the magnitude of the corrections. Normally information about tail effectiveness is determined from wind-tunnel tests where lift, drag, and pitching moment are measured for discrete horizontal tail settings ($\delta_H$) and plotted as shown in Fig. IV.39.

Typically the wind-tunnel data are plotted for the moments about the reference CG position. If constant angle-of-attack ($\alpha$) lines are superimposed on the moment plot, the tail setting required as well as the increment in lift for trimming to zero moment can be determined at the various $\alpha$, zero moment points. Constant $\alpha$ lines can be superimposed on the drag polar plot ($C_D$ vs $C_L$, for a range of tail settings), and the required tail angle to trim, as determined from the moment plot, can be interpolated along these lines. These interpolated points graphically describe the trimmed drag polar. This routine is readily computerized using nonlinear curve fitting. The horizontal tail effectiveness for the airplane in flight is influenced by flexibility and should be adjusted accordingly. The FLEXSTAB[7] computer program was used during the TACT program to evaluate these flexibility influences.

Lash and Sims[6] describe an approach to determine stability information from dynamic flight maneuvers which, if properly used, may provide appropriate data to correct for center-of-gravity shifts for the flight vehicle. Good

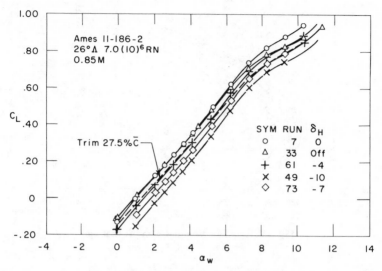

Fig. IV.39a   Longitudinal trim lift curve.

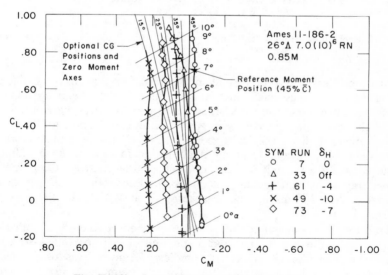

Fig. IV.39b   Longitudinal trim lift vs moment.

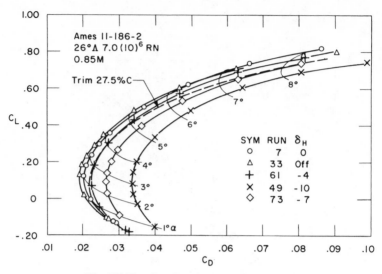

Fig. IV.39c   Longitudinal trim lift vs drag.

test planning can minimize the trim corrections by assuring that the actual flight center of gravity is close to the reference value.

## Wind-Tunnel Model Data Adjustments

*Aeroelasticity*

Airframe flexibility produces geometric variances that are very difficult to reconcile. This is especially true on an aircraft that has large flight and maneuver envelopes. The model rigid shape can (usually) represent, at most, one flight condition. Any change in speed, altitude, or load factor during flight will produce, in effect, a different geometry. This change in geometry, primarily wing shape, can have an effect on all aerodynamic coefficients, including trim requirements. Fortunately the magnitude of the drag change is often small, and some drag increments actually act to offset each other (see, for example, Paterson, MacWilkinson, and Blackerby[109] and Nelson, Gornstein, and Dornfeld[104]).

To estimate the magnitude of the aeroelastic effects requires detailed information about the structural characteristics and the aerodynamic load distribution. The FLEXSTAB aeroelastic prediction routine was used effectively during the TACT correlation effort. This program calculates aerodynamic coefficients (longitudinal, lateral, and directional) based on the input configuration geometry and a Woodward-type aerodynamic prediction procedure. The resulting airloads are coupled with the input structural influence coefficients and, through an interactive process, the aerodynamic coefficients for the deflected shape and the ratios for the rigid to flexible aerodynamics are established. The FLEXSTAB aerodynamic procedure is inviscid and, except for induced drag, does not give accurate drag ratios.

*Scaling Effects*

All model-related discrepancies are, in a sense, scaling effects; however, the term is usually reserved for viscous-related phenomena that are caused or aggravated by the relative Reynolds number difference between the model and the full scale vehicle. Under this heading, such topics as boundary-layer transition, skin-friction corrections, and separated flows will be discussed. The fact that a new high Reynolds number cryogenic tunnel (the National Transonic Facility) is being built specifically to study viscous-related effects on models is strong evidence of the uncertainties in scaling effect corrections.

For most correlations, total airplane drag can be represented as the sum of minimum drag, plus drag due to lift, plus drag due to trim. The major adjustments made to the wind-tunnel data are minimum drag corrections. The drag due to lift of the wind-tunnel model (without wing separation) is generally accepted as being representative of the full scale drag due to lift. The drag due to trim based on wind-tunnel tail effectiveness (as mentioned earlier) and adjusted for aeroelastic effects provides a reasonable estimate of full scale trim drag increments.

Turbulent boundary layers are known to exist over the surface of most flight vehicles. It is a generally accepted test procedure to use some form of distributed roughness (grit) to force the laminar boundary layer that normally exists on scaled models to transition to a turbulent boundary layer.[15] For subsonic flow, the transition strips are usually located near the leading edge of each model component (fuselage, wing, tail, etc.). Properly chosen grit, size, and distribution patterns can effectively stimulate transition to a turbulent boundary layer without creating a "grit drag" penalty.[14] This "proper" choice, however, is not as straightforward as one would hope. Reynolds number and Mach number obviously must be considered, but there are also more subtle factors. The turbulence level or "noise" in the tunnel can promote transition. Also some airfoil shapes seem more prone to transition than others. Model surface finish can influence the tendency to transition and, in fact, model surface deterioration can change the transition characteristics during the course of a test. Figure IV.40 shows the transition patterns used on the TACT model. Suffice it to say that a prudent wind-tunnel/flight-test correlation should include a series of transition trip size and location studies and careful attention to detail.

At transonic speeds where shock waves begin to form on the airfoil, it is generally agreed that the transition strip should not be arbitrarily placed near the airfoil leading edge. Blackwell addresses this subject in Ref. 8 with rationale for locating the transition strip so that the model trailing-edge boundary-layer thickness-to-chord ratio matches that of flight test (Fig. IV.41). Blackwell's approach is widely used; however, there is no general agreement as to exactly where the transition strip should be. The location would probably have to change with each change in Reynolds number, Mach number, and angle of attack to exactly represent each flight condition. Paterson,

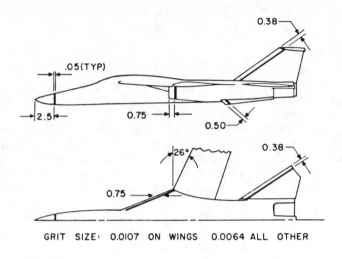

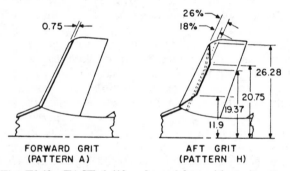

GRIT SIZE: 0.0107 ON WINGS     0.0064 ALL OTHER

Fig. IV.40   TACT, 1/12 scale-model transition grit patterns.

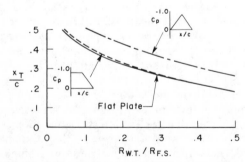

Fig. IV.41   Effect of Reynolds number on theoretical transition location, $M = 0.7$, $R_{W.T.} = 3 \times 10^6$ (Ref. 8).

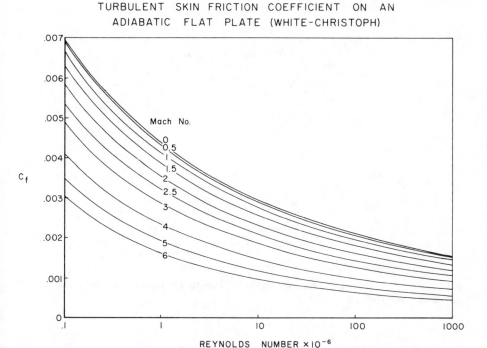

TURBULENT  SKIN  FRICTION  COEFFICIENT  ON  AN
ADIABATIC  FLAT  PLATE  (WHITE-CHRISTOPH)

Fig. IV.42     Turbulent skin-friction coefficient on an adiabatic flat plate (Ref. 153).

MacWilkinson, and Blackerby[109] give a good discussion of this problem in their report on the C-5A. Assuming that drag of the grit particles themselves is avoided, the effects of improper transition location are manifest in poor simulation of induced drag and/or trim drag and possibly wave drag from improperly simulated shocks.

One of the major wind-tunnel minimum drag data corrections is for the variations in skin friction between the relatively low Reynolds number flow on the model and the relatively high Reynolds number flow on the flight vehicle. Basically this incremental drag coefficient embodies the classical, turbulent, flat-plate skin-friction variation from model to full scale Reynolds number. Figure IV.42 shows the White-Christoph[153] variation of skin friction with Reynolds number, and Ref. 109 discusses several different sources, with Fig. IV.43 showing the variation in these sources. Reference 122 gives a step-by-step procedure for calculating the drag variation. The model configuration can be arbitrarily sectioned so that representative lengths can be established for each section and an average $RN$ can be calculated for each of these sections. With these $RN$, the skin friction (based on wetted area) can be established from Fig. IV.42, converted to the reference area, and summed for the total model friction drag. Normally the form (FF) and interference (IF) factors are empirically estimated and included with the friction drag estimate to obtain the minimum

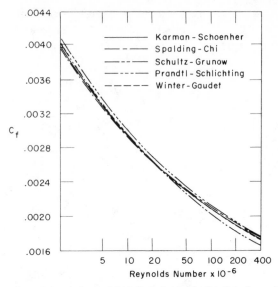

**Fig. IV.43   Comparison of empirical flat-plate skin-friction formulae for incompressible turbulent flow (Ref. 109).**

profile drag of the model at its test Reynolds number.

$$C_D = C_f \frac{A_{\text{wet}}}{A_{\text{ref}}} * \text{FF} * \text{IF}$$

Empirical values of form and interference drag factors are given in Ref. 122. Table IV.3 shows typical skin-friction and roughness drag estimates for the F-111 TACT airplane. Skin-friction estimates for the wind-tunnel model at its test Reynolds number are also made, and the difference between these estimates is the Reynolds number correction applied to the wind-tunnel minimum drag. It should be noted that the minimum wind-tunnel profile drag may be higher than the estimated skin-friction, form and interference drag. This is normally due to additional interference drag and can only be determined by wind-tunnel tests.

In estimating the full scale airplane skin-friction drag, a roughness cutoff for the variation with Reynolds number is typically used (Fig. IV.44). This surface roughness is described in Ref. 122. A typical roughness value for fighter aircraft finishes is $K = 0.0004$ in. This number is sometimes referred to as "sand grain roughness." Wind-tunnel model surfaces are assumed to be aerodynamically smooth (no roughness).

It is generally true that turbulent flow exists from the leading edge aft in full scale flight and that laminar flow exists from the leading edge aft to the transition strip on the wind-tunnel model. This requires that the wind-tunnel drag data be adjusted for that portion of the model forward of the transition

Table IV.3  Skin-friction drag/roughness
F-111/TACT airplane $M = 0.90$, $q = 825$ psf (10,000 ft)

| Component | $A_{wet}$, ft² | $\frac{A_{wet}}{A_{REF}}$ | $\ell$, ft | $RN_\ell$, ×10⁶ | $C_{f_{SMOOTH}}$ | FFᵃ | IF | $C_D$ | $K_l\left(\frac{\ell}{K}\right)^{1.0489}$ ×10⁶ | $C_{f_{ROUGH}}$ | FF*IF | $C_D$ | $\Delta C_{D_{ROUGH}}$ |
|---|---|---|---|---|---|---|---|---|---|---|---|---|---|
| Fuselage | 1020 | 1.689 | 70.8 | 341.2 | 0.00169 | 1.065 | 1.01 | 0.00307 | 98.5 | 0.00199 | 1.076 | 0.00362 | |
| Nacelle | 443 | 0.733 | 33.5 | 161.4 | 0.00185 | 1.040 | 1.30 | 0.00184 | 44.9 | 0.00222 | 1.352 | 0.00220 | |
| Strakes | 50 | 0.083 | 14.0 | 67.5 | 0.00207 | — | 1.25 | 0.00021 | 18.0 | 0.00255 | 1.250 | 0.00026 | |
| Wing | 845 | 1.400 | 10.2 | 49.1 | 0.00220 | 1.180 | 1.00 | 0.00363 | 12.9 | 0.00269 | 1.180 | 0.00444 | |
| Horizontal tail | 334 | 0.533 | 10.3 | 49.6 | 0.00220 | 1.080 | 1.15 | 0.00145 | 13.0 | 0.00269 | 1.242 | 0.00178 | |
| Vertical tail | 223 | 0.369 | 13.3 | 64.1 | 0.00210 | 1.050 | 1.15 | 0.00093 | 17.0 | 0.00259 | 1.207 | 0.00115 | |
| Glove-up | 260 | 0.430 | 37.1 | 178.8 | 0.00183 | 1.130 | 1.15 | 0.00023 | 50.0 | 0.00220 | 1.299 | 0.00123 | |
| Glove-lower | 53 | 0.088 | 8.6 | 41.4 | 0.00225 | 1.130 | 1.15 | 0.00026 | 10.8 | 0.00276 | 1.299 | 0.00031 | |
| Splitter plate | 62 | 0.103 | 4.8 | 23.1 | 0.00245 | — | 1.30 | 0.00032 | 5.86 | 0.00305 | 1.300 | 0.00041 | |
| Total | 3290 | 5.428 | | | | | | 0.01194 | | | | 0.0154 | 0.00346 |

$K_l = 47.2$ (Ref. 122).
$K = 0.0008$ in. (Ref. 64 and F-111 miscellaneous items).
ᵃCompressibility not included.
N.B.: Roughness will not normally be a factor at higher altitudes where Reynolds number is lower.

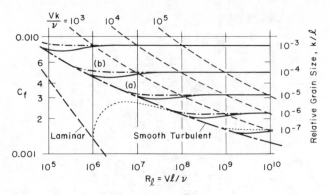

Fig. IV.44   Skin-drag coefficient of sand-rough surfaces (Ref. 64).

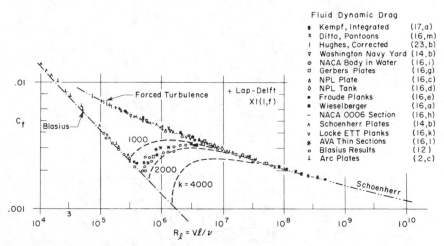

Fig. IV.45   Average or total skin-friction drag coefficient of smooth and plane surfaces (in compressible flow) in air and in water (Ref. 64).

strip where laminar flow exists. A typical variation in flat-plate skin friction for laminar and turbulent flow is shown in Fig. IV.45. The Blasius[9] curve ($C_f = 1.328\sqrt{RN}$) is generally accepted for laminar flow. The flat-plate skin friction for turbulent flow as shown in Ref. 122 accounts for various percentages of the surface being in laminar flow and for compressibility.

Most of the above discussion assumes attached flow over the surface of the model. Flight configurations with large curvatures on aft-facing surfaces, such as fuselage "boattails" or blunt bases, often experience flow separation at some flight conditions. There is no reason, however, to expect the model to exhibit the exact same separation characteristics at the simulated wind-tunnel conditions. This is particularly true for wings at large angles of attack (typically 8–12 deg and above). Also the transition grit patterns that worked well for the cruise conditions may be all wrong for high angles of attack where separation is eminent or already started. Good correlation between wind-tunnel and flight data at test conditions where separated flow is prevalent is, in general, beyond the present state of the art. The new high Reynolds number National Transonic Facility cryogenic tunnel may provide some insight into this problem area.

*Geometric Difference*

Protuberances that exist on the airplane are very difficult to simulate on the wind-tunnel model. It is doubtful that their drag measured at low Reynolds number would be representative of the full scale value. The drag of these protuberances (lights, air vents, angle of attack/yaw probes, etc.) is usually estimated and added to the clean aircraft wind-tunnel data. Hoerner[64] is an excellent source of data for these individual drags. Paterson, MacWilkinson, and Blackerby[109] (Table IV.4) include an extensive list of similar items under the heading "roughness drag" and estimate a drag increment of $\Delta C_D = 0.0009$ for the C-5A ($M = 0.7$, $C_L = 0.45$). It should be noted that protuberance drag and Reynolds number cutoff roughness drag are separate items; however, protuberances that are difficult to analyze are often empirically included in the roughness cutoff Reynolds number. Protuberance drag for the F-111/TACT airplane[6] was evaluated by analysis of the individual protuberances using Hoerner.[64] The subsonic drag increment was estimated to be $\Delta C_D = 0.00065$. Young, Patterson, and Jones[170] have a good discussion on this subject.

The simulation of all auxiliary air inlets and ducts is not often accomplished on the force and moment wind-tunnel model. The source of this air is usually from the engine inlet or inlet boundary-layer diverters. As such, this airflow will influence the spillage drag of the airplane. The auxiliary air duct losses or drag can be determined from the change in momentum between entrance and exit airflow. Design of the model inlet diverter should consider the fact that the model boundary-layer thickness at the diverter lip may differ from that of the full scale airplane.

The most obvious geometric difference between a model and the corresponding flight vehicle is the support mechanism. The F-111 TACT model[6] had significant modifications to the rear base area of the fuselage to accommodate a "sting" support (Fig. IV.46). The distorted aft-end lines of a force and moment model with flow-through ducts can be made to eliminate most aft-end airflow separation by continuing the fuselage and nacelle lines parallel to the fuselage reference plane. All aft-facing areas on the modified geometry should be 90 deg to the zero angle-of-attack position, and all edges should be sharp to ensure immediate separation and constant static pressure over the total aft-facing area. This model base drag can then be removed from the balance data by removing the aft-facing pressure-area force term. Most wind-tunnel tests use a variation of the sting or blade support methods. The C-5A study,[109] for example, favored the forward blade-mount approach. Regardless of the method of supporting the model, there has to be some influence of the support on the total model drag. An accurate procedure for estimating the effect of the support on the aircraft drag must be developed before drag estimates can be obtained. In the TACT program, as mentioned earlier, a blade-mounted model

Table IV.4   Estimated and target roughness drag increments on C5-A (Ref. 109)

| Item | $\Delta C_D$ (Counts) Estimate | Target | Item | $\Delta C_D$ (Counts) Estimate | Target |
|---|---|---|---|---|---|
| *General* | | | *Fuselage* | | |
| Antennas | 0.535 | | Doors | 0.250 | |
| Anticollision light | 0.010 | | Waviness | 0.046 | |
| CDPIR installation | 0.024 | | Air conditioning I/O system | 1.350 | |
| Windshield wiper | 0.160 | | Nose radome and vent | 0.115 | |
| Total | 0.729 | 0.500 | Leakage (pressurization) | 0.500 | |
| | | | Negative pressure relief vents | 0.200 | |
| *Wing* | | | Wheel well fairing vents | 0.056 | |
| Skin joint steps and gaps | 0.291 | | Wheel well fairing slip joints | 0.020 | |
| Aileron spacer gaps | 0.245 | | Wheel well fairing surface | 0.072 | |
| Step at slat trailing edge | 0.150 | | Auxilliary power unit exhaust outlet | 0.020 | |
| Discontinuities at slat segment ends | 0.500 | | Fuselage doubler | 0.250 | |
| Slat actuator doors | 0.010 | | Wing root fairing slip joints | 0.029 | |
| Step at spoiler trailing edge | 0.210 | | Visor door-fuselage joint | 0.186 | |
| Spoiler gaps | 0.340 | | Kneeling blisters | 0.200 | |
| Wing tip lights | 0.002 | | Total | 3.294 | 3.100 |
| Static discharge wicks | 0.060 | | | | |
| Anti-icing air exhaust | 0.022 | | *Empennage* | | |
| Steps and gaps around flaps | 0.200 | | Skin joints | 0.094 | |
| Exposed flap tracks | 1.600 | | Spacer gaps | 0.452 | |
| Exposed bearings | 0.250 | | Bullet base | 0.007 | |
| Gear boxes | 0.200 | | Total | 0.553 | 0.500 |
| Flap trolley bumps | 0.400 | | | | |
| Total | 4.480 | 2.878 | *Pylons* | | |
| | | | Steps and gaps | 0.022 | |
| | | | Total | 0.022 | 0.022 |
| | | | TOTAL | 9.075 | 7.000 |

Fig. IV.46   TACT model in Ames 11-×11-ft transonic tunnel.

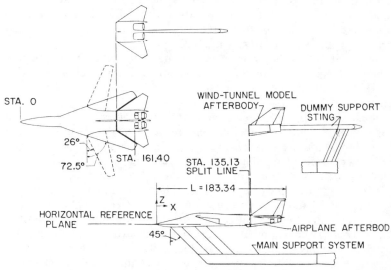

Fig. IV.47   Sketch of wind-tunnel propulsion and aero models showing metric and nonmetric sections (Ref. 6).

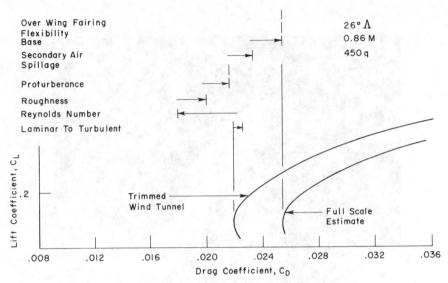

Fig. IV.48   TACT adjustments to wind-tunnel drag.

with a metric aft end (propulsion model) was used to measure the difference in drag of the actual aircraft aft end shape (with and without nozzle exhaust flow simulation) and the modified model aft end with a "dummy" sting (Fig. IV.47). The static pressure in the sting cavity and base area should be measured during all sting-mounted testing so that the balance axial force can be corrected to a reference pressure value of freestream static.

Figure IV.48 shows the order of magnitude of the adjustments made to the TACT model wind-tunnel data to determine the full scale estimated flight data. Table IV.5 summarizes the drag estimation procedures used for the TACT program correlation.

A much more insidious geometric problem comes from manufacturing tolerances, alignments, installation and removal inconsistencies, and model deterioration. Accurate simulation of complex inlet shapes on a small scale model is very difficult. Spillage drag and internal drag may be significantly affected by improper representation of the inlet geometry. Geometric discrepancies, while hopefully kept to a minimum by competent manufacturers and good model handling, can cause annoying tunnel-to-flight, tunnel-to-tunnel, and even entry-to-entry differences that are never fully explained.

The effects of stores (bombs, missiles, fuel tanks) will not be discussed except to say that, in general, a carefully conducted wind-tunnel test can reasonably predict the incremental effect of store drag over the clean aircraft configuration.[118]

*Tunnel Influence*

The effects of the wind-tunnel environment on the model, and hence the measured loads and moments, could easily be the subject of an entire book.

Table IV.5   Summary of drag estimation procedures

| Property | Method/procedure | Limits |
|---|---|---|
| $C_{D_{min}}$ | | |
| $\quad C_D$ Friction | Classical flat-plate skin friction vs $RN$, Ref. 122 | Adequate for all Mach Nos. |
| $\quad C_D$ Protuberance | Individual item analysis, Ref. 64 or empirical statistics | Subsonic analysis adequate, shock effects—uncertain |
| $\quad C_D$ Spillage | Wind-tunnel model with flow-through nacelle, variation of mass flow | Adequate for all Mach Nos., requires accurate inlet modeling |
| $\quad C_D$ Auxiliary air | Momentum loss determined from ground mock-up | Adequate for all Mach Nos. |
| $\quad C_D$ Base | Powered jet-effects wind-tunnel model | Adequate for all Mach Nos., requires elaborate model and test, uncertain $RN$ effects |
| $\quad C_D$ Flexibility | CFD program for induced drag changes | Influence on viscous and shock drag not available |
| $C_D$ Induced | Analysis, Ref. 122, and wind-tunnel tests | Adequate up to moderate angles of attack |
| $C_D$ Trim | Wind-tunnel model tests with various horizontal surface incidence | Adequate for all Mach Nos. |

The most that will be attempted here is to discuss some of the variables having the greatest influence on drag estimates. Tunnel type (continuous, blowdown, etc.), speed (subsonic, transonic, supersonic), test section size and wall construction (solid, slotted, slanted holes), and turbulence levels all impact the model, test techniques, and corrections that are appropriate. Ideally one would select the largest possible tunnel and the largest practical model, keeping in mind the tradeoff between scale effects and wall effects.

The ratio of model maximum cross-sectional area (at zero lift) to the tunnel test section area is referred to as tunnel blockage. To minimize the effects on the test data, it is generally accepted that this blockage should be kept below 1%. Wind tunnels are usually calibrated to determine the freestream buoyancy (longitudinal pressure variation along the tunnel centerline), and the test data are normally adjusted for this freestream buoyancy. Buckner and Webb,[18] however, encountered additional buoyancy blockage due to the presence of the model in the tunnel. Table IV.6 shows the percent blockage of three different scale models in two wind tunnels, and Fig. IV.49 shows the pressure variation along the centerline of one of the models in these two different tunnels.

Adjustments for flow angularity are normally made to the wind-tunnel measured angle of attack ($\alpha$). The transfer of the recorded internal force balance, normal force coefficient ($C_N$), and axial force coefficient ($C_A$), to lift and drag coefficients ($C_L = C_N \cos \alpha - C_A \sin \alpha$, $C_D = C_A \cos \alpha + C_N \sin \alpha$) requires that $\alpha$ be very accurately known. The tunnel airflow that is not parallel to the model zero $\alpha$ line must be accounted for. This is normally accomplished by initially testing the model upright and inverted, and applying the average

Table IV.6   Model/wind-tunnel geometric data (Ref. 18)

| Model | Wing area, ft$^2$ | Span, in. | MAC, in. | $\Lambda_{LE}$, deg |
|---|---|---|---|---|
| 1/9th YF-16 | 3.46 | 38.8 | 14.582 | 40 |
| 1/15th YF-16 | 1.245 | 23.3 | 8.749 | 40 |
| 0.0266 C-5A | 3.174 | 59.516 | 8.399 | 28.48 |

| Model | % Blockage | | Span/width | |
|---|---|---|---|---|
| | Ames 11-TWT | Calspan 8-TWT | Ames 11-TWT | Calspan 8-TWT |
| 1/9th YF-16 | 0.24 | 0.45 | 0.29 | 0.40 |
| 1/15th YF-16 | 0.08 | 0.16 | 0.17 | 0.24 |
| 0.0266 C-5A | 0.41 | 0.77 | 0.45 | 0.62 |

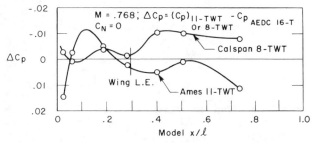

Fig. IV.49   Measured pressure differences between facilities on C-5A fuselage (Ref. 18).

difference in $\alpha$, at a fixed $C_L$, to the $\alpha$ during the test. These flow angularity corrections are generally around 0.01 deg.

Wall corrections in solid wall tunnels can get quite large at high lift coefficients and/or high subsonic speeds, and such conditions should be avoided. The primary model size consideration in a supersonic tunnel is that the reflected shock waves do not impinge on the model. Wind-tunnel wall interference effects in transonic tunnels are becoming more important with increasing interest in testing at Mach numbers approaching 1 and at high lift conditions. The procedure for correcting for these effects by making small adjustments in incidence and Mach number is probably not adequate. Considerable research is being done in this area.[151] However, for the near term, the proper wall corrections for transonic tunnels are best left to the experience of the tunnel operators unless the test program can afford the luxury of running a series of calibration tests. Ground-plane test, inlet or nozzle test, stores-release, and all other special-purpose tests have their own particular problems and are beyond the scope of this book.

**Flight Variances**

The first order of business in evaluating flight-test drag is accurately determining in-flight net thrust. As the saying goes, this must be done "very carefully." This problem has been addressed in Chap. III. Atmospheric conditions change with time and location and must be considered. This section will examine the difficulties of obtaining the exact flight conditions desired.

Many atmospheric condition problems are solved by simple avoidance. Flight testing should be avoided during periods of wind gust and in air with high moisture content. Variations in temperature and atmospheric pressure are automatically accounted for when the data are reduced to coefficient form; however, accurate measurement of Mach number and dynamic pressure are essential. All instrumentation must be calibrated to give true readings regardless of speed or attitude. An excellent report on flight-test procedures is provided by Cooper, Hughes, and Rawlings,[33] who describe the F-111/TACT flight-test program.

Wind-tunnel tests are conducted in a very well-controlled environment. Mach number, Reynolds number, and model attitude can all be established and maintained quite accurately. Similarly the configuration (wing sweep, control surface setting) is fixed and all test conditions are essentially steady-state. Flight testing, however, is quite different. While it may be possible to measure Mach number, airspeed, angle of attack, yaw angle, etc., with acceptable accuracy, it is very difficult to establish and maintain set values. Control surface positions must be monitored to assess their input to the data being correlated (lift, drag, and pitching moment).

Drag-polar data have been obtained from quasi-static (push over/pull up and wind-up turns) flight maneuvers; however, the pitch rate must be kept low (less than 3 deg/s), and $C_{m_{\dot{\alpha}}}$ and $C_{m_q}$ should be accounted for if trimmed data are desired. The engine must be in a stabilized power condition. Angle-of-attack calculations are quite complicated, and again the reader is referred to Cooper, Hughes, and Rawlings.[33]

In the TACT program, the lift, drag, angle-of-attack, and tail-setting results from each flight (for a given configuration and dynamic pressure) were faired and plotted vs Mach number. These fairings were used to establish a data bank from which a computer could interpolate results for any specified flight conditions. It is these results, then, that are compared to the adjusted model data in the report by Baldwin,[6] and later in this chapter.

## Correlation Results

The most complete and concise wind-tunnel/flight correlation effort found in the open literature is given by Paterson, MacWilkinson, and Blackerby[109] for the C-5A. MacWilkinson, Blackerby, and Paterson[86] complete the C-141 correlation in their report. While the correlation is quite good (Fig. IV.50), the Mach number and $C_L$ range are limited ($0.6 \le M \le 0.8$; $0.2 \le C_L \le 0.6$). Rooney[118] undertakes a much more ambitious task in his F-14 correlation. The variable-sweep configuration and high Mach number capability add signifi-

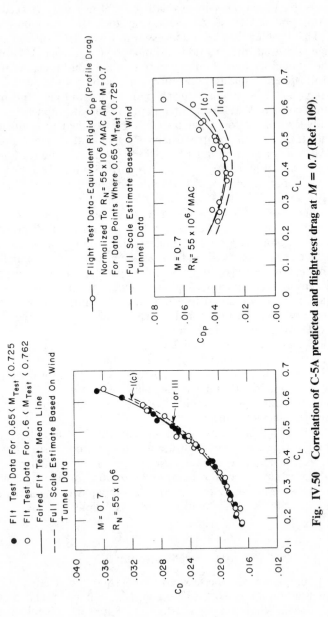

Fig. IV.50   Correlation of C-5A predicted and flight-test drag at $M = 0.7$ (Ref. 109).

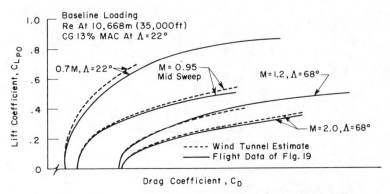

Fig. IV.51   Induced drag variation with Mach number (Ref. 118).

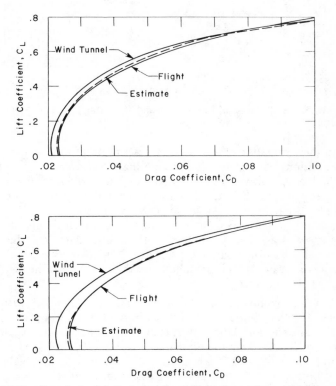

Fig. IV.52   Correlation of wind-tunnel and flight-drag polars at 26 deg sweep.

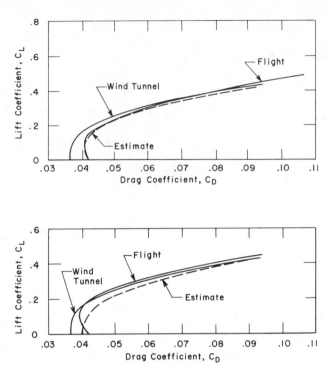

Fig. IV.53   Correlation of wind-tunnel and flight-drag polars at 58 deg sweep.

cantly to the correlation effort. A good description of the airplane and engine instrumentation required to get accurate flight-test drag data is presented in Ref. 118. Rooney also gives a good discussion on flight maneuvers used to develop drag polars. Unfortunately his discussion on corrections to the model data is somewhat limited. At subsonic Mach numbers, the $C_{D_{min}}$ agreement between wind tunnel and flight is good; whereas at high lift the flight data indicate more drag than predicted by the tunnel data. At Mach 0.95 and above, the shape of the flight and predicted drag polars is good; the minimum drag, however, was not well predicted and the plots shown in Fig. IV.51 have been adjusted to match $C_{D_{min}}$.

The TACT program correlated results on an F-111 modified with supercritical wings. Details of the flight test are provided by Cooper, Hughes, and Rawlings[33] and the correlation effort is discussed by Baldwin.[6] Again, low wing sweep (26 deg), subsonic, low dynamic pressure (300–500 psf) correlation is very good up to $C_L \leq 0.6$. At higher Mach numbers and higher dynamic pressure, both $C_{D_{min}}$ and the polar shapes from flight data deviate from the estimates based on wind-tunnel data (Figs. IV.52 and IV.53). It is apparent that the flexibility for this configuration with aft wing-sweep angles at high Mach numbers produce large changes in the drag polar.

Buckner and Webb[18] show some correlation with preliminary data on the YF-16 (Fig. IV.54). The polar shape and minimum drag levels appear to agree

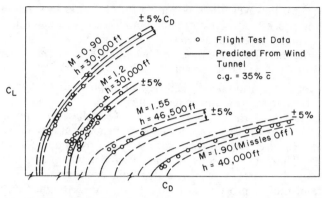

**Fig. IV.54  Preliminary flight data show good agreement with wind-tunnel polar shape predictions (Ref. 18).**

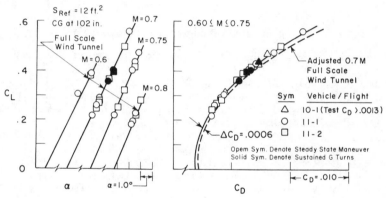

**Fig. IV.55  Wind-tunnel / flight-lift / drag / angle-of-attack correlation Tomahawk Missile (Ref. 120).**

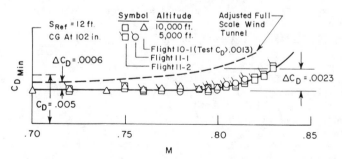

**Fig. IV.56  Flight-test minimum drag variation with Mach number Tomahawk Missile (Ref. 120).**

well (within 5%); however, there is quite a lot of scatter in some of the flight results. In the process of their correlation analysis, it was determined that the model free tunnel longitudinal pressure distribution was influenced by the presence of the model and the modified buoyancy correction accounted for as much as 4% of $C_{D_{min}}$ at 0.6 Mach number. Peterson's[113] correlation attempt on the XB-70 suffers from too few data points over too wide a Mach number and $C_L$ range to be able to draw any general conclusions.

Rooney, Craig, and Lauer's[120] correlation of a full scale Tomahawk cruise missile with flight-test data showed a good correlation at low subsonic speeds (Fig. IV.55). However, at transonic speeds, the correlation of drag was poor (Fig. IV.56). The main reason for this discrepancy appears to be tunnel blockage, which becomes more critical at the higher transonic Mach numbers.

The state-of-the-art procedures used to estimate aerodynamic data in the subsonic/transonic, moderate $\alpha$ range appear adequate. However, care and attention to detail in all areas are required. Additional effort is necessary to fully understand strong compressibility, aeroelastic and viscous effects.

### Annotated Bibliography for Chapter IV

[1]Abercrombie, J. M., "Flight Test Verification of F-15 Performance Predictions," AGARD CP-242, 1977, pp. 17-1–17-13.

[2]Aidala, P. V., Davis, W. H., and Mason, W. H., "Smart Aerodynamic Optimization," AIAA Paper 83-1863, Applied Aerodynamics Conference, Danvers, Mass., July 1983.

[3]Arnaiz, H. H., "Flight-Measured Lift and Drag Characteristics of a Large, Flexible, High Supersonic Cruise Airplane," NASA TM X-3532, May 1977.

[4]Axelson, J. A., "AEROX—Computer Program for Transonic Aircraft Aerodynamics to High Angles of Attack," Vols. I–III, NASA TM X-73,208, Feb. 1977.

[5]Baals, D. D., Robins, A. W., and Harris, R. V. Jr., "Aerodynamic Design Integration of Supersonic Aircraft," Journal of Aircraft, Sept.–Oct. 1970.

[6]Baldwin, A. W., ed., "Symposium on Transonic Aircraft Technology (TACT)," AFFDL-TR-78-100, Aug. 1978.

[7]Bills, G. R., Hink, G. R., and Dornfeld, G. M., "A Method for Predicting the Stability Characteristics of an Elastic Airplane," AFFDL-TR-77-55, 1977.

[8]Blackwell, J. A. Jr., "Preliminary Study of Effect of Reynolds Number and Boundary Layer Transition Location on Shock Induced Separation," NASA TN D-5003, 1969.

[9]Blasius, H., "Boundary Layers in Fluids of Small Viscosity," Dissertation, Göttingen, FRG, 1907.

[10]Bonner, E., "Expanding Role of Potential Theory in Supersonic Aircraft Design," Journal of Aircraft, May 1971, pp. 347–353.

[11]Bonner, E., Clever, W., and Dunn, K., "Aerodynamic Preliminary Analysis II, Part I—Theory," NASA CR 165627, April 1981.

[12]Bonner, E. and Gingrich, P., "Supersonic Cruise/Transonic Maneuver Wing Section Development Study," AFWAL-TR-80-3047, June 1980.

[13]Bowes, G. M., "Aircraft Lift and Drag Prediction and Measurement," AGARD-LS-67, May 1974, pp. 4-1–4-44.

[14]Braslow, A. L., "Use of Grit-Type Boundary-Layer-Transition Trips on Wind Tunnel Models," NASA TN D-3579, Sept. 1966.

[15]Braslow, A. L. and Knox, E. C., "Simplified Method for Determination of Critical

Height of Distributed Roughness Particles for Boundary Layer Transition at Mach Numbers from 0 to 5," NACA TN-4363, 1958.

[16] Bristow, D. R. and Hawk, J. D., "Subsonic Panel Method for the Efficient Analysis of Multiple Geometry Perturbations," NASA CR-3528, March 1982.

[17] Brown, S. W. and Bradley, D., "Uncertainty Analysis for Flight Test Performance Calculations," AIAA Paper 81-2390, AIAA/SETP/SAE Flight Testing Conference, Las Vegas, Nev., Nov. 1981.

[18] Buckner, J. K. and Webb, J. B., "Selected Results from the YF-16 Wind Tunnel Test Program," AIAA Paper 74-619, AIAA 8th Aerodynamic Testing Conference, Bethesda, Md., July 1974.

[19] Butler, S. F. J., "Aircraft Drag Prediction for Project Appraisal and Performance Estimation," AGARD CP-124, 1973, pp. 6-1–6-50, with Appendix.

[20] Butler, S. F. J., "Technical Evaluation Report," AGARD CP-124, 1973, pp. vii-xv.

[21] Carlson, H. W. and Mack, R. J., "Estimation of Wing Nonlinear Aerodynamic Characteristics at Supersonic Speeds," NASA TP-1718, Nov. 1980.

[22] Carlson, H. W., Mack, R. J., and Barger, R. L., "Estimation of Attainable Leading-Edge Thrust for Wings at Subsonic and Supersonic Speeds," NASA TP 1500, Oct. 1979.

[23] Carlson, H. W. and Miller, D. S., "Numerical Methods for the Design and Analysis of Wings at Supersonic Speed," NASA TN D-7713, Dec. 1974.

[24] Carlson, H. W. and Walkey, K. B., "A Computer Program for Wing Subsonic Aerodynamic Performance Estimates Including Attainable Thrust and Vortex Lift Effect," NASA CR-3515, March 1982.

[25] Carmichael, R. L. and Woodward, F. A., "An Integrated Approach to the Analysis and Design of Wings and Wing-Body Combinations in Supersonic Flow," NASA TN D-3685, Oct. 1966.

[26] Carmichael, R. L. and Erickson, L. L., "PAN AIR: A Higher-Order Panel Method for Predicting Subsonic or Supersonic Linear Potential Flows About Arbitrary Configurations," AIAA Paper 81-1255, June 1981.

[27] Caughey, D. A., "A (Limited) Perspective on Computational Aerodynamics," FDA-80-07, Fluid Dynamics and Aerodynamics Program—Cornell University, July 1980.

[28] Cenko, A., "PAN AIR Applications to Complex Configurations," AIAA Paper 83-0007, Aerospace Sciences Meeting, Reno, Nev., Jan. 1983.

[29] Chapman, D. R., Mark, H., and Pirtle, M. W., "Computers vs Wind Tunnels for Aerodynamic Flow Simulations," *Progress in Astronautics and Aeronautics*, AIAA, New York, April 1975, pp. 21–30.

[30] Chapman, D. R., "Computational Aerodynamics Development and Outlook," AIAA Paper 79-0129R, 17th Aerospace Sciences Meeting, Jan. 1979.

[31] Chen, A. W. and Tinoco, E. N., "PAN AIR Applications to Aero-Propulsion Integration," AIAA Paper 83-1368, 19th Joint Propulsion Conference, Seattle, Wash., June 1983.

[32] Coakley, T. J., "Turbulence Modeling Methods for the Compressible Navier-Stokes Equations," AIAA Paper 83-1693, 16th Fluid and Plasma Dynamics Conference, Danvers, Mass., July 1983.

[33] Cooper, J. M., Hughes, D. L., and Rawlings, K., "Transonic Aircraft Technology —Flight-Derived Lift and Drag Characteristics," AFFTC-TR-12, July 1977.

[34] Craig, R. E. and Reich, R. J., "Flight Test Aerodynamic Drag Characteristics Development and Assessment of Inflight Propulsion Analysis Methods for AGM-109 Cruise Missile," AIAA Paper 81-2423, AIAA/SETP/SFTE/SEA/ITEA/IEEE 1st Flight Testing Conference, Las Vegas, Nev., Nov. 1981.

[35] Crosthwait, E. L., Kennon, I. G. Jr., and Roland, H. L. et al., "Preliminary Design

Methodology for Air-Induction Systems," SEG-TR-67-1, Jan. 1967.

[36] Daugherty, J. C., "Wind Tunnel/Flight Correlation Study of Aerodynamic Characteristics of a Large Flexible Supersonic Cruise Airplane (SB-70-1)," NASA TP-1514, Nov. 1979.

[37] DeLaurier, J., "Drag of Wings with Cambered Airfoils and Partial Leading-Edge Suction," *Journal of Aircraft*, Oct. 1983, pp. 882–886.

[38] Dvorak, F. A., Woodward, F. A., and Maskew, B., "A Three-Dimensional Viscous/Potential Flow Interaction Analysis Method for Multi-Element Wings," NASA CR-152012, 1977.

[39] Erickson, L. L. and Strande, S. M., "PAN AIR Evolution and New Directions…Transonics for Arbitrary Configurations," AIAA Paper 83-1831, Applied Aerodynamics Conference, Danvers, Mass., July 1983.

[40] Firmin, M. C. P. and Potter, J. L., "Wind Tunnel Compatibility Related to Test Sections, Cryogenics, and Computer-Wind Tunnel Integration," AGARD AR-174, April 1982.

[41] Fornasier, L., "Calculation of Supersonic Flow Over Realistic Configurations by an Updated Low-Order Panel Method," AIAA Paper 83-0010, 21st Aerospace Sciences Meeting, Jan. 1983.

[42] "Future Computer Requirements for Computational Aerodynamics," NASA Workshop Series—Mixed Papers, NASA CP-2032, Oct. 1977.

[43] Gaudet, L. and Winter, K. G., "Measurements of the Drag of Some Characteristic Aircraft Excrescences Immersed in Turbulent Boundary Layers," AGARD CP-124, 1973, pp. 4-1–4-12.

[44] Geddes, J. P., "Taking an Airliner from Certification to Airline Acceptance—the DC-10-30," *Interavia*, June 1973, pp. 609–611.

[45] Gilmer, B. R. and Bristow, D. R., "Analysis of Stalled Airfoils by Simultaneous Perturbations to Viscous and Inviscid Equations," AIAA Paper 81-023, 14th Fluid and Plasma Dynamics Conference, Palo Alto, Calif., June 1981.

[46] Gingrich, P. B. and Bonner, E., "Wing Design for Supersonic Cruise/Transonic Maneuver Aircraft," ICAS-82-5.8.2, 13th Congress of the International Council of the Aeronautical Sciences, Seattle, Wash., Aug. 1982.

[47] Glauert, H., *The Elements of Aerofoil and Airscrew Theory*, 2nd ed., Cambridge University Press, New York, 1959.

[48] Graves, R. A., "Computational Fluid Dynamics—The Coming Revolution," *Progress in Astronautics and Aeronautics*, AIAA, New York, March 1982, pp. 20–28.

[49] Grellmann, H. W., "YF-17 Full Scale Minimum Drag Prediction," AGARD CP-242, 1977, pp. 18-1–18-12.

[50] Hahn, M., Brune, G. W., Rubbert, P. E., and Nark, T. C., "Drag Measurements of Upswept Afterbodies and Analytical Study on Axisymmetric Separation," AFFDL-TR-73-153, Feb. 1974.

[51] Haines, A. B., "Subsonic Aircraft Drag: An Appreciation of Present Standards," *The Aeronautical Journal of the Royal Aeronautical Society*, Vol. 72, March 1968, pp. 253–266.

[52] Haines, A. B., "Turbulence Modelling," *Aeronautical Journal*, Aug./Sept. 1982, pp. 269–277.

[53] Hall, M. G., "Computational Fluid Dynamics—A Revolutionary Force in Aerodynamics," AIAA Paper 81-1014, 1981.

[54] Harris, R. V., "An Analysis and Correlation of Aircraft Wave Drag," NASA TM X-947, 1964.

[55] Harris, R. V. and Landram, E. J., "Drag Characteristics of a Series of Low-Drag Bodies of Revolution at Mach Numbers from 0.6 to 4.0," NASA TN D-3163, Dec. 1965.

[56] Henne, P. A., "Inverse Transonic Wing Design Method," *Journal of Aircraft*, Feb. 1981.

[57] Henne, P. A., Dahlin, J. A., and Peavey, C. C., "Applied Computational Transonics—Capabilities and Limitations," Douglas Paper 7025, *Transonic Perspective Symposium*, *NASA Ames*, Feb. 1981.

[58] Hess, J. L., "A Higher Order Panel Method for Three-Dimensional Potential Flow," Naval Air Development Center, Rept. No. NADC 77166-30, June 1979.

[59] Hess, J. L. and Friedman, D. M., "An Improved Higher Order Panel Method for Three-Dimensional Lifting Potential Flow," Naval Air Development Center, Final Rept. NADC-79277-60, Dec. 1981.

[60] Hicks, R. M., "Transonic Wing Design Using Potential-Flow Codes—Successes and Failures," SAE 810565, 1981.

[61] "High Angle of Attack Aerodynamics," AGARD CP-247, 1973.

[62] Hitzel, S. M. and Schmidt, W., "Slender Wings and Leading-Edge Vortex Separation—A Challenge for Panel-Methods and Euler-Codes," AIAA Paper 83-0562, 21st Aerospace Sciences Meeting, Reno, Nev., Jan. 1983.

[63] Hoak, D. E., *USAF Stability and Control DATCOM*, Rev., April 1978.

[64] Hoerner, S. F., *Fluid Dynamic Drag*, published by the author, 1965.

[65] Hopps, R. H. and Danforth, E. C. B., "Correlation of Wind Tunnel and Flight Test Data for the Lockheed L-1011 Tristar Airplane," AGARD CP-242, 1977, pp. 21-1–21-12.

[66] Hunt, B. L. and Gowadia, N. S., "Determination of Thrust and Throttle-Dependent Drag for Fighter Aircraft," AIAA Paper 81-1692, Aircraft Systems and Technology Conference, Dayton, Ohio, Aug. 1981.

[67] Jackson, C. M. Jr. and Smith, R. S., "A Method for Determining the Total Drag of a Pointed Body of Revolution in Supersonic Flow with Turbulent Boundary Layer," NASA TN D-5046, March 1969.

[68] James, R. M., "An Investigation of Induced Drag for Field Calculation Methods for Planar Distributions," MDC J7046, Dec. 1975.

[69] Johnson, K. J., "Accuracies and Limitations of a Statistical Method for Computing Lift and Drag Due to Lift of Fighter Type Wing-Body Configurations," AFFDL/FXM-TM-73-46, Aug. 1973.

[70] Kjelgaard, S. O., "Evaluation of a Surface Panel Method Coupled with Several Boundary Layer Analyses," AIAA Paper 83-0011, 21st Aerospace Sciences Meeting, Reno, Nev., Jan. 1983.

[71] Knaus, A., "A Technique to Determine Lift and Drag Polars in Flight," *Journal of Aircraft*, Vol. 20, July 1983, pp. 587–593.

[72] Kolesar, C. E. and May, F. W., "An Aftbody Drag Prediction Technique for Military Airlifters," AIAA Paper 83-1787, Applied Aerodynamics Conference, Danvers, Mass., July 1983.

[73] Korkegi, R. H., "The Impact of CFD on Development Test Facilities—A National Research Council Projection," AIAA Paper 83-1764, 16th Fluid and Plasma Dynamics Conference, Danvers, Mass., July 1983.

[74] Kuhlman, J. M. and Ku, T. J., "Numerical Optimization Techniques for Bound Circulation Distribution for Minimum Induced Drag of Nonplanar Wings: Computer Program Documentation," NASA CR-3458, Aug. 1981.

[75] Kulfan, R. M., "Application of Hypersonic Favorable Aerodynamic Interference Concepts to Supersonic Aircraft," AIAA Paper 78-1458, Aircraft Systems and Technology Conference, Aug. 1978.

[76] Kulfan, R. M., "Wing Geometry Effects on Leading-Edge Vortices," AIAA Paper 79-1872, Aircraft Systems and Technology Meeting, New York, Aug. 1979.

[77] Kulfan, R. M. and Sigalla, A., "Real Flow Limitations in Supersonic Airplane

Design," AIAA Paper 78-147, Jan. 1978.

[78] Kulfan, R. M., Yoshihara, H., Lord, B. J., and Friebel, G. O., "Application of Supersonic Favorable Aerodynamic Interference to Fighter Type Aircraft," AFFDL-TR-78-33, April 1978.

[79] Kutler, P., "A Perspective of Theoretical and Applied Computational Fluid Dynamics," AIAA Paper 83-0037, 21st Aerospace Sciences Meeting, Reno, Nev., Jan. 1983.

[80] Landrum, E. J. and Miller, D. S., "Assessment of Analytic Methods for the Prediction of Supersonic Flow Over Bodies," *AIAA Journal*, Feb. 1981.

[81] Leyman, C. S. and Markham, T., "Prediction of Supersonic Aircraft Aerodynamic Characteristics," AGARD-LS-67, May 1974, pp. 5-1-5-52.

[82] Liepmann, H. W., "The Rise and Fall of Ideas in Turbulence," *American Scientist*, Vol. 67, March-April 1979, pp. 221-228.

[83] Lomax, H., "Some Prospects for the Future of Computational Fluid Dynamics," *AIAA Journal*, Vol. 20, Aug. 1982.

[84] Longo, J., Schmidt, W., and Jameson, A., "Viscous Transonic Airfoil Flow Simulation," *Zeitschrift Flugwissenschaft Wettraumforsh. 7*, Vol. 1, 1983.

[85] Lotz, M. and Friedel, H., "Analysis of Error Sources in Predicted Flight Performance," *Performance Prediction Methods*, AGARD-CP-242, 1977, pp. 14-1-14-11.

[86] MacWilkinson, D. G., Blackerby, W. T., and Paterson, J. H., "Correlation of Full Scale Drag Prediction with Flight Measurements on the C-141A Aircraft—Phase II, Wind Tunnel Test, Analysis, and Prediction Techniques. Vol. I—Drag Predictions, Wind Tunnel Data Analysis and Correlation," NASA CR-2333, Feb. 1974.

[87] Marvin, J. G., "Turbulence Modeling for Computational Aerodynamics," AIAA Paper 82-0164, 20th Aerospace Sciences Meeting, Orlando, Fla., Jan. 1982.

[88] Maskew, B., "Prediction of Subsonic Aerodynamic Characteristics—A Case for Low-Order Panel Methods," AIAA Paper 81-0252, Jan. 1981.

[89] Maskew, B., Rao, B. M., and Dvorak, F. M., "Prediction of Aerodynamic Characteristics for Wings with Extensive Separations," AGARD CP-291, Feb. 1981, pp. 31-1-31-15.

[90] Mason, W. H., "Wing-Canard Aerodynamics at Transonic Speeds—Fundamental Considerations on Minimum Drag Span Loads," AIAA Paper 82-0097, 20th Aerospace Sciences Meeting, Jan. 1982.

[91] Mason, W. H., "SC3, A Wing Concept for Supersonic Maneuvering," AIAA Paper 83-1858, July 1983.

[92] Mason, W. H. and Rosen, B. S., "The COREL and W12SC3 Computer Programs for Supersonic Wing Design and Analysis," NASA CR-3676, 1983.

[93] McCluney, B. and Marshall, J., "Drag Development of the Belfast," *Aircraft Engineering*, Oct. 1967, pp. 33-37.

[94] McGeer, T. and Shevell, R. S., "A Method for Estimating the Compressibility Drag of an Airplane," SUDAAR 535, Jan. 1983.

[95] McKinney, L. W., ed., *Wind Tunnel/Flight Correlation—1981*, NASA Miniworkshop Series, NASA Pub. 2225, Nov. 1981.

[96] Middleton, W. D. and Carlson, H. W., "Numerical Method of Estimating and Optimizing Supersonic Aerodynamic Characteristics of Arbitrary Planform Wings," *Journal of Aircraft*, July-Aug. 1965.

[97] Middleton, W. D. and Lundry, J. L., "A System for Aerodynamic Design and Analysis of Supersonic Aircraft," NASA CR-3351, Dec. 1980.

[98] Miller, S. G. and Youngblood, D. B., "Applications of USSAERO-B and the PAN AIR Production Code to the CDAF Model—A Canard/Wing Configuration," AIAA Paper 83-1829, AIAA Applied Aerodynamics Conference, Danvers, Mass., July 1983.

[99] Miranda, L. R., "A Perspective of Computational Aerodynamics from the View-

point of Airplane Design Applications," AIAA Paper 82-0018, Orlando, Fla., Jan. 1982.

[100] Miranda, L. R., Elliott, R. D., and Baker, W. M., "A Generalized Vortex Lattice Method for Subsonic and Supersonic Flow Applications," NASA CR-2865, Dec. 1977.

[101] Morris, S. J., Neums, W. P., and Bailey, R. O., "A Simplified Analysis of Propulsion Installation Losses for Computerized Aircraft Design," NASA TM X-73, 136, Aug. 1976.

[102] Murillo, L. E. and McMasters, J. H., "A Method for Predicting Low-Speed Aerodynamic Characteristics of Transport Aircraft," AIAA Paper 83-1845, Applied Aerodynamics Conference, Danvers, Mass., July 1983.

[103] Murman, E. M. and Cole, J. D., "Inviscid Drag at Transonic Speeds," AIAA Paper 74-540, 7th Fluid and Plasma Dynamics Conference, Palo Alto, Calif., June 1974.

[104] Nelson, D. W., Gornstein, R. J., and Dornfeld, G. M., "Prediction of Maneuvering Drag Polars Including Elasticity Effects for AFTI-F-111," AIAA Paper 81-1658, Aug. 1981.

[105] Nixon, D., *Progress in Astronautics and Aeronautics: Transonic Aerodynamics*, Vol. 81, AIAA, New York, 1982.

[106] Norton, D. A., "Airplane Drag Prediction," *Annals New York Academy of Sciences*, 1968, pp. 306–328.

[107] O'Conner, W. M., "Lift and Drag Prediction in Computer Aided Design," Vols. I and II, ASD/XR-73-8, April 1973.

[108] Paterson, J. H., "Scaling Effects on Drag Prediction," AGARD-LS-37-10, June 1970, pp. 4-1–4-12.

[109] Paterson, J. H., MacWilkinson, D. G., and Blackerby, W. T., "A Survey of Drag Prediction Techniques Applicable to Subsonic and Transonic Aircraft Design," AGARD-CP-124, 1973, pp. 1-1–1-38.

[110] Perrier, P., "Computational Fluid Dynamics Around Complete Aircraft Configurations," ICAS Paper 82.6.11, 13th Congress of the International Council of the Aeronautical Sciences, Seattle, Wash., Aug. 1982.

[111] Paynter, G. C., Vaidyanathan, T. S., Maskew, B., and Dvorak, F. A., "Experience with Hybrid Aerodynamic Methods," AIAA Paper 83-1819, Applied Aerodynamics Conference, Danvers, Mass., July 1983.

[112] Peterson, J. B., "A Comparison of Experimental and Theoretical Results for the Compressible Turbulent-Boundary-Layer Skin Friction with Zero Pressure Gradient," NASA TN D-1795, March 1963.

[113] Peterson, J. B. Jr., "Wind Tunnel/Flight Correlation Program on XB-70-1," NASA CP-2225, Nov. 1981.

[114] Pittman, J. L., "Preliminary Supersonic Analysis Methods Including High Angle of Attack," AIAA Paper 82-0938, AIAA/ASME 3rd Joint Thermophysics, Fluids, Plasma, and Heat Transfer Conference, St. Louis, Mo., June 1982; *Journal of Aircraft*, Vol. 20, Sept. 1983, pp. 784–790.

[115] Poisson-Quinton, P., "From Wind Tunnel to Flight, the Role of the Laboratory in Aerospace Design," *Journal of Aircraft*, May–June 1968, pp. 193–214.

[116] Robins, A. W., Carlson, H. W., and Mack, R. J., "Supersonic Wings with Significant Leading-Edge Thrust at Cruise," NASA TP-1632, April 1980.

[117] Rodi, W., "Progress in Turbulence Modeling for Incompressible Flows," AIAA Paper 81-0045, 19th Aerospace Sciences Meeting, Jan. 1981.

[118] Rooney, E. C., "Development of Techniques to Measure In-Flight Drag of a U.S. Navy Fighter Airplane and Correlation of Flight Measured Drag with Wind Tunnel Data," AGARD-CP-124, Oct. 1973.

[119] Rooney, E. C. and Craig, R. E., "Development of Techniques and Correlation of

Results to Accurately Establish the Lift/Drag Characteristics of an Air Breathing Missile from Analytical Predictions, Subscale and Full Scale Wind Tunnel Tests and Flight Tests," *Performance Prediction Methods*, AGARD CP-242, 1977, pp. 16-1–16-18.

[120] Rooney, E. C., Craig, R. E., and Lauer, R. F., "Correlation of Full Scale Wind Tunnel and Flight Measured Aerodynamic Drag," AIAA Paper 77-996, AIAA/SAE 13th Propulsion Conference, Orlando, Fla., July 1977.

[121] Rubbert, P. E. and Tinoco, E. N., "Impact of Computational Methods on Aircraft Design," AIAA Paper 83-2060, Atmospheric Flight Mechanics Conference, Gatlinburg, Tenn., Aug. 1983.

[122] Schemensky, R. T., "Development of an Empirically Based Computer Program to Predict the Aerodynamic Characteristics of Aircraft," Vols. I and II, AFFDL-TR-73-144, Nov. 1973.

[123] Schlichting, H., *Boundary-Layer Theory*, 6th Ed., McGraw-Hill, New York, N. Y., 1968.

[124] Schmidt, W., "Advanced Numerical Methods for Analysis and Design in Aircraft Aerodynamics," *International Journal of Vehicle Design*, Technological Advances in Vehicle Design Series, SP6, Applications of New Techniques for Analysis and Design of Aircraft Structures, 1983, pp. 2–37.

[125] Schmidt, W. and Jameson, A., "Euler Solvers as an Analysis Tool for Aircraft Aerodynamics," *Recent Advances in Numerical Methods in Fluids*, Vol. 4, 1983.

[126] Sears, W. R., "On Projectiles of Minimum Wave Drag," *Quarterly Journal of Applied Mathematics*, Jan. 1947, pp. 361–366.

[127] Shankar, V. and Goebel, T., "A Numerical Transformation Solution Procedure for Closely Coupled Canard-Wing Transonic Flows," AIAA Paper 83-0502, 21st Aerospace Sciences Meeting, Reno, Nev., Jan. 1983.

[128] Shrout, B. L., "Aerodynamic Characteristics at Mach 2.03 of a Series of Curved-Leading-Edge Wings Employing Various Degrees of Twist and Camber," NASA TN D-3827, Feb. 1967.

[129] Shrout, B. L. and Covell, P. F., "Aerodynamic Characteristics of a Series of Bodies with Variations in Nose Camber," NASA TP-2206, Sept. 1983.

[130] Sidwell, K. W., Baruah, P. K., and Bussoletti, J. E., "PAN AIR—A Computer Program for Predicting Subsonic or Supersonic Linear Potential Flows About Arbitrary Configurations Using a Higher Order Panel Method," NASA CR-3252, May 1980.

[131] Simon, W. E., Ely, W. L., Niedling, L. G., and Voda, J. J., "Prediction of Aircraft Drag Due to Lift," AFFDL-TR-71-84, July 1971.

[132] Smelt, R., "The Role of Wind Tunnels in Future Aircraft Development," *The Aeronautical Journal*, Nov. 1978, pp. 467–475.

[133] Smith, A. M. O., "The Boundary Layer and I," *AIAA Journal*, Nov. 1981.

[134] Smith, K. G., "Methods and Charts for Estimating Skin Friction Drag in Wind Tunnel Tests with Zero Heat Transfer," Technical Note No. AERO 2980, Royal Aircraft Establishment, Ministry of Aviation, London, Aug. 1964.

[135] Snodgrass, R. R., "A Computerized Method for Predicting Drag and Lift for Aeronautical Systems," ASD-ENF-TM-76-1, Oct. 1976.

[136] Sommer, S. C. and Short, B. J., "Free-Flight Measurements of Turbulent-Boundary-Layer Skin Friction in the Presence of Severe Aerodynamic Heating at Mach Numbers from 2.8 to 7.0," NACA TN-3391, March 1955.

[137] Sorrells, R. B., Jackson, M. W., and Czarnecki, K. R., "Measurement by Wake Momentum Surveys at Mach 1.61 and 2.01 of Turbulent Boundary-Layer Skin Friction on Five Swept Wings," NASA TN D-3764, Dec. 1966.

[138] Stallings, R. L. Jr. and Lamb, M., "Wind Tunnel Measurements and Comparison with Flight of the Boundary Layer and Heat Transfer on a Hollow Cylinder at Mach 3," NASA TP 1789, Dec. 1980.

[139]Stancil, R. T., "Improved Wave Drag Predictions Using Modified Linear Theory," *Journal of Aircraft*, Jan. 1979, pp. 41–46.

[140]Sytsma, H. S., Hewitt, B. L., and Rubbert, P. E., "A Comparison of Panel Methods for Subsonic Flow Computation," AGARD AG-241, Feb. 1979.

[141]Tendeland, T., "Effects of Mach Number and Wall Temperature Ratio on Turbulent Heat Transfer at Mach Numbers from 3 to 5," NACA TN 4236, April 1958.

[142]Thomas, J. L. and Miller, D. S., "Numerical Comparisons of Panel Methods at Subsonic and Supersonic Speeds," AIAA Paper 79-0404, 17th Aerospace Sciences Meeting, New Orleans, La., Jan. 1979.

[143]Thomas, J. L., Luckring, J. M., and Sellers, W. L. III, "Evaluation of Factors Determining the Accuracy of Linearized Subsonic Panel Methods," AIAA Paper 83-1821, Applied Aerodynamics Conference, Danvers, Mass., July 1983.

[144]Tinoco, E. N., Johnson, F. T., and Freeman, L. M., "Application of a Higher Order Panel Method to Realistic Supersonic Configurations," AIAA Paper 79-0274R, Jan. 1980.

[145]Tinoco, E. N. and Rubbert, P. E., "Experience, Issues, and Opportunities in Steady Transonics," AERO-B8111-P82-006, *Computational Methods in Potential Aerodynamics (ICTS) Short Course*, Amalfi, Italy, May–June 1982.

[146]Tinoco, E. N., and Rubbert, P. E., "Panel Methods: PAN AIR," AERO-B8111-P82-005, *Computational Methods in Potential Aerodynamics (ICTS) Short Course*, Amalfi, Italy, May–June 1982.

[147]Tjonneland, E., Paynter, G., and May, F., "Progress Toward Airframe/Propulsion Design Analysis," Paper 8, *AGARD Flight Mechanics Panel Symposium on Sustained Supersonic Cruise/Maneuver*, Brussels, Belgium, Oct. 1983.

[148]Tomasetti, T. A., "Statistical Determination of the Lift and Drag of Fighter Aircraft," GAM/AE/73-1, June 1972.

[149]Towne, M. C. et al., "PAN AIR Modeling Studies," AIAA Paper 83-1830, Applied Aerodynamics Conference, Danvers, Mass., July 1983.

[150]Waggoner, E. G., "Computational Transonic Analysis for a Supercritical Transport Wing-Body Configuration," AIAA Paper 80-0129, 18th Aerospace Sciences Meeting, Pasadena, Calif., Jan. 1980.

[151]*Wall Interference in Wind Tunnels*, AGARD CP-335, May 1982.

[152]Webb, T. S., Kent, D. R., and Webb, J. B., "Correlation of F-16 Aerodynamics and Performance Predictions with Early Flight Test Results," AGARD CP-242, 1977, pp. 19-1–19-17.

[153]White, F. M. and Christoph, G. H., "A Simple New Analysis of Compressible Turbulent Two-Dimensional Skin Friction Under Arbitrary Conditions," AFFDL-TR-70-133, Feb. 1971.

[154]White, F. M. and Christoph, G. H., "Rapid Engineering Calculation of Two-Dimensional Turbulent Skin Friction," AFFDL-TR-72-136, Sup. 1, Nov. 1972.

[155]White, F. M., Lessman, R. C., and Christoph, G. H., "A Simplified Approach to the Analysis of Turbulent Boundary Layers in Two and Three Dimensions," AFFDL-TR-136, Nov. 1972.

[156]Williams, J., "Ground/Flight Test Techniques and Correlation," AGARD CP-339, Oct. 1982.

[157]Williams, J., "Technical Evaluation Report on the Flight Mechanics Panel Symposium on Ground/Flight Test Techniques and Correlation," AGARD AR-191, June 1983; also AGARD CP-339, Oct. 1982.

[158]Williams, J., "Synthesis of Responses to AGARD-FMP Questionnaire on Prediction Techniques and Flight Correlation," AGARD CP-339, 1982, pp. 24-1–24-30.

[159]Williams, J., "Aircraft Performance—Prediction Methods and Optimization," AGARD LS-56, London.

[160] Williams, J., "Prediction Methods for Aircraft Aerodynamic Characteristics," AGARD LS-67, May 1974.

[161] Wood, R. A., "Brief Prepared Remarks for Session II, Performance Correlation," AGARD CP-339, 1982, pp. 9A-1–9A-4.

[162] Wood, R. M., Dollyhigh, S. M., and Miller, D. S., "An Initial Look at the Supersonic Aerodynamics of Twin-Fuselage Aircraft Concepts," ICAS Paper 82-1.8.3, 13th Congress of the International Council of the Aeronautical Sciences with AIAA Aircraft Systems and Technology Conference, Seattle, Wash., Aug. 1982.

[163] Wood, R. M. and Miller, D. S., "Experimental Investigation of Leading-Edge Thrust at Supersonic Speeds," NASA TP-2204, Sept. 1983.

[164] Wood, R. M., Miller, D. S., Hahne, D. E., Niedling, L., and Klein, J., "Status Review of a Supersonically Biased Fighter Wing-Design Study," AIAA Paper 83-1857, Applied Aerodynamics Conference, Danvers, Mass., July 1983.

[165] Woodward, F. A., "Development of the Triplet Singularity for the Analysis of Wings and Bodies in Supersonic Flow," NASA CR-3466, Sept. 1981.

[166] Wu, J. C., Hackett, J. E., and Lilley, D. E., "A Generalized Wake-Integral Approach for Drag Determination in Three-Dimensional Flows," AIAA Paper 79-0279, Aerospace Sciences Meeting, New Orleans, La., Jan. 1979.

[167] Yoshihara, H., "Special Course on Subsonic/Transonic Aerodynamic Interference for Aircraft," AGARD R-712, May 1983.

[168] Yoshihara, H. and Spee, B. M., "Applied Computational Transonic Aerodynamics," AGARD AG-266, Aug. 1982.

[169] Young, A. D., "The Calculation of the Total and Skin Friction Drags of Bodies of Revolution at Zero Incidence," ARC R&M 1874, April 1939.

[170] Young, A. D., Patterson, J. H., and Jones, J. L., "Aircraft Excrescense Drag," AGARD AG-264, July 1981.

[171] Youngren, H. H., Bouchard, E. E., Coopersmith, R. M., and Miranda, L. R., "Comparison of Panel Method Formulation and Its Influence on the Development of QUADPAN, an Advanced Low Order Method," AIAA Paper 83-1827, Applied Aerodynamics Conference, Danvers, Mass., July 1983.

[172] Yu, N. J., Chen, H. C., Samant, S. S., and Rubbert, P. E., "Inviscid Drag Calculations for Transonic Flows," AIAA Paper 83-1928, Computational Fluid Dynamics Conference.

# Chapter V

# Throttle-Dependent Forces

Douglas L. Bowers* and Gordon Tamplin*
*Flight Dynamics Laboratory, Wright Patterson Air Force Base, Ohio*

## Introduction

$A$IRCRAFT propulsion systems are sized to satisfy critical mission points, such as subsonic cruise or high-altitude takeoff for transports and maximum acceleration or high-altitude dash for tactical aircraft. This sizing determines the inlet and nozzle areas required, and any variable geometry to be incorporated. Any time the aircraft flies at other than the design point, the propulsion system is operating in an off-design condition with attendant performance penalties. Throttle-dependent forces are a product of changes in engine power setting and the subsequent inlet flow and nozzle geometry variations. The issue of throttle-dependent forces becomes more critical as aircraft are required to operate over wider Mach number ranges, as is the case for tactical aircraft, or to finer performance tolerances, as is the case for transports. The impact of these forces on specific fuel consumption, shown in Fig. V.1, is large for high-performance fighters and has a significant, but diminished, impact in the case of subsonic transport.

For tactical aircraft, Ref. 9 points out that increasing the maximum-design Mach number leads to larger inlet capture areas and therefore greater off-design losses when operating at subsonic cruise Mach numbers. In addition, the required nozzle area is increased, contributing to large drag penalties as a result of high nozzle boattail angles at reduced Mach numbers. Two examples of required inlet/nozzle area variability for a Mach 2.5 tactical aircraft are shown in Figs. V.2 and V.3. The engine requirement can range from 60 to 100% of that air which can be brought into the propulsion system. When operating at capture areas that are substantially less than 1.0, the incurred penalty can be substantial. The nozzle area variation, as a function of Mach number and engine flow, is shown in Fig. V.3. For efficient internal perfor-

---

*Aerospace Engineer.

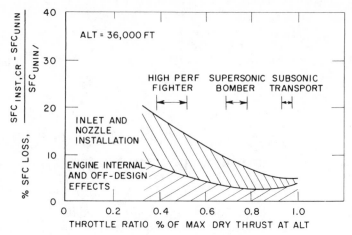

Fig. V.1   Installation penalties at subsonic cruise.

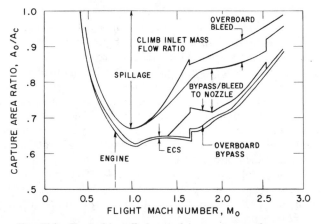

Fig. V.2   Typical installed propulsion system performance.

mance across the aircraft mission profile, the nozzle exit-to-throat area ratio must vary from 1.0 to 2.6. The associated external geometry variations result in drag penalties during subsonic cruise. The combined impact (inlet flow and nozzle geometry) of these penalties on specific fuel consumption for a Mach 2.5 aircraft at Mach 0.9 cruise can be as large as 22% (Fig. V.4).

Although not required to operate over the wide Mach number range required for tactical aircraft, the subsonic transport is driven by a requirement for finely tuned performance. The subsonic inlet is sized for cruise altitude and Mach number and generally requires no variable geometry other than blow-in doors for additional inlet area at takeoff. For this class of installation, the

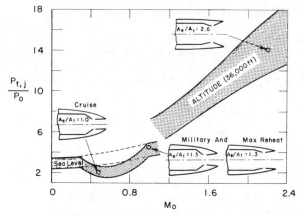

**Fig. V.3    Required variation of nozzle geometry.**

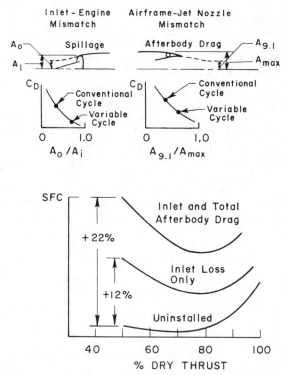

**Fig. V.4    Representative installation losses for a Mach 2.5 aircraft with a mixed flow turbofan engine.**

inlet-related throttle-dependent forces at off-design operation can be 10 to 100 times greater than at cruise.[3]

The importance attributed to throttle-dependent forces, for both tactical and subsonic transport aircraft, is demonstrated by the many efforts to determine their impact across the flight regime. To aid in the understanding of this critical area, this chapter will provide basic definitions, examples at various flight points, and an examination of techniques for determining these forces.

## Basic Concepts and Examples

Throttle-dependent forces include all internal and external forces acting on the aircraft that change with engine power setting. The engine power setting changes affect not only the forces on the engine streamtube but also the external forces on the aircraft related to the inlet and nozzle flowfields. A complete set of basic definitions relevant to throttle-dependent forces is included at the end of this chapter. This section will present concepts and examples relating to inlet and nozzle throttle-dependent forces.

### Supersonic/Transonic Aircraft Inlets

An aircraft inlet is used to capture freestream air and reduce its velocity so the engine can process it in a stable, efficient manner. In order to minimize compressor work, inlet diffusion should be accomplished with a minimum of total pressure loss. The inlet must also deliver the working fluid with minimum distortion and a uniform velocity profile—all over a wide range of Mach number, angle of attack, yaw angle, and engine demand. The supersonic inlet for a tactical aircraft must also be able to provide the maximum-demand airflow which usually occurs at maneuver or acceleration conditions. When the aircraft is at a subsonic cruise condition at 0.9 Mach number, the engine is processing a mass flow associated with 40–60% maximum dry thrust.[12] The inlet, however, is still attempting to process a larger mass flow closer to maximum demand. Figure V.5 shows the excess airflow the inlet provides subsonically for a Mach 3.0 design aircraft. The forces resulting from handling

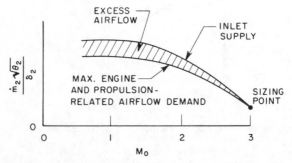

Fig. V.5   Inlet/engine flow matching for a supersonic cruise sizing point.

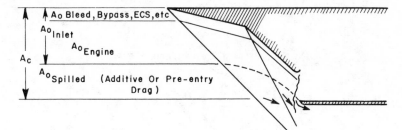

$A_c$ — $A_{0}$ Bleed, Bypass, ECS, etc
$A_{0}$ Inlet
$A_{0}$ Engine
$A_{0}$ Spilled  (Additive Or Pre-entry Drag)

Fig. V.6   Inlet spillage flow schematic.

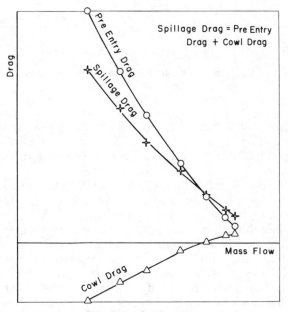

Drag

Pre Entry Drag
Spillage Drag

Spillage Drag = Pre Entry
Drag + Cowl Drag

Mass Flow

Cowl Drag

Fig. V.7   Spillage drag.

this excess air are referred to as spillage drag and bypass drag. The general flow conditions for a supersonic inlet is shown in Fig. V.6. Spillage drag, the result of the inlet operating on freestream air not demanded by the engine, consists of additive drag and lip suction components. Additive drag (also referred to as pre-entry drag) is a calculated force used in thrust-drag book-keeping procedures and is the loss in momentum from freestream to the inlet entrance of that flow influenced by the inlet capture area $A_c$. For a Mach 3.0 cruise aircraft, the additive drag can be as much as 20% of the total drag at subsonic cruise.[5] Lip suction, sometimes referred to as cowl drag or cowl suction, is the result of the spilled inlet airflow moving along the external inlet cowl surface. This term may be a positive or negative axial force depending on

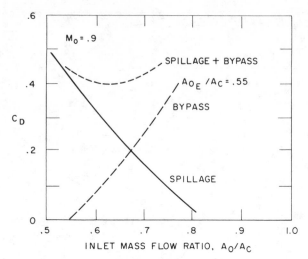

**Fig. V.8   Tradeoff of spillage and bypass drag.**

the flow characteristics and surface contours. Spillage drag is the sum of additive drag and lip suction and is a function of inlet mass flow (or throttle setting). Generally the spillage drag decreases as engine demand mass flow increases (Fig. V.7).

A portion of the air taken through the inlet may not be used to produce engine thrust. This excess airflow, or bypass air, is brought onboard the aircraft, worked by the inlet compression system, and then dumped upstream of the engine face through doors or slots or routed to the nozzle. This results in a force due to the loss in momentum in the flight direction between freestream and exit conditions and the change in local external forces on surfaces washed by the bypassed air. As will be seen in the examples, this bypass flow can affect not only lift and drag but also aircraft stability and control. The bypass drag and spillage drag are related and can be traded off against each other. Both are a function of inlet mass flow (Fig. V.8) and are included as a portion of inlet drag. Advanced engine cycles such as the variable-cycle engine or high-airflow engine may result in reduced spillage and bypass drag.

The control of the shock-wave position of flow separation in the inlet can be accomplished by bleeding boundary air from the inlet ramps, cowls, or sidewalls and dumping that flow overboard. This produces forces similar to the bypass flow which must be considered in supersonic inlet throttle-dependent forces.

### Supersonic/Transonic Aircraft Nozzles

Traditionally the nozzle functions as an engine control valve and as a device to accelerate the engine flow converting the thermal energy of the flow to jet kinetic energy. These functions require that the nozzle geometry be responsive

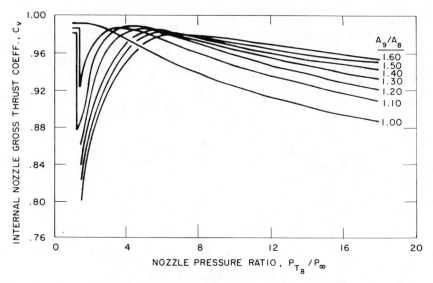

**Fig. V.9   Nozzle internal performance map = 1.35.**

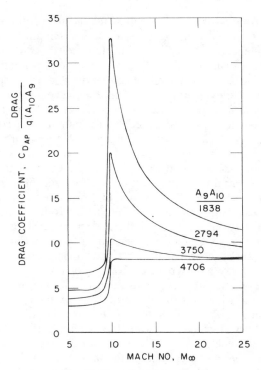

**Fig. V.10   Aft end drag map.**

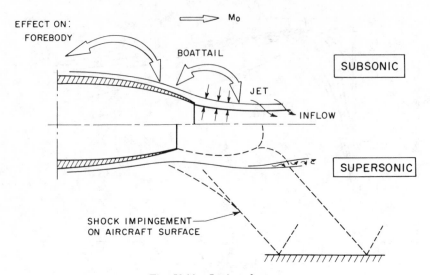

**Fig. V.11   Jet interference.**

to changing throttle positions. The forces related to throttle position changes are gross thrust (internal) and external nozzle boattail and base drag. Another related throttle-dependent force is jet interference or aerodynamic interaction with nearby aircraft surfaces. Nozzle pressure ratio is primarily a function of Mach number and is indirectly related to throttle setting (especially at afterburning power and above), and shows a dependence on engine airflow in much the same way mass flow ratio relates inlet performance to engine airflow.

Typical nozzle internal performance, as a function of nozzle pressure ratio and nozzle area ratio, is shown in Fig. V.9. No one area ratio provides optimum thrust across the nozzle pressure ratio range. This leads to variable area nozzles and geometry changes which create external throttle-dependent forces. Typical external drag coefficients as a function of nozzle exit area are shown in Fig. V.10. As nozzle airflow and exit pressure vary, both plume shape and drag change. In subsonic flow, the plume turns and entrains the external flow, producing a strong upstream influence. In supersonic flow, the "solid" plume produces a shock system that can impact adjacent aircraft surfaces. In addition, jet exhaust can propagate for a short distance upstream through the subsonic boundary layer on separated flow regions. Figure V.11 illustrates the influence of the plume for these freestream conditions. Boattail drag results from external air flowing over the afterbody and producing a nonpotential flow pressure distribution. This flow is affected by boundary-layer displacement on the boattail surface which can be large due to an adverse pressure gradient at the nozzle/plume junction. In the extreme case, a large boattail angle or a strong shock at the nozzle/plume junction results in flow separation and increased drag.

The base drag of the nozzle is dependent on the nozzle pressure ratio. It should be noted that most modern aircraft have extremely small base areas.

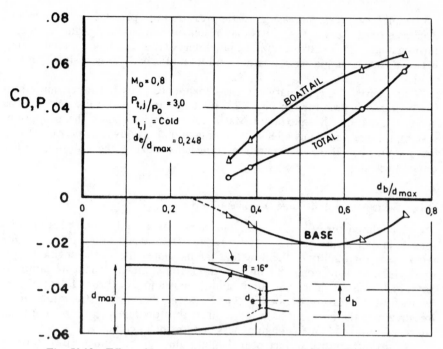

Fig. V.12    Effect of base size on boattail-, base-, and total pressure drag.

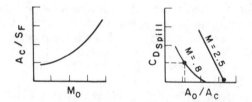

Inlet capture areas become larger as maximum Mach
number is increased, leading to higher spillage
drag at subsonic cruise or the requirement for
variable capture area (matching problems).

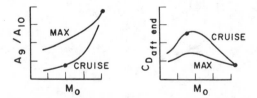

Exhaust nozzle maximum exit areas become larger as
maximum Mach number is increased, resulting in higher
nozzle/afterbody drag at subsonic cruise.

Fig. V.13    Inlet/nozzle throttle-dependent forces.

For small base areas, increasing nozzle pressure ratio pressurizes the area and decreases drag. Larger base areas may result in increased drag, due to steeper turning angles at the boattail/plume junction. A typical functional relationship between base area and base, boattail, and total afterbody drag is shown in Fig. V.12.

A summary of the primary throttle-dependent forces for supersonic/transonic aircraft is presented in Fig. V.13. Inlet capture area and nozzle exit areas become larger as the maximum Mach number, which sizes the propulsion system, is increased. Thus inlet spillage drag and afterbody drag become strong factors during off-design operation, especially during the subsonic cruise mission leg.

### Other Tactical Aircraft Throttle-Dependent Forces

Other related throttle-dependent considerations are inlet total pressure distortion and pressure recovery, tail interference, and jet-induced effects. One measure of the quality of the flow that the inlet provides to the engine face is total pressure distortion. This phenomenon, sometimes transient in nature, is not a factor unless the engine surges and there is a loss of thrust. Inlet pressure recovery, the measure of total pressure loss of the engine streamtube from freestream to the engine face, is not a direct throttle-dependent force but is an indirect result of changing engine mass flow or inlet geometry.

Any aircraft control surface near the jet plume can be influenced by throttle position. An example of the change in normal force on a horizontal tail, as a function of angle-of-attack and jet on/off, is presented in Fig. V.14. The high tail shows no jet influence, while the low tail is strongly influenced by the presence of the jet. Further indirect throttle-dependent forces can result from lift generated on inlets and nozzles and from induced lift and moments on adjacent aircraft lifting surfaces. If trim changes are required to balance these forces, a throttle-dependent trim drag will occur. A graphical presentation of the change in pressure distribution on a wing with a close-coupled jet plume is shown in Fig. V.15. This change in pressure distribution results in an increment of total aircraft lift, drag, and pitching moment which is dependent on

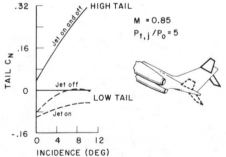

Fig. V.14   Jet effect on tail normal force.

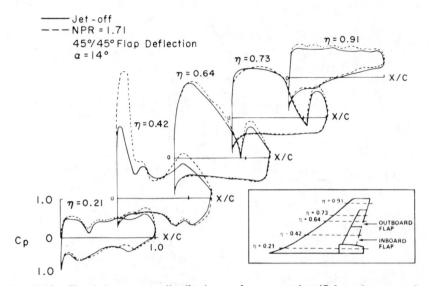

Fig. V.15    Chordwise pressure distributions—thrust vectoring 45 deg primary nozzle.

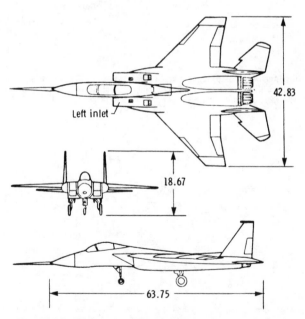

Fig. V.16    Three-view drawing (dimensions are in feet).

INLET SYSTEM

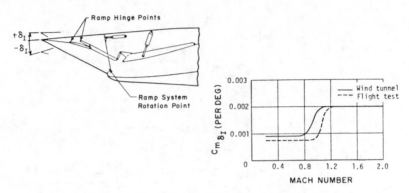

**Fig. V.17    Inlet longitudinal control effectiveness.**

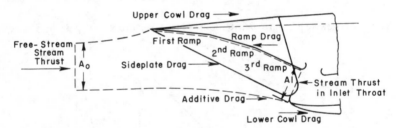

Inlet drag = Additive drag + Upper cowl drag + Lower cowl drag + Sideplate drag *where*
Additive drag = Drag on three ramps + Stream thrust in inlet throat − Freestream stream thrust
(a) Relationships used in the calculation of pressure integrated inlet drag.

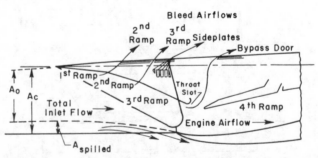

(b) Diagram showing sum of inlet captured stream tube area ($A_0$) and geometric capture area ($A_c$).

**Fig. V.18    General terms for pressure-integrated inlet drags and inlet capture ratios.**

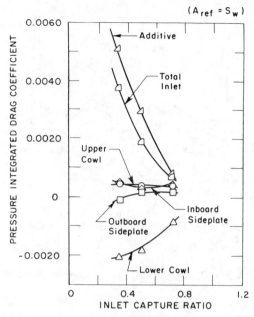

Fig. V.19    Typical pressure-integrated inlet drag components from F-15 flight data.

engine throttle setting. This influence is well documented, and advanced aircraft are currently being configured to take advantage of this effect.

*Examples — Supersonic/Transonic Aircraft Throttle-Dependent Forces*

*F-15.* The F-15 aircraft, shown in Fig. V.16, has variable inlets and exhaust nozzles and offers several good examples of throttle-dependent forces. The inlet (Fig. V.17) has a variable second and third ramp and a variable capture area by rotating about a transverse hinge point at the lower cowl lip. This surface, which is well forward of the center of gravity, has a significant effect on aircraft longitudinal stability and control. The impact of inlet rotation on aircraft longitudinal control is significant across the subsonic/transonic Mach number spectrum of the F-15 (Fig. V.17). The components of throttle-dependent forces, spillage drag, and bleed/bypass drag for the F-15 are evident in Fig. V.18. The resulting forces and their contribution to the total inlet drag as a function of inlet capture ratio are presented in Fig. V.19. Note, as discussed earlier, that the cowl drag partially offsets the additive drag. The F-15 nozzle, required to pass engine airflows for flight up to 2.5 Mach number, can have 8–18 deg boattail angles, depending on flight condition. The resulting nozzle drag variation (Fig. V.20) across the flight regime can be from 2 to 44 drag counts as boattail angle increases. The flow separation on the nozzle/afterbody is very unsteady; in fact, the external nozzle leaves have experienced acoustic structural failure, and have been removed from operational F-15s.

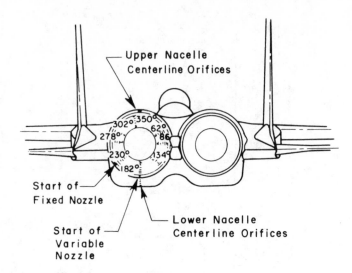

Fig. V.20a    Left nozzle surface pressure instrumentation.

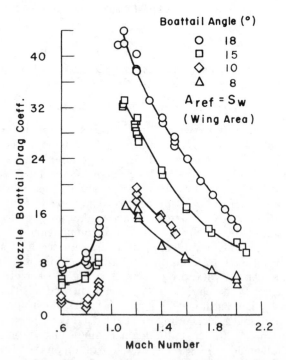

Fig. V.20b    Nozzle boattail drag coefficient vs Mach number.

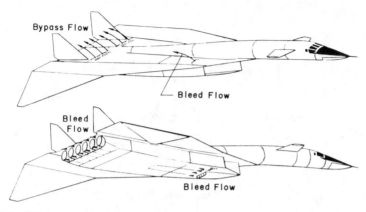

**Fig. V.21   Schematic drawing showing the overboard discharge exits for the inlet bypass and boundary-layer bleed flows.**

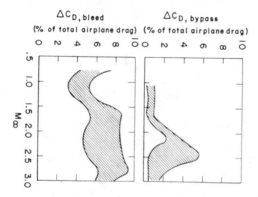

**Fig. V.22   Variation in measured inlet bypass and bleed drag with Mach number for typical 1g flight condition for the XB-70 airplane.**

*XB-70.* For large supersonic/transonic aircraft with high-airflow propulsion systems, the throttle-dependent forces are reflected in total drag and aircraft stability and control. The XB-70, with six engines, has dumped bleed and bypass flow at various locations around the aircraft (Fig. V.21). As a result, as much as 10% of total airplane drag is attributed to these flows in supersonic flight (Fig. V.22). The bypass operation also affected aircraft stability and control as shown in Fig. V.23. Both rolling moment and pitching moment were strong functions of bypass door opening. The overall aircraft drag change with power lever angle at two Mach numbers is shown in Fig. V.24. Note that nearly one-third of the change is base drag from this six-engine installation.

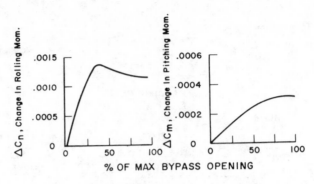

Fig. V.23  Effect of bypass operation on aircraft stability.

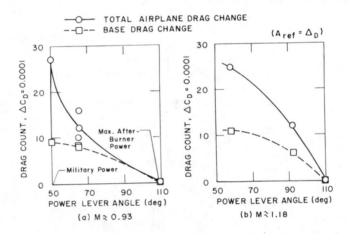

Fig. V.24  Changes in drag with engine power lever angle (averaged for six engines). Drag increments referenced to maximum afterburner power condition and corrected for lift effects.

## Subsonic Aircraft Inlets/Nozzles

The propulsion system for subsonic aircraft, while not required to respond to the large engine mass flow changes in supersonic aircraft, is required to produce high levels of performance in sometimes very complicated aircraft integrations. The engine is usually podded under a wing or on the fuselage to take advantage of high-energy freestream flow and is usually designed to satisfy the maximum airflow requirement with the smallest possible inlet capture area. Further driven by weight considerations, the subsonic inlet will not likely have bleed/bypass provisions and the problem of off-design operation increases. The inlet may incorporate blow-in doors to handle the extra airflow required for takeoff (compared with cruise airflow) and avoid an

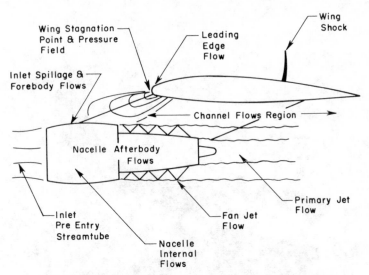

**Fig. V.25   Major elements of wing/pylon/nacelle flowfield.**

oversized inlet for cruise. A typical wing pylon/nacelle installation (Fig. V.25) shows the potential for throttle-dependent forces around the inlet, the nozzle/afterbody, the pylon, and the wing. For an aircraft with short takeoff and landing (STOL) requirements, configurations such as the Boeing YC-14 with upper surface blowing have been proposed. The engine/aircraft interaction is obvious (Fig. V.26).

The subsonic inlet generally has rounded cowl lips and is sized for a capture area ratio ($A_0/A_i$) of 1 at cruise. Figure V.27 presents the subsonic inlet flow character for a range of mass flows. Note the changing stagnation point $S$, which produces both components of spillage drag, i.e., additive drag and cowl lip suction. With a velocity ratio greater than 1, the flow stagnates outside the cowl and part of the captured flow experiences rapid acceleration around the cowl. This can lead to flow separation on inside cowl surfaces resulting in reduced engine face total pressure recovery and thus reduced thrust. This problem is especially acute for STOL aircraft, which have climb-out angles typically around 40–50 deg.

Changes in transport nozzle flow can affect the engine cowl, the pylon, and the wing lower surface pressure distribution. Reference 1 attributes up to 2.5% of the total aircraft drag to this interaction. This is a significant factor for transports. For upper surface blowing aircraft, the jet obviously changes the scrubbed wing forces and both conventional and advanced installations will experience throttle-dependent lift and drag changes.

*Examples — Subsonic Aircraft Throttle-Dependent Forces*

*Airbus A300B.* The Airbus A300B nacelle, shown in schematic in Fig. V.28, is a good example of a transport propulsion installation with throttle-dependent

Fig. V.26   Boeing YC-14.

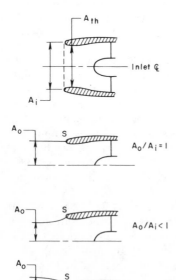

Fig. V.27   Definition of inlet capture area $A_i$ and characterization of flow incidence approaching inlet via mass flow ratio $A_0/A_i$.

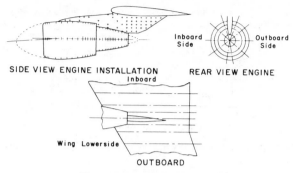

Fig. V.28   A300B nacelle.

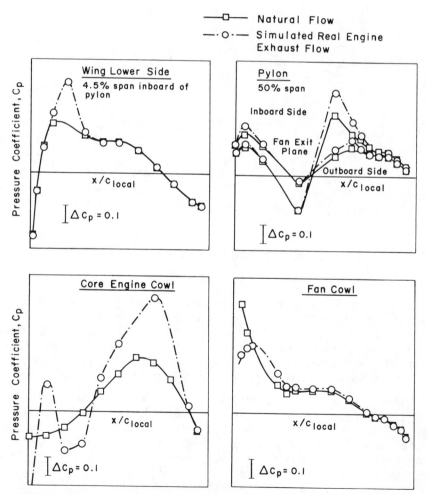

Fig. V.29   Pressure distribution on transport wing lower surface, pylon, core engine cowl, and fan cowl.

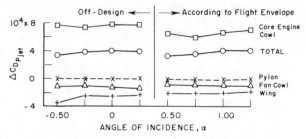

Fig. V.30    $\Delta C_{D_P}$ jet vs $\alpha$ for wing, pylon, fan cowl, and core engine cowl.

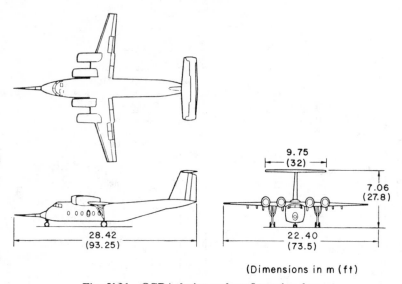

(Dimensions in m (ft)

Fig. V.31    QSRA design and configuration data.

forces. The changes in pressure distribution on the wing lower surface, pylon, core engine cowl, and fan cowl as the jet is changed from flow-through to a real exhaust flow simulation are presented in Fig. V.29. As expected, the largest changes are on the core engine cowl, with lesser changes on fan cowl and pylon. These pressure changes translate into drag components (Fig. V.30), with the core engine cowl drag being somewhat offset by the other contributors.

*QSRA.* The NASA Quiet Short-Haul Research Aircraft (QSRA), Fig. V.31, is typical of propulsive lift transports with strong throttle-dependent effects. It has a four-engine nacelle-over-wing propulsion installation similar to other upper surface blowing concepts. The engine exhaust flows over the upper wing and flaps, and so changes in throttle setting affect aircraft lift, drag, and pitching moment. Total aircraft lift coefficient at takeoff (Fig. V.32) shows the lift dependence on engine rpm. Curve A, at 89% fan rpm (100% QSRA thrust),

Upper Surface Blowing

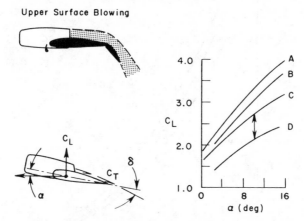

Graphic representation of USB showing thrust and lift vectors. Graph is plot of lift coefficient vs angle of attack for takeoff configuration. Curve A is total lift at 89% fan rpm, curve B is total lift at 70% fan rpm. Curve C is A and B less thrust vector $C_L - C_T \sin(\delta + \alpha)$. Curve D is approximate basic wing lift (all engines at idle).

**Fig. V.32 Upper surface blowing aerodynamics.**

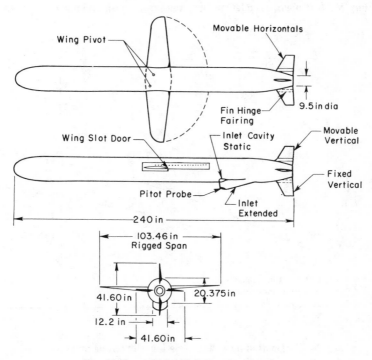

**Fig. V.33 AGM-109 three-view drawing.**

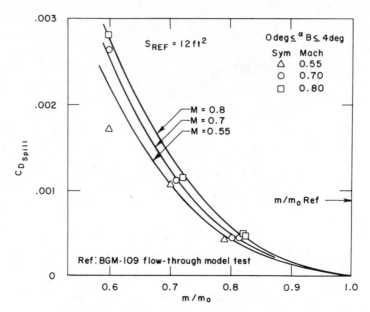

Note: $C_D$ scale adjusted (by capture area ratio) to account for enlarged inlet on AGM-109

**Fig. V.34   Subcritical inlet spillage drag variation with inlet mass flow ratio.**

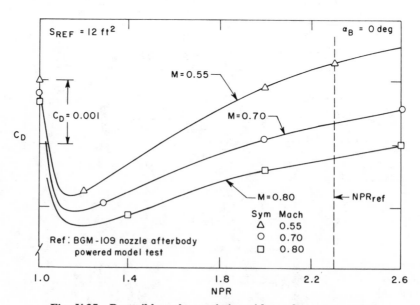

**Fig. V.35   Boattail-base drag variation with nozzle pressure ratio.**

is appreciably greater than curve B, at 70% fan rpm (60% thrust). Curve C is total lift generated with the direct thrust component removed from both curves A and B. This curve is approximately 17% higher across the angle-of-attack range than the basic wing lift curve D. Clearly the throttle-dependent interaction of the propulsive flow and the airframe is a large contributor to the mission performance of this aircraft class.

*AGM-109.* The throttle-dependent forces for transonic cruise missiles follow the trends seen in larger aircraft. For the AGM-109 cruise missile (Fig. V.33), spillage drag increases with decreasing mass flow (Fig. V.34), and missile boattail/base drag first decreases and then increases with nozzle pressure ratio (Fig. V.35).

### Summary

The primary throttle-dependent forces are inlet spillage drag and nozzle boattail drag. Additional inlet factors may be inlet bypass/bleed flows, pressure recovery, and distortion. Additional nozzle forces are tail interference, entrainment and induced effects, and plume scrubbing. All of these factors can impact the total aircraft lift, drag, and pitching moment.

## Determination of Throttle-Dependent Forces

Since the measurement of throttle-dependent forces in flight is generally not possible, experimental results from wind tunnels and analytical techniques are used. This section will present tools and techniques for the experimental and analytical development of these forces.

### Experimental Techniques

The experimental approach for determining these forces is based on the thrust/drag bookkeeping system (Chap. II) typically established for aircraft systems. Three different wind tunnel models are typically used: an aerodynamic force and moment model, an inlet model, and a nozzle (jet-effects) model (Fig. V.36). The aerodynamic force and moment model is the aerodynamic reference configuration which usually does not have accurate inlet and nozzle geometry representation. This model is typically 4–7% scale for fighter aircraft, usually sting supported, fully metric, and uses a flow-through propulsion system simulation.

The inlet spillage model, or inlet drag model, usually simulates only a partial aircraft. The fuselage forebody and inlet are present, and the fuselage is represented to approximately three intake heights downstream.[3] At least partial span wings should be present to the extent needed to account for all areas affected by the inlet flowfield. This model is larger than the aerodynamic force and moment model (5–20% scale); is equipped with suction at low speed or relies on ram air for inlet flow; is instrumented with pressures (static and total) at the engine face and on cowl surfaces; accurately represents the inlet

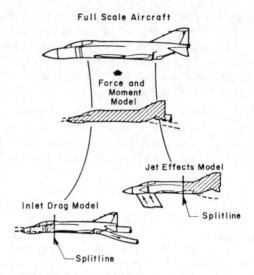

Fig. V.36    Projection of full scale aircraft performance.

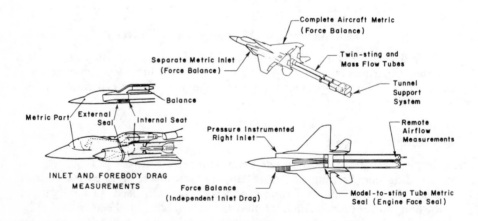

Fig. V.37    Inlet model force balance arrangement.

geometry; incorporates simulation of bleed, bypass, and auxiliary doors if necessary; and measures total forces with a force balance.

The accuracy of the mass flow through the inlet is especially important with typical accuracies of mass flow quoted to $\pm 0.5\%$.[1] The spillage drag, as well as bleed and bypass forces, are determined from pressures or the force balance. Representative inlet spillage models are shown in Fig. V.37. The F-15 model in this example is supported by two flow-through sting tubes and is instrumented with a force balance and surface pressures.

The jet-effects model is used to determine throttle-dependent forces due to the exhaust plume and the variable geometry nozzle boattail. The model

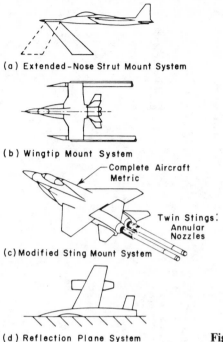

(a) Extended-Nose Strut Mount System

(b) Wingtip Mount System

Complete Aircraft
Metric

Twin Stings:
Annular
Nozzles

(c) Modified Sting Mount System

(d) Reflection Plane System

**Fig. V.38   Jet effects model support systems.**

generally duplicates aircraft lines although usually only the aftbody is metric. High-pressure cold air at the correct nozzle pressure ratio is generally used to represent the nozzle flow. The model may be supported in a number of ways (Fig. V.38), with the wing-tip support often favored. Parameters include nozzle pressure ratio, nozzle boattail angle, nozzle area ratio, angle of attack, and Mach number. Instrumentation for this model generally consists of force balances that measure aftbody drag or thrust-minus-drag, and pressures that may be integrated to determine the nozzle forces.

While the experimental determination of throttle-dependent forces for subsonic transports uses an approach similar to that used for tactical aircraft, the requirement for guaranteed aircraft performance puts increased emphasis on accuracy. Propulsion forces, that depend on nacelle shape, pylon placement, and jet exhaust interactions, require careful calibration of mass flow and controlled testing. On a short nacelle the simultaneous simulation of the inlet and nozzle flow often requires use of propulsion simulators, adding to model complexity. Typical model techniques for transport throttle-dependent forces are shown in Fig. V.39.

The selection of model hardware to determine throttle-dependent forces is driven by factors present in all wind tunnel tests. The resources available limit the overall test objectives which define the model complexity. This in turn determines the test approach. Issues include: model size vs accuracy, use of force balances vs pressure area integration, extent of aircraft and propulsion stream simulation, Reynolds number simulation, available hardware, facility,

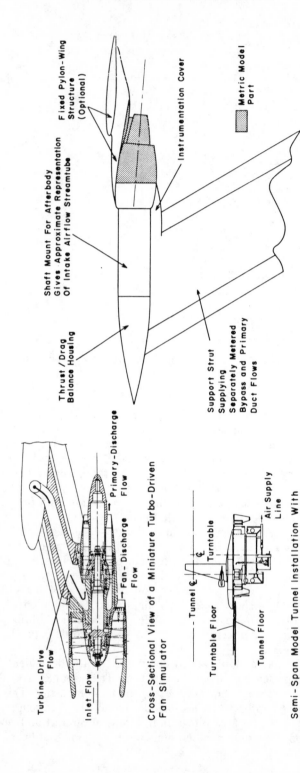

Fixed Pylon-Wing Structure (Optional)

Instrumentation Cover

Metric Model Part

Shaft Mount For Afterbody Gives Approximate Representation Of Intake Airflow Streamtube

Thrust/Drag Balance Housing

Support Strut Supplying Separately Metered Bypass and Primary Duct Flows

Primary-Discharge Flow

Fan-Discharge Flow

Turbine-Drive Flow

Inlet Flow

Cross-Sectional View of a Miniature Turbo-Driven Fan Simulator

Tunnel ₵

Turntable ₵

Turntable

Air Supply Line

Turntable Floor

Tunnel Floor

Semi-Span Model Tunnel Installation With Powered Nacelle

Fig. V.39   Transport testing techniques.

and cost. Each test presents individual problems in sizing, model internal space, data corrections, and other items that impact the desired and achievable accuracy.[32] All test variables involved must be closely examined and compared so that the most effective compromises can be made between what is desired and what is feasible. The model is then designed based on local options, past experience, and cost. Some of these factors are explored in the next section.

### Difficulties in Experimental Testing

Reference 9 identifies difficulties in wind tunnel determination of inlet and nozzle throttle-dependent forces, including errors in parameter matching, inadequate flowfield simulation, model support influence, and tunnel effects such as blockage and buoyancy. Problems also arise in the wind tunnel with propulsion flow simulation, model support testing options, measurement and determination of forces, and determination of accuracy and repeatability.

The simulation of the propulsion stream has been and is currently the subject of a great deal of research and debate. Current techniques used in wind tunnel testing are presented in Fig. V.40. For the inlet spillage model, the inlet flow must be correctly simulated to give proper levels of spillage drag, cowl drag, and inlet flow/aircraft interactions. The location of the stagnation point on a transport nacelle, for example, must match between wind tunnel and flight to provide comparable inlet flow. This flow can be simulated by active suction for low-speed conditions or rely on ram air and a variable-position mass flow plug for high Mach number. For systems that have independent inlet and nozzle flowfields—for example, tactical aircraft—these techniques are adequate. For closely coupled inlet/nozzle flows, typically encountered in transports, special techniques may be required.

The exhaust nozzle flow from a jet engine has characteristics related to nozzle pressure ratio, temperature, gas combustion products, total pressure and temperature distortions, swirl, and turbulence. Perhaps because it is not possible to simulate all parameters in a scaled wind-tunnel model, one source states[32]: "the jet should be simulated as simply as possible." This has led to a variety of techniques for jet plume simulation, the most common being the use of high-pressure cold air. Such a simulation is relatively easy to control and incorporate into a wind-tunnel model. This technique, however, accounts for only the nozzle jet pressure ratio parameter. There is also the solid plume simulator, with different fairings to simulate nozzle pressure ratio changes. This technique partially accounts for the nozzle pressure ratio effect using a calculated plume shape to fabricate plume contours but does not provide jet entrainment/mixing effects. A hybrid jet simulation which can double as a model support system is an annular jet. The model is supported by a sting(s) through the nozzle, and high-pressure air is ducted around the sting (Fig. V.41). This technique offers promise but has limited angle of attack and jet simulation capability.

In experiments where hot and cold exhaust plumes have been utilized to determine nozzle boattail forces, the cold jet consistently had higher boattail drag values than a hot jet at similar nozzle pressure ratios. This is due to a

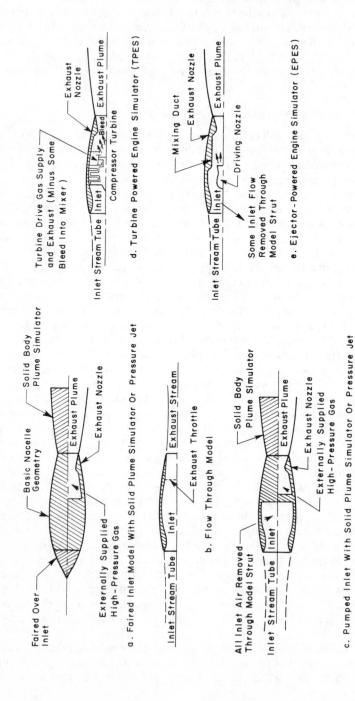

Fig. V.40   Current engine simulation techniques for wind tunnel testing.

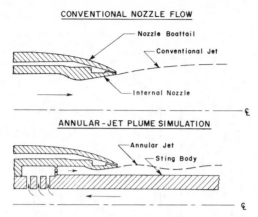

CONVENTIONAL NOZZLE FLOW

Nozzle Boattail

Conventional Jet

Internal Nozzle

℄

ANNULAR-JET PLUME SIMULATION

Annular Jet

Sting Body

℄

Fig. V.41 Concept of annular-jet plume simulation with sting support of a jet effects model.

different plume shape and entrainment character, with the largest differences occurring at high transonic Mach numbers and with large boattail angles. Exhaust temperature has been simulated using gases having different ratios of specific heats, by burning ethylene in the nozzle plenum, and by using real engines. This adds cost and complexity to an already difficult model. An approximate correction for temperature effects can be achieved by increasing the cold jet nozzle pressure ratio to produce the maximum plume diameter for the corresponding hot jet.

The assumption that the inlet and nozzle flowfields are independent allows separate determination of inlet and nozzle throttle-dependent forces. Simultaneous simulation of the inlet/nozzle flow is limited by available wind-tunnel testing techniques. Also, if the inlet is faired over, as in a jet-effects model, the entire flow is spilled and can cause changes in the aerodynamic character of the aftbody/nozzle.[23] Figure V.42 demonstrates the impact of inlet fairing, illustrating the changes in nozzle/afterbody drag for two aircraft models. Further, for closely coupled nacelles, typically found on transports or on fighter configurations which have a canard/inlet/wing/nozzle in close proximity, the inlet/nozzle flow should be provided simultaneously. The solution to the simultaneous simulation of inlet/nozzle flows has been primarily the use of turbine simulators and, to a limited extent, ejector simulators. Turbine simulators are small compressors with high-pressure air-tip turbine drive or electric motor drive. This technique has been utilized with success on transport nacelles and is being investigated for close-coupled fighter aircraft. The turbine simulator partially matches inlet and nozzle flow conditions but can be mechanically complex and does not account for temperature. The ejector simulators use ejector pumping action to control the inlet/nozzle flow. Although primarily emphasized in the European technical community, ejector simulators have been investigated at Arnold Engineering Development Center, with best results achieved at transonic/supersonic Mach numbers. An array of

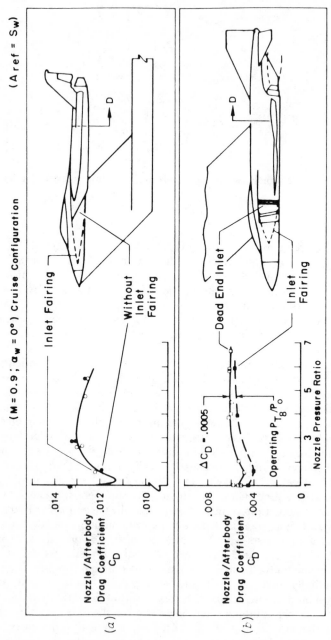

Fig. V.42   Inlet fairing effects on afterbody drag.

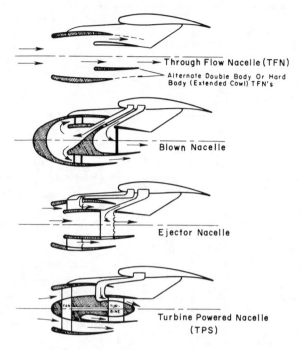

Fig. V.43   Types of nacelle simulators.

propulsion stream simulation techniques for jet transport aircraft is shown in Figs. V.43 and V.44. These approaches are generally applicable for fighter aircraft as well. The correct simulation of the inlet and nozzle flows must be considered when throttle-dependent forces are determined.

One reason it is not possible to test an accurate scale version of the complete aircraft is that wind-tunnel models have support systems which interfere with the flowfield and model and compromise the data. The quest for interference-free support systems has led to a number of concepts for both transport and fighter aircraft. Figure V.45 shows stings, plates, half-models, and wing-tip supports all tested for transports. Representative support techniques for one fighter aircraft model are presented in Fig. V.46. Although each support system has its own set of deficiencies, all have a strong transonic influence. Based on Glidewell and Fanning[23] and Kennedy,[36] a strut support can be properly designed for subsonic or supersonic use; however, such a support can be a large contributor to blockage and interference transonically. The wing-tip support minimizes blockage and interference but distorts aircraft wing lines on the outer span and limits model metric arrangements. The dual-sting and annular-sting support system provides minimum blockage and interference. Regardless of the support system, the support influence could be determined either by calculating local pressures near the support system or by special support testing where an alternate support is used to hold the model and the

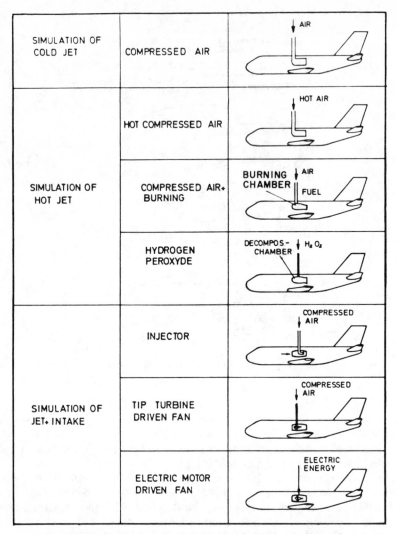

Fig. V.44   Principles of jet engine simulation in wind tunnels.

primary support is tested on and off to examine the support interference. The choice of a model support system depends on the test objective, test facility, existing hardware, and cost. For throttle-dependent force determination, the support system should be tailored for minimum interference.

The experimental techniques previously described require use of a force balance and pressure measurements. A typical jet-effects model with a nonmetric section and two separate metric sections is shown in Fig. V.47. The use of either force balance or pressure integration or both techniques is a much discussed and controversial area. How much, if any, of the aircraft should be metric? Can surface pressures and pressure area integration accurately de-

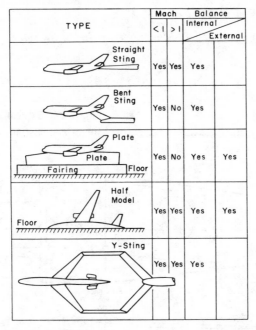

| TYPE | Mach | | Balance | |
|---|---|---|---|---|
| | < 1 | > 1 | Internal | External |
| Straight Sting | Yes | Yes | Yes | |
| Bent Sting | Yes | No | Yes | |
| Plate / Plate / Fairing / Floor | Yes | No | Yes | Yes |
| Half Model / Floor | Yes | Yes | Yes | Yes |
| Y-Sting | Yes | Yes | Yes | |

**Fig. V.45   Examples of wind tunnel model mounting techniques.**

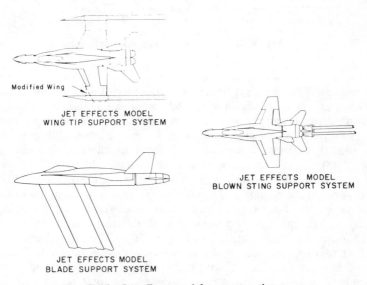

Modified Wing

JET EFFECTS MODEL
WING TIP SUPPORT SYSTEM

JET EFFECTS MODEL
BLOWN STING SUPPORT SYSTEM

JET EFFECTS MODEL
BLADE SUPPORT SYSTEM

**Fig. V.46   Jet effects model support options.**

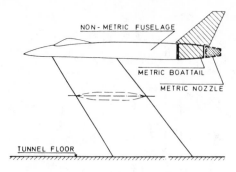

Fig. V.47   Typical aftbody test rig with subdivided boattail.

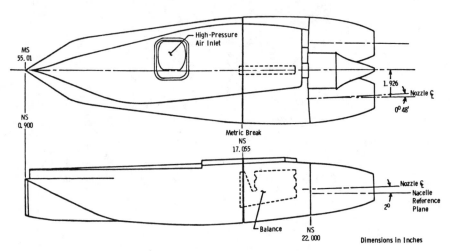

Fig. V.48   Force balance nacelle configuration.

termine throttle-dependent forces? The presence of a metric break on a model can have an impact on the flow downstream, creating discontinuities evident in pressure distributions. In addition, the presence of a physical gap in the model necessitates additional data corrections for metric break seal forces and cavity pressures. Reference 36 presents data for a nacelle with a metric afterbody (Fig. V.48). In this test arrangement the balance loads and the corrections (metric seal and cavity pressure) are of the same magnitude at subsonic and supersonic Mach numbers (Fig. V.49). The uncertainty of these corrections impacts overall data accuracy. The issue of close-coupled inlet/nozzle flows has opened the question of how much of the aircraft must be modeled and how much must be metric. Data such as Fig. V.50 indicate that the throttle-dependent change in aircraft afterbody pressure distribution with nozzle power setting and nozzle pressure ratio can continue upstream ahead of the metric section. Obviously the entire throttle-dependent force is not being measured since the pressure measurement at the metric break does not show a zero pressure coefficient and in fact varies with power setting. The current guidance

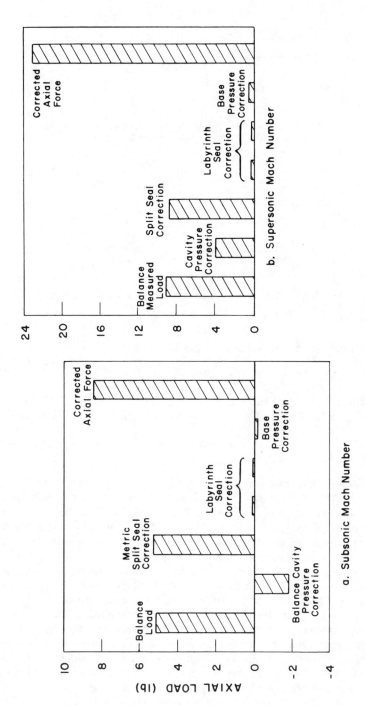

Fig. V.49  Nacelle axial-force components.

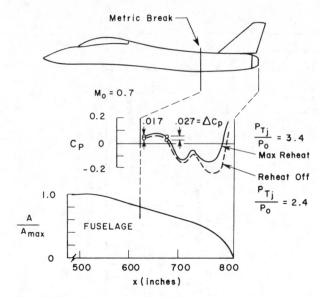

**Fig. V.50   Effect of nozzle position on pressure distribution.**

from a number of sources on metric break placement is:

1) Isolate the smallest piece to satisfy the test objectives with greatest accuracy.[2]

2) If, however, disturbances from inlet/nozzle flow carry far upstream or downstream, a large part of the aircraft must not only be simulated but also must be metric.

3) Further, other sources[9,22] recommend as much of the aircraft as possible be on a force balance with the entire aircraft metric if allowable.

Other factors that complicate experimental determination of throttle-dependent forces are accuracy and repeatability of data and wind tunnel effects. Accuracy, as defined by Ref. 4, is the uncertainty of a value due to systematic or bias error. Accuracy can be improved by careful calibration and controlled test techniques, however, Jaarsma[32] notes that accuracy is difficult to assess because overall accuracy includes a combination of the effects of many instruments such as mass flow meters, pressure transducers, thermocouples, and force balances. As stated previously, each model test apparatus presents individual problems in sizing, restricted internal space, pressure corrections, metric break seal restraints, thermal expansion, clearances, and other items which make any general statement of achievable accuracy difficult.[32]

Repeatability variations, defined as the uncertainty of measurement due to the presence of random errors, can be high. Reference 3 quotes $\pm 1\%$ aircraft drag repeatability for a technique that measures aircraft transonic spillage drag of a supersonic aircraft. Both high repeatability and reduced bias error should be the goals for throttle-dependent force measurements.

Additional factors in experimental throttle-dependent force determination are wind-tunnel effects, such as buoyancy, blockage, flow quality, and Reynolds number simulation. These items are discussed in Chap. IV.

### Qualitative Techniques to Assess Throttle-Dependent Forces

The nature and extent of the influence of the propulsion stream on the inlet/nozzle and adjacent aircraft surfaces can be determined qualitatively by various flow visualization techniques. Wind-tunnel techniques include the use of tufts, dye, and oil flow, all of which indicate surface flow direction, presence and location of shock waves, areas of separation, and location of vortices. Schlieren and shadowgraph techniques can indicate the presence of density gradients, primarily shock waves, in supersonic tunnel flow. A technique applicable to low-speed wind tunnel work is the use of smoke to trace streamlines influenced by the propulsion stream. Dye streams and bubbles are often used to indicate propulsion stream influence in low-speed water tunnels. All of these techniques can provide the investigation further insight into throttle-dependent effects.

### Analytical Techniques

In addition to experimental techniques, analytical or computational techniques can be used to predict throttle-dependent forces. In general, due to the complex nature of the interaction of the propulsion stream with the external flow and adjacent aircraft surfaces, throttle-dependent forces are difficult to predict. Analytical techniques vary from empirical approaches to potential flow solutions, from quasi-time-dependent to time-dependent, inviscid/viscous patch methods to full Navier-Stokes equation solutions. These techniques can play a useful role in evaluating and screening concepts relative to throttle-dependent forces. It should be noted that these methods, in general, predict surface pressure distributions. Throttle-dependent forces are then determined from pressure area integration. The success of these techniques varies relative to the level of accuracy required and level of resources available to support the prediction. The extent of the aircraft simulated also determines the success of these techniques, but this is limited by the storage size of available computers. In this section, representative techniques will be discussed for throttle-dependent drag primarily for inlets and nozzles for supersonic/transonic aircraft and for nacelle installations for transport aircraft.

*Inlet Analytical Techniques*

Empirical procedures are correlations of experimental data which indicate trends and approximate force levels for a class of inlets. Figure V.51 shows a spillage drag correlation compared with experimental data. The procedure provides an adequate trend but does not indicate the absolute value. These methods can be useful in conceptual and preliminary design procedures. The success of all empirical techniques depends on the data base and the degree of similarity of the configuration being analyzed to those in the data base. If

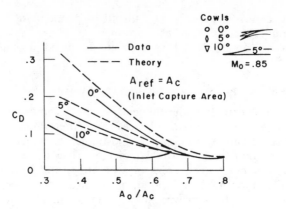

Fig. V.51    Inlet drag correlations.

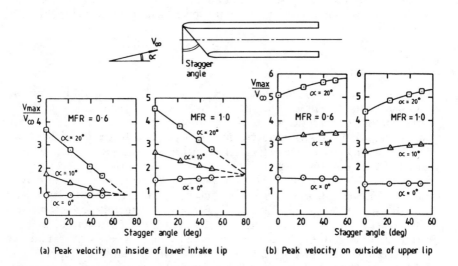

(a) Peak velocity on inside of lower intake lip          (b) Peak velocity on outside of upper lip

Fig. V.52    Theoretical peak velocities on lips of a two-dimensional staggered intake.

carefully restricted and applied, these techniques can be useful tools for determining throttle-dependent forces.

Inlet surface pressures are often calculated by two-dimensional (2-D) potential flow solutions over a range of angles-of-attack and mass flow. Boundary-layer analysis is sometimes added to account for viscous effects. These techniques are moderately successful if no separation is present. Local velocity profiles on a supersonic inlet lip at different angles-of-attack and mass flow ratios calculated by a 2-D incompressible technique are shown in Fig. V.52. These methods have been extended to three-dimensional (3-D) potential and viscous approaches. The effect of inlet flow simulation on the engine nacelle pressure distribution, as calculated by one of these techniques, is

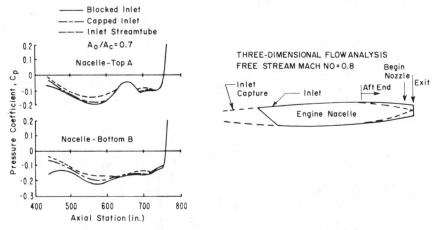

Fig. V.53   Predicted twin-jet nacelle pressure distributions.

presented in Fig. V.53. Note that the effect of inlet flow condition extends considerably downstream of this inlet station, indicating the extent of the nacelle to be simulated to measure the entire inlet mass flow effect.

Although not an explicit example of a throttle-dependent force calculation, the solution of a 2-D mixed compression inlet by a Navier-Stokes solution (Fig. V.54) indicates the potential for very detailed inlet analysis. This method[47] is described as a MacCormack's explicit finite difference algorithm with modified treatment of the viscous sublayer and of the turbulent boundary layer. The method provides good agreement with experimental data and, for the case shown, ran for 1.6 h on a CYBER 175 computer. Advanced inlet analysis is being developed for 3-D viscous flow Navier-Stokes solutions. Some trades using these complex methods are discussed in the next section on nozzle computational techniques.

*Afterbody/Nozzle Analytical Techniques*

References 38, 46, and 49 all point out the difficulty involved in modeling and predicting the flow over the aftbody boattail region. Consequently the throttle-dependent forces due to changes in nozzle boattail angle and jet plume effects are also difficult to predict. The influence of the jet is difficult to represent relative to its effect on the freestream flow turning and jet entrainment. Quoting Ref. 49, "Flow over nozzle boattail configurations at subsonic and transonic speeds is dominated by strong viscous/inviscid interaction effects." As illustrated on Fig. V.55, this flow is characterized by gradients which may cause the boundary layer to separate, a jet exhaust plume that may grow or diminish in size depending on nozzle pressure ratio, and a large viscous mixing region between the jet and external flow. If the flow is transonic or supersonic, shock waves will also be present. Because of these strong viscous/inviscid interactions, inviscid theoretical methods have been found to be inadequate for predicting nozzle/afterbody flow. In general, nozzle surface

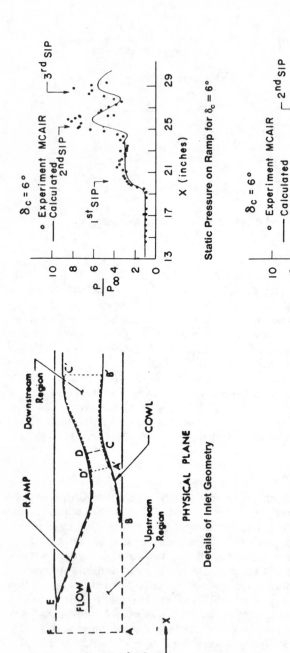

Fig. V.54  Supersonic inlet pressure distributions, calculated and experiment.

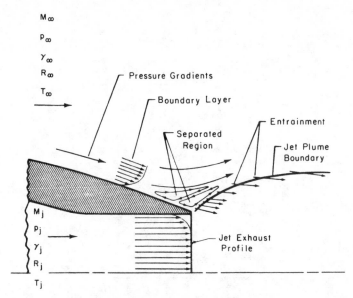

Fig. V.55 Schematic of nozzle aftbody flow.

pressures are difficult to predict, especially at transonic speeds. Trends can be predicted with more success than absolute values. As with the inlet, the full range of techniques, from empirical correlations to 3-D Navier-Stokes solutions, has been applied to predict afterbody/nozzle forces.

Empirical force prediction techniques for exhaust nozzles are derived from parametric afterbody/nozzle data. Important geometric parameters include nozzle boattail angle, base area, initial plume angle, overall aftbody closure, nozzle type and spacing, interfairing type, empennage location, and the proximity of lifting surfaces. In general, the geometric effects and jet effects are handled separately. The geometry parameters can be characterized by the integral mean slope (IMS) correlation. This parameter is an area-weighted average slope of the nondimensional cross-sectional area plot and is proportional to the ideal pressure area integral of the afterbody/nozzle. The formulation and a correlation of the parameter for one Mach number are shown in Fig. V.56. The procedure provides a good prediction of aftbody pressure drag if the area is free of separation and is "clean," i.e., with minimum nearby empennage surfaces. This correlation is combined with a plume correlation parameter to account for geometric as well as jet-induced effects.

Subsonic viscous flow and patch methods which combine the inviscid regions with a boundary-layer calculation and an exhaust simulation yield more rigorous subsonic/transonic solutions of the nozzle boattail flowfield. These techniques include representations of the effects of skin friction, axial pressure gradients, jet entrainment, separated flow, jet mixing, and jet temperature. Two examples of these techniques are given in Refs. 12 and 49. Bower[67] was able to predict the surface pressure distribution of an axisymmetric 15-deg

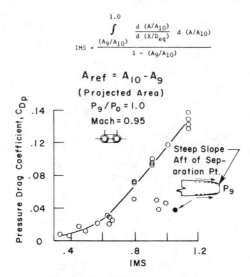

$$IMS = \frac{\displaystyle\int_{(A_9/A_{10})}^{1.0} \frac{d\ (A/A_{10})}{d\ (X/D_{eq})}\ d\ (A/A_{10})}{1 - (A_9/A_{10})}$$

$$A_{ref} = A_{10} - A_9$$

(Projected Area)

$$P_9/P_0 = 1.0$$

Mach = 0.95

**Fig. V.56    IMS data correlation.**

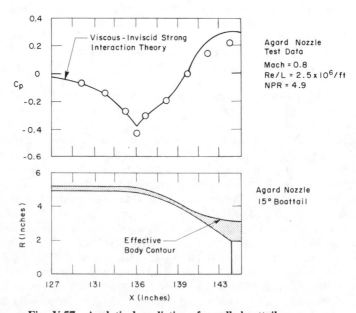

**Fig. V.57    Analytical prediction of nacelle boattail pressures.**

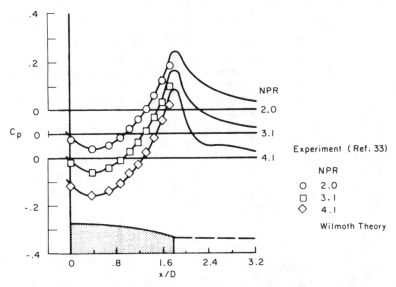

Fig. V.58   Comparison of predictions of patched method with experiment.

boattail with a strong viscous/inviscid interaction method (Fig. V.57). The method of Wilmoth, described in Ref. 49, is a successful patch method which combines an inviscid flowfield, a boundary layer, a mixing layer between the jet and freestream, and a displacement thickness corrected for jet plume entrainment and growth. A comparison of predicted boattail pressures at different nozzle pressure ratios compared to experimental data is shown in Fig. V.58. This method provides a good prediction of pressure distribution and is capable of accounting for jet temperature, composition, and chemistry effects. This is an interactive procedure that uses approximately 30 s of processing time on a CYBER 175 computer. More work is needed to refine the code for use in describing regions of boundary-layer separation. An extension of the method to 3-D would be difficult.

Strictly supersonic external nozzle boattail flow calculations can be approached by small-perturbation potential flow solutions. Special consideration must be given to the viscous portion of the flowfield. This can be accomplished by a Method-of-Characteristics solution or by a time-dependent finite-difference solution to the compressible Navier-Stokes equations for turbulent separating flows.

The calculation of the afterbody/nozzle flowfield and surface pressure distribution using the Navier-Stokes equations is the focus of many research efforts in industry and government. Three-dimensional, time-dependent Navier-Stokes solutions including a turbulence model are the goal of many researchers investigating this complicated flowfield. These solutions, requiring large computers and elaborate computational grid generation techniques, are being supplanted in the near term with approximate solutions to these equations. Good prediction across the Mach number range of the surface pressures

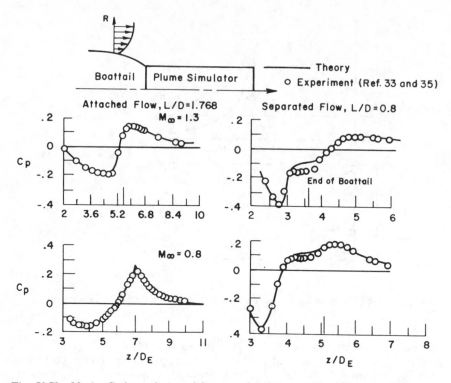

Fig. V.59  Navier-Stokes solution of Swanson for flow over circular arc boattail nozzles with solid plume simulators.

of an axisymmetric afterbody is shown in a calculation by Putnam and Mace.[49] The results (Fig. V.59) required 5 min of computing time for a supersonic case and approximately 2 h for a subsonic case on a CYBER 203 computer. A velocity splitting technique for solving the steady-state 3-D Navier-Stokes equations at transonic speeds for general bodies has been developed by Cosner.[50] This method has shown good agreement between theory and experiment for clean afterbodies with approximately $\frac{1}{2}$-h running times on a CYBER 175 computer.[50]

*Transport Nacelle Analytical Techniques*

The predominantly subsonic flowfield of the transport nacelle has been approached primarily by potential flow codes and by limited axisymmetric Navier-Stokes solutions. Incompressible, inviscid techniques, corrected for compressibility and iterated to a solution, are successful for entirely subsonic flowfields. Results of a low-speed inlet calculation are shown in Fig. V.60. The incompressible solution with the compressibility and boundary-layer corrections agrees reasonably well with the data. Mixed flowfields require a transonic relaxation potential flow technique with an iterative boundary-layer solution. The application of this technique to a transonic cowl is shown in Fig. V.61.

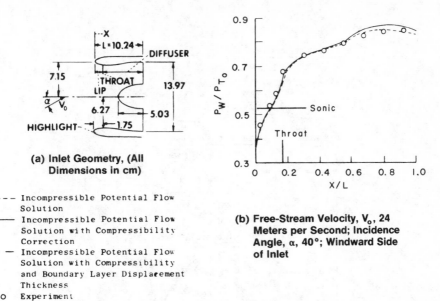

(a) Inlet Geometry, (All
Dimensions in cm)

- - - - Incompressible Potential Flow
      Solution
─────── Incompressible Potential Flow
      Solution with Compressibility
      Correction
─ ─ ─ Incompressible Potential Flow
      Solution with Compressibility
      and Boundary Layer Displacement
      Thickness
  o   Experiment

(b) Free-Stream Velocity, $V_o$, 24
Meters per Second; Incidence
Angle, $\alpha$, 40°; Windward Side
of Inlet

Fig. V.60  Comparison of calculated and measured internal surface pressure distributions for a translating centerbody inlet in the unchoked mode.

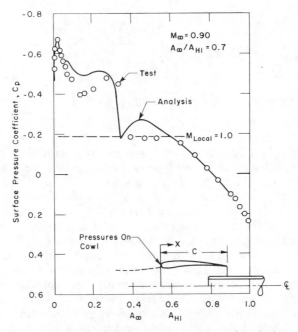

Fig. V.61  Transonic cowl pressures comparison of analysis and test.

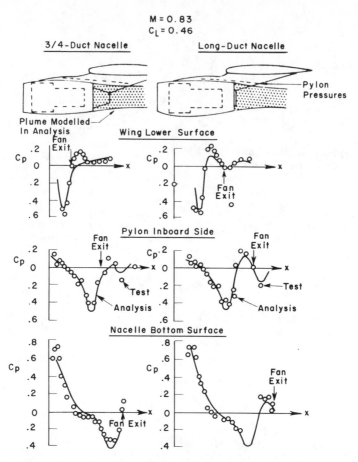

**Fig. V.62   Comparison of analytical vs measured pressures for installed powered nacelle.**

Moderately good agreement is seen between test data and the analysis. Three-dimensional geometries, such as a nacelle pylon, are also solved by linearized compressible potential flow methods. A flowfield of a powered nacelle, with the plume represented as a solid body, is predicted with some success (Fig. V.62). Areas of concern in this solution are the wing lower surface, the pylon inboard side, and the nacelle bottom surface.

Panel methods, 3-D linearized compressible potential flow solutions, are often utilized for the flow around high-bypass engines. The surface of complex configurations and, to some extent, the inlet and nozzle flows are represented by flat panels. A semiempirical compressibility term for the highly curved regions at the inlet leading edge may be employed.[39] A typical application of this approach is presented in Ref. 42. The nacelle representation is comprised of 260 panels and includes inlet suction, jet spreading, and flow entrainment. The wing/fuselage is represented by 160 panels. The wing pressures with the

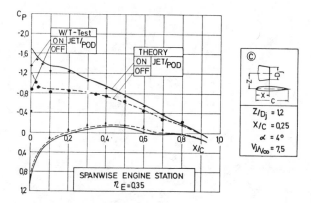

Fig. V.63    Wing pressure at engine position.

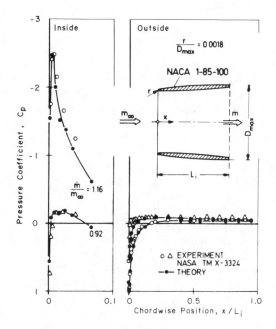

Fig. V.64    Pressure distribution at an axisymmetric inlet NACA 1-85-100 for different mass flow rates.

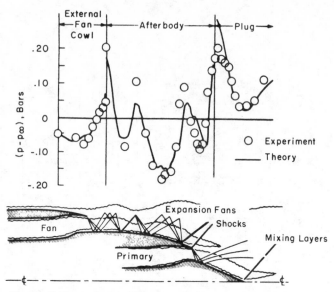

**Fig. V.65   Navier-Stokes solution of Peery and Forester for multistream exhaust nozzle flow; NPR = 2.6 and $M_\infty$ = 0.9.**

jet on and off show good agreement between theory and wind tunnel test (Fig. V.63). Another application is from Haberland et al.[48] where an axisymmetric panel method predicted the inlet cowl pressures for two mass flow ratios. For the most part, the measured and predicted pressures agree (Fig. V.64). Leakage, a common problem for panel methods caused by a difference in calculated and prescribed mass flow rates at the engine face, can be reduced by careful selection and distribution of panels around the cowl.

Limited Navier-Stokes solutions have been reported for transport nacelles. A mixed flow solution for a JT9D turbofan engine is summarized in Ref. 49. At a Mach number of 0.9 and a nozzle pressure ratio of 2.6, the pressure distribution on the external fan cowl, afterbody, and plug shows good agreement between theory and experiment (Fig. V.65).

Some conclusions can be drawn from this brief review of analytical techniques for the complex flowfield associated with throttle-dependent forces. Empirical techniques are useful as predictors in preliminary design if care is exercised in their application and use. Two-dimensional and three-dimensional potential flow methods with boundary-layer and compressibility corrections are moderately successful in predicting surface pressure distributions around inlets/nozzles and nacelles. Solutions for separated regions and areas of mixed flow are difficult to obtain. Patched methods can provide a good prediction at reasonable cost. The solutions of the Navier-Stokes equations are limited by the lack of a good turbulence model, by available data/computational storage, and by cost. All of the methods vary in success relative to the level of accuracy

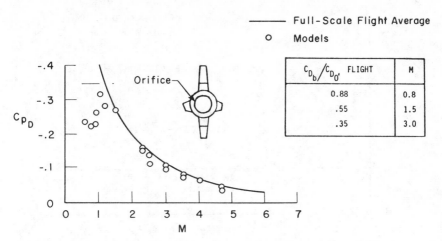

Fig. V.66   X-15 fuselage base pressure coefficients (power off).

required and the level of resources available. Analytical methods of all levels of complexity can play a role in the prediction of throttle-dependent forces.

### Throttle-Dependent Forces in Flight

The dynamic nature of an aircraft in flight makes determination of throttle-dependent forces difficult, if not impossible. The aircraft is rarely instrumented to determine all the parameters required, such as mass flow ratio, nozzle total pressure and exit static pressure, and inlet/nozzle surface pressures. Further, the aircraft can vary mass flow ratio or nozzle pressure ratio at a fixed Mach number only over a small range which precludes a determination of spillage or boattail drag at off-design conditions. The alternative to flight determination of throttle-dependent forces is to use wind tunnel data or analytical techniques, predict the throttle-dependent forces, and then verify these predictions by correlating the surface pressure distribution predicted or measured with those measured in flight. Plots such as Figs. V.66 and V.67 for the X-15 and XB-70, respectively, show examples of correlations of base pressures measured in flight with wind tunnel and analytical predicted values. This section will describe difficulties and discrepancies in using these correlations and will then present examples of wind tunnel/analytical pressure distributions compared to pressure distributions measured in flight.

### Sources of Error—Wind Tunnel/Analysis to Flight

Hunt and Gowadia[9] state that separate determination of inlet and nozzle throttle-dependent forces is not possible in flight. Further, when relying on a wind tunnel model to determine these forces, it is not possible to test an accurate scale version of the aircraft due to wind tunnel support systems and

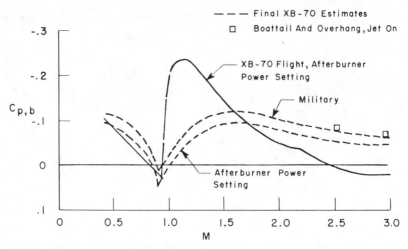

**Fig. V.67  Comparison of flight-measured XB-70 average base pressure coefficient with predicted values.**

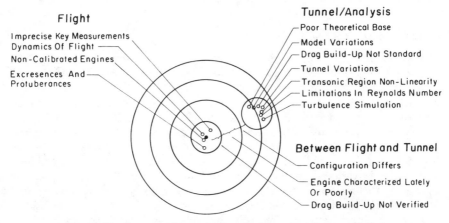

**Fig. V.68  Sources of error in throttle-dependent drag predictions.**

their interference. Also, there is no well-established method of reproducing the inlet and nozzle flows simultaneously in the model in order to measure their interference effects. Ayers[60] notes that prediction of full scale aircraft forces has been historically hampered by the inability to extrapolate nozzle boattail and base effects. This is primarily attributed to model support effects and inadequate Reynolds number simulation. Due to wind tunnel limitations, model data are rarely acquired at flight Reynolds numbers, thus for flow areas with strong viscous effects and separated regions, such as an afterbody/nozzle, the wind tunnel simulation can be questionable. A compilation of possible discrepancies between wind tunnel and flight data developed from Refs. 27, 30, and 46 includes: test techniques, wall and support effects, inadequate duplication of inlet/nozzle flow, wind tunnel flow quality, model detail, aeroelastic deformation in flight, scaling and Reynolds number effects, and incorrect assessment of inflight inlet/engine/nozzle operating characteristics.

All these possible sources of error, plus those graphically presented in Fig. V.68 for determining throttle-dependent forces in the wind tunnel or by analysis, do not indicate that the task is impossible but rather point out the special care and considerations required to simulate these propulsion/airframe interactions. The next section will present five examples of the correlation of wind tunnel data and analysis with flight data where different levels of success have been achieved.

*Wind Tunnel/Analysis to Flight Correlation — Examples*

Since throttle-dependent forces are very difficult to determine in flight, wind tunnel/analytical assessments of these forces must be deduced from surface pressure distributions. The accuracy of these methods is verified by comparing experimentally or analytically predicted pressure distributions with flight data. These comparisons have been performed on several aircraft systems. The B-1 and F-15 were the subjects of extensive wind tunnel/flight correlations, especially with regard to the inlet and nozzle, and lesser comparisons have been completed for the YF-17, Tornado, and A-7 aircraft.

*B-1.* A correlation between wind tunnel and flight inlet and nozzle pressure data was performed with the objective of eliminating as many of the discrepancies between the data sets as possible. The wind tunnel data were taken after the flight test to ensure matching of conditions. The surface pressure orifice locations were also matched. Support interference was determined in a series of wind tunnel tests to remove this parameter. Factors that could not be simulated were wing flexibility and the resulting wing gap between the propulsion nacelle and the lower wing surface, environmental control system (ECS) purge air mass flow and pressure, and absolute Reynolds number. The model

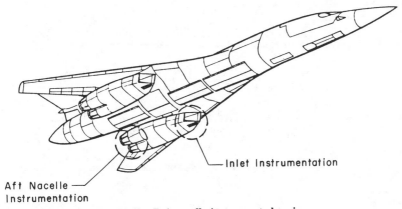

Aft Nacelle
Instrumentation

Inlet Instrumentation

Fig. V.69   B-1 nacelle instrumented regions.

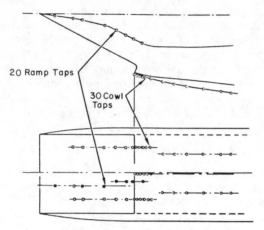

**Fig. V.70   Inlet surface pressure instrumentation common to the wind tunnel models and aircraft.**

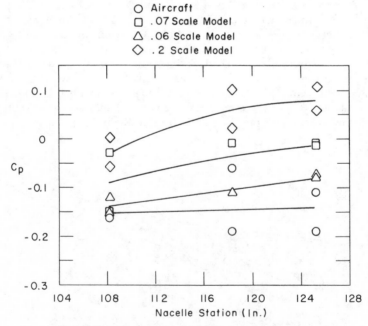

**Fig. V.71   Aircraft and model inlet cowl pressures.**

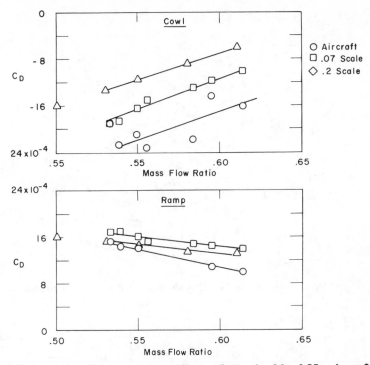

**Fig. V.72   Drag coefficient variation with mass flow ratio; $M = 0.85$ and $\alpha = 2.4$ deg.**

wing was tested to determine the appropriate wing gap effect, and the ECS air was varied to determine its effect, but an absolute simulation of these parameters was not possible. The Reynolds number was 20–40 times greater on the aircraft than the 6% propulsion model. In addition to the flight vehicle and the propulsion wind tunnel model, data were also obtained on a 20% inlet/forebody model with bypass and bleed and a 7% inlet force model to determine inlet ramp and cowl forces. Some 200 pressures were matched on the left-hand inlet and aft nacelle of the B-1 aircraft and wind tunnel models (Fig. V.69). In general, the pressures showed excellent agreement between wind tunnel and flight-test data. The agreement was better in the nozzle region than the inlet area and individual differences in local pressures tended to cancel, yielding pressure area integrated drag estimates which correlated to within approximately 10 drag counts. A comparison of the inlet cowl pressures, shown installed in Fig. V.70 and as pressure coefficients in Fig. V.71, shows better agreement for the 6% aircraft model than the partial aircraft simulation of the 20% model. The 20% model pressure coefficient deviation was attributed to problems encountered with high pressures feeding forward from the large cross-sectional area of the mass flow metering system and the downstream model support. The change of inlet drag with mass flow is shown in Fig. V.72.

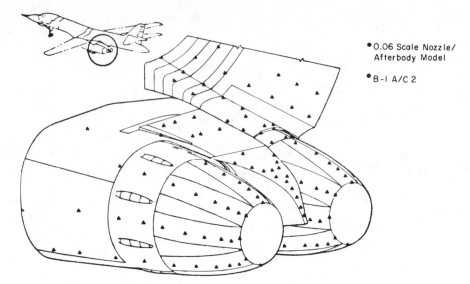

• 0.06 Scale Nozzle/
Afterbody Model

• B-1 A/C 2

**Fig. V.73  Aft nacelle isometric showing measurement locations.**

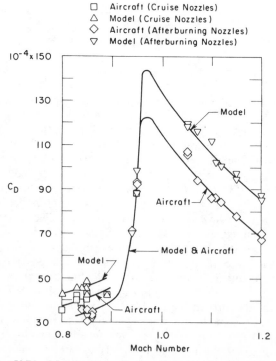

☐  Aircraft (Cruise Nozzles)
△  Model (Cruise Nozzles)
◇  Aircraft (Afterburning Nozzles)
▽  Model (Afterburning Nozzles)

**Fig. V.74  Model aft nacelle drag is higher than aircraft drag.**

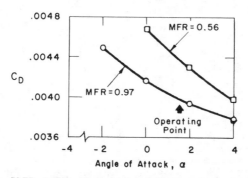

**Fig. V.75   MFR effect on aft nacelle/nozzle pressures.**

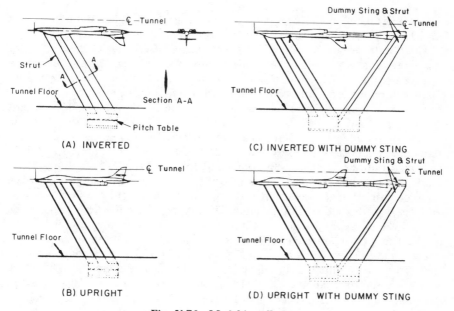

**Fig. V.76   Model installations.**

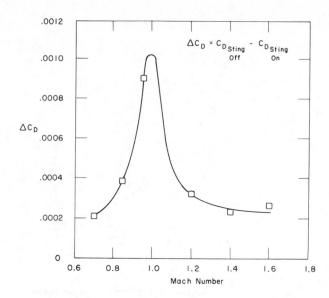

Fig. V.77   Sting effects (nozzle/afterbody tests).

The wind tunnel model drag values are consistently higher than flight test (3–10 drag counts) but show similar trends with mass flow.

The nozzle surface pressure distribution in the afterbody/nozzle region for the flight vehicle and wind tunnel model are shown in Fig. V.73. Comparison of pressure data yielded results similar to those for the inlet area: the pressure distributions were roughly comparable, with some local pressures higher and some lower. The integrated pressure drag values are within 8 drag counts subsonically and 15 counts supersonically (Fig. V.74). As before, the data trends are parallel. These differences are attributed to nozzle cross flow, some remaining support system interference, inlet fairing effects, nonuniform pressure distributions, incorrect environmental control system air simulation, and inadequate simulation of separation regions on the wind tunnel model. The issue of independent assessment of inlet and nozzle propulsion flow was also addressed in this study. For this nacelle, varying inlet mass flow changes the drag of the reference nozzles by 4 drag counts at the operating condition (Fig. V.75). The support interference studies provided insight into the magnitude of sting and strut effects. As shown in Fig. V.76, the model was tested upright and inverted to determine a strut effect, estimated at 3 drag counts at 0.85 Mach number and 6 drag counts at 1.1 Mach number. The model was also tested to determine a sting interference effect for corrections to the aerodynamic model. Figure V.77 demonstrates the effect on the aftbody/nozzle drag as 2–10 drag counts, depending on Mach number. This correlation effort for the highly interactive flow on the B-1 aircraft nacelle pointed out the difficult technical problems surrounding determination of throttle-dependent forces.

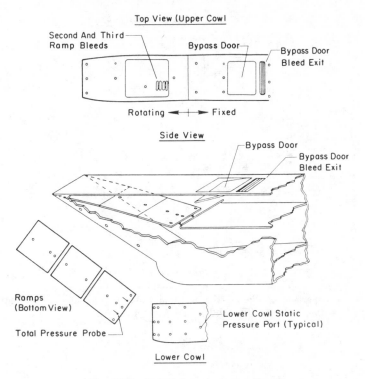

Top View (Upper Cowl)

Second And Third Ramp Bleeds

Bypass Door

Bypass Door Bleed Exit

Rotating ◀ ─┼─ ▶ Fixed

Side View

Bypass Door

Bypass Door Bleed Exit

Ramps (Bottom View)

Total Pressure Probe

Lower Cowl Static Pressure Port (Typical)

Lower Cowl

**Fig. V.78    F-15 inlet pressure orifices.**

*F-15.* The F-15 wind tunnel/flight-test data correlation effort is well documented. Typical references include Refs. 16, 58, and 59. The 7.5% wind tunnel model and flight vehicle were instrumented with approximately 300 static surface pressures in the inlet and nozzle region. The wind tunnel model was tested at 0.6, 0.9, and 1.2 Mach numbers over a range of angles-of-attack and mass flows. Not simulated were a variable bypass door, scaled inlet throat probes, or hot exhaust jets. The pressure coefficient uncertainty targets were $\pm 0.005$ for the wind tunnel and $\pm 0.03$ for the flight test. Flight tolerances desired were: angle-of-attack $\pm 0.25$, angle-of-yaw $\pm 0.25$, and Mach number $\pm 0.01$. The wind tunnel unit Reynolds numbers were usually 12 million compared to a flight Reynolds number of 150–280 million. Inlet ramp and cowl pressure instrumentation locations are shown in Fig. V.78. The upper cowl pressure data show good agreement between wind tunnel and flight tests, except downstream of the throat slot bleed/bypass exit (Fig. V.79). Similar agreement was evident on the lower cowl and the wing fairing (Fig. V.80). Comparison of the pressure area integrated drag from the wind tunnel and flight vehicle and the wind tunnel force balance measurements shows that the drag correlates well at 0.6, 0.9, and 1.2 Mach numbers across the range of inlet capture ratios except at higher capture ratios for 0.6 Mach number (Fig. V.81).

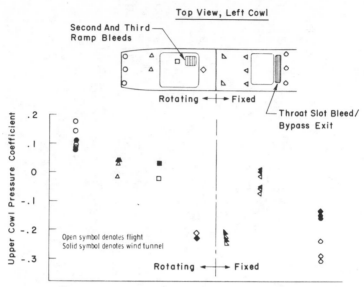

Fig. V.79   Comparison of flight and wind tunnel pressure coefficients for upper cowl surface; $M_\infty = 0.9$, $\alpha = 3$ deg, $\beta = 0$ deg, and $\rho = 7$ deg.

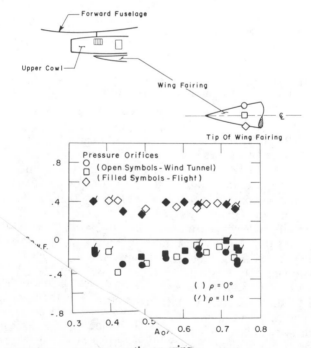

Fig. V.80   Effect of capture area ratio on wing ... ing pressure coefficients; $M = 0.9$ and $\alpha = 0$ deg.

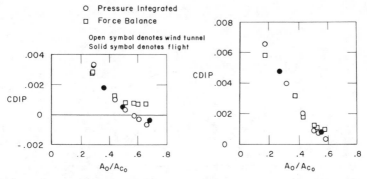

Fig. V.81a  Comparison of wind tunnel and flight pressure-integrated drag; $M_\infty = 0.6$, $\alpha = 0$ deg, $\rho = 0$ deg, and altitude = 20,000 ft.

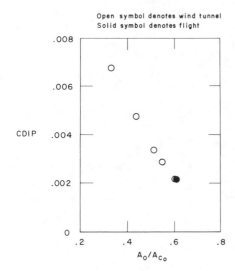

Fig. V.81b  $M_\infty = 1.2$, $\alpha = 0$ deg, $\rho = 0$ deg, and altitude = 30,000 ft.

This discrepancy between the pressure area integration wind tunnel and flight and the force balance from wind tunnel test is attributed to the large flow angularity at the inlet lip which the pressure area integration cannot account for. Additional work is prescribed for improving the measurement technique near the inlet lip.

The pressure instrumentation on the nozzle boattail is described in Fig. V.82. The nozzle pressures correlate well except for the pressures on the nozzle sides near the interfairing or tail booms. One other point of interest concerning throttle-dependent forces is shown in Fig. V.83. Changing the inlet rotation angle produces the expected upper cowl pressure changes, but also unexpectedly changes the mid-fuselage and upper nozzle pressures. This could result from the inlet rotation changing the aircraft trim, necessitating a

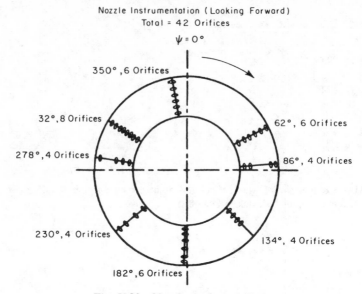

**Fig. V.82   Nozzle pressure orifices.**

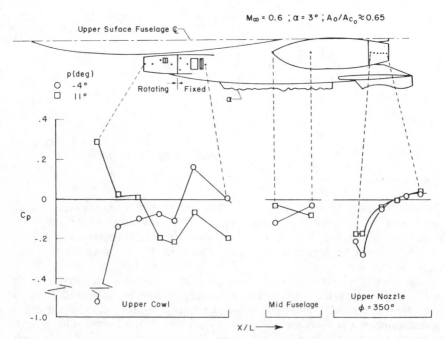

**Fig. V.83   Upper fuselage pressure coefficient.**

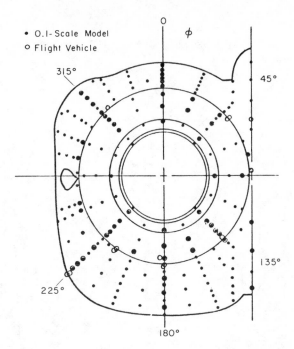

**Fig. V.84 Pressure orifice locations—left engine nacelle and nozzle.**

horizontal tail incremental deflection, resulting in this pressure change. The common assumption that the inlet and nozzle propulsion streams are independent is therefore subject to question.

*YF-17.* Flight-test surface pressure data correlations, primarily in the nozzle boattail area (Fig. V.84), were performed using the YF-17 aircraft.[62] Two wind tunnel models, 10 and 20% scale, were tested at Arnold Engineering Development Center from 0.6 to 1.5 Mach number over a range of angles of attack and nozzle pressure ratio. Model-to-flight sources of error are support interference, cold jet simulation, faired inlet, engine bay purge, variable exhaust nozzle, speed brake, and Reynolds number. Wind tunnel repeatability was ±0.0005 for pressure coefficient, ±0.1 deg angle-of-attack, and ±0.10 nozzle pressure ratio. Flight-test pressure coefficient uncertainties were ±0.002 to 0.011, depending on altitude and Mach number. To evaluate support interference, the model was held by a sting support with and without a dummy wing-tip support (Fig. V.85). These corrections were incorporated where possible. For both 0.6 and 0.9 Mach numbers, the wind tunnel and flight pressures correlated well. Some differences were present near the metric break on the wind tunnel models. Pressure comparisons at 0.6 Mach number are presented in Fig. V.86. Data for 1.2 Mach number (Fig. V.87) indicate that more expansion and recompression were present in the flight data. This is attributed to the thinner flight boundary layer at higher Reynolds numbers and to the

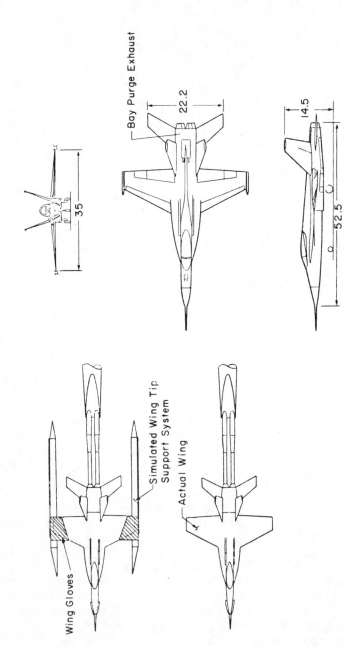

**Fig. V.85   YF-17 wind tunnel models and aircraft.**

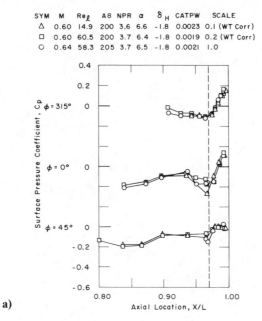

a)

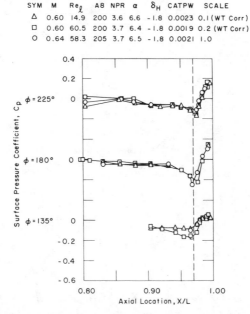

b)

Fig. V.86 Comparison of wind tunnel and flight vehicle surface pressure distribution at $M = 0.6$, $A8 = 200$ in.$^2$, $\alpha \sim 6.5$ deg, and NPR $\sim 3.7$; a) lee-side pressures, b) windward-side pressures.

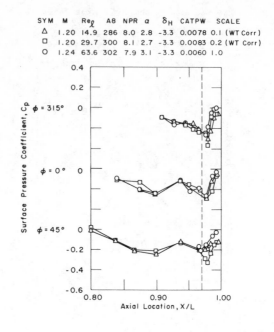

a)

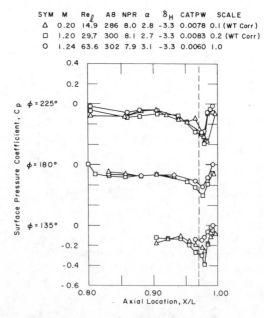

b)

Fig. V.87  Comparison of wind tunnel and flight vehicle surface pressure distributions at $M = 1.2$, $A8 \sim 300$ in.$^2$, $\alpha \sim 2.7$ deg, and NPR $\sim 8.0$; a) lee-side pressures, b) windward-side pressures.

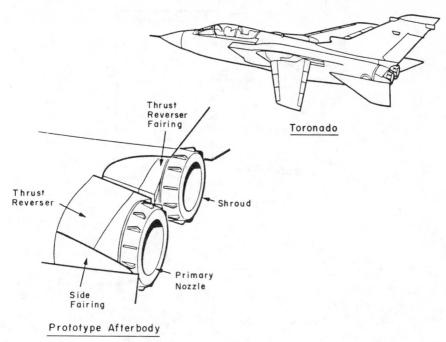

Fig. V.88   Tornado prototype afterbody.

difference in nozzle construction from a solid model to a flight article with boattail leaves and leakage. In general, the support system interference is not significant subsonically and improves the correlation if accounted for supersonically.

*Tornado.* The multinational Tornado aircraft also was the subject of a general correlation of wind tunnel and flight-test afterbody pressures. The aircraft and afterbody/nozzle (Fig. V.88) were pressure-instrumented in the boattail and base regions to compare with data from the wind tunnel model (Fig. V.89). In general, there was agreement between wind tunnel and flight-test boattail and base pressure data. The upper and lower surface boattail pressure data are compared to flight-test pressures at subsonic speeds in Fig. V.90. The pressure data on the wind tunnel model show good agreement with flight-test data, especially on the lower surface. The base pressures, especially on this nozzle, were good correlating parameters for drag[56] and were used to monitor drag changes in flight. The base pressures in wind tunnel and flight tests are at approximately the same level, especially above 1.1 Mach number (Fig. V.91). An unsteady throttle-dependent force was also identified in this effort. A fix to an afterbody flow instability problem, besides reducing the prototype aircraft drag, also identified the usefulness of unsteady pressure measurements as an indication of aftbody flow quality and separation.

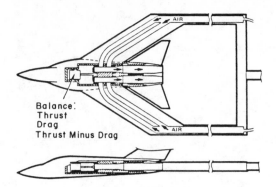

Fig. V.89   Tornado afterbody model.

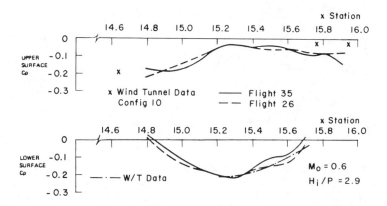

Fig. V.90   Boattail surface static pressure distributions.

## Summary

The success of a wind tunnel/analytical correlation of flight throttle-dependent forces is determined by the care and consideration directed to eliminating or correcting the discrepancies created by the differences between the wind tunnel model and the flight vehicle. Support effects can be corrected with some success. Reynolds number mismatches, especially important in the nozzle/aftbody region where separated flow can perhaps be present, can be resolved in high Reynolds number facilities. Simultaneous duplication of inlet and nozzle flows is critical if the inlet and nozzle flows are not independent. Extensive pressure instrumentation is required over the inlet and nozzle regions and all other aircraft surfaces influenced by the propulsion stream. The task is not impossible but requires use of much engineering skill and resources.

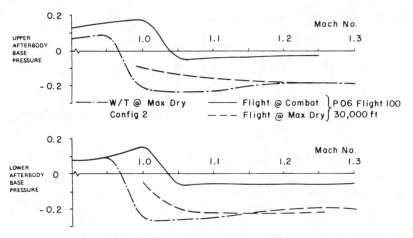

**Fig. V.91  Comparison of inflight and wind tunnel base pressures at supersonic speeds for production afterbody.**

## Final Remarks

Throttle-dependent forces are all forces, internal and external, acting on the engine streamtube and aircraft surfaces that change with engine power setting. These forces can be significant for tactical aircraft which are required to perform a mission over a wide Mach number range and for transport aircraft which have demanding cruise performance requirements. The primary inlet throttle-dependent forces are spillage and bypass drag. For the nozzle, these forces are boattail and base drag and jet interference and entrainment. These forces can be determined with varying success by analytical techniques, wind tunnel tests, and in flight test. The computational methods and test techniques have been and will continue to be developed.

## Appendix V.1

### Basic Definitions

*Additive drag* ($D_{\text{ADD}}$)—Static pressure force exerted, in the wind direction, on the inlet streamtube, between freestream conditions and the inlet stagnation point, with the inlet operating at zero external bleed flow. Alternatively it may be looked upon as the total momentum change of the inlet air from freestream to the inlet stagnation point with the inlet operating at zero external bleed flow.

*Additive drag correction factor* ($K_{\text{ADD}}$)—The change in spillage drag from the mass flow ratio at which the additive drag is zero to an operating mass flow ratio, divided by the theoretical additive drag:

$$K_{ADD} = (D_{ADD} - \Delta D_{Lip})/D_{ADD} \text{ (theoretical)}$$

If it is assumed that the correction factor is not a function of mass flow ratio (a common assumption), the additive drag correction factor is the change in spillage drag divided by the change in theoretical additive drag between any two mass flow conditions.

$$K_{ADD} = \Delta D_{Spill}/\Delta D_{ADD} \text{ (theoretical)} \left[ \text{if } K_{ADD} \neq f \text{ (MFR)} \right]$$

*Aerodynamic reference configuration* (*reference configuration*)—Configuration tested on external aerodynamics model. Usually a fixed configuration.

*Aerodynamic reference model*—Wind-tunnel model used to determine the forces and moments of the reference aircraft configuration.

*Afterbody*—Total fuselage and/or nacelle from the maximum cross section to the nozzle exit stations, including base area.

*Afterbody drag*—Boattail drag plus base drag.

*Base area*—Cross-sectional area projected normal to the fuselage reference line of any geometrically definable area aft of the maximum cross-section station that is known to be totally in a separated flow region.

*Base drag*—Drag attributable to a base area.

*Bleed drag* ($D_{bl}$)—Wind direction component of total momentum loss from freestream to exit station of the bleed air, plus the incremental change in external drag, at constant inlet airflow, from no bleed to the operating bleed. Note that in the case of external bleed drag (bleed air taken off ahead of the inlet lip station), although the definition is identical, there is an important difference. In this case the bleed flow alters the inlet stagnation streamtube shape, even at a constant inlet airflow. Therefore the incremental change in external drag includes a change in the "spillage drag" in addition to the change in external drag due to the perturbations at the bleed exit station.

*Boattail*—Amount of fuselage and/or nacelle closure between two stations. Measured by planar cuts normal to the fuselage reference line through various aircraft sections (less any base closure between the stations).

*Boattail drag*—Total drag (pressure plus friction) on the fuselage afterbody minus base drag (does not include tail surfaces).

*Bypass drag* ($D_{by}$)—Wind direction component of total momentum loss from freestream to exit station of the bypass air plus the incremental change in external drag, at constant inlet airflow, from no bypass to the operating bypass.

*Capture area, design* ($A_{Des}$ or $A_{ref}$)—For variable capture area inlets, this is the capture area setting at which the inlet contours are designed.

*Capture area, inlet* ($A_c$)—Area enclosed by the locus of most forward points (measured normal to the inlet reference line) on the inlet cowl, sideplates, and first ramp or boundary-layer diverter, projected in the freestream direction.

*Cowl length* ($l_{max}$)—Length from cowl lip station to the fuselage or nacelle maximum cross-section station measured along the fuselage reference line.

*Datum configuration*—Configuration used as a basis of comparison with other configurations of the same type.

*Drag polar*—A plot of drag vs lift. For bookkeeping purposes it is presented for the operating reference configuration.

*Drag-polar exhaust increments*—Force and moment differences between the aerodynamic reference and operating reference exhaust system configurations which are allocated to the aircraft drag polars.

*Drag-polar inlet increments*—Force and moment differences between aerodynamic reference and operating reference inlet configurations which are allocated to the aircraft drag polar.

*Engine air*—All air entering the engine compressor face including all air removed from or injected into the engine airstream aft of the compressor face. This will include, for example, compressor bleed air, fan bleed air, nozzle bleed air, and secondary air injected into the nozzle flow. The lone exception to this definition will be tertiary nozzle air which is defined as an external flow stream.

*Exhaust system force model*—Wind tunnel model used to determine the relative forces and moments of the aerodynamic reference, operating reference, and operating exhaust system configurations.

*Exit station* ($e$)—The most downstream station, measured along the fuselage reference line, at which the internal airflow is completely confined by solid surfaces.

*External drag* ($D_{ext}$)—Pressure plus friction force in the wind direction on the external surface of the aircraft, minus the static force on all external surfaces affected by the propulsion airstreams.

*External surfaces*—All aerodynamic surfaces not included in the internal surfaces.

*Friction force, calculated*—Skin-friction force obtained from skin-friction coefficient correlations and model local flow conditions.

*Fuselage boattail*—Closure of the fuselage and/or nacelle from maximum cross section to customer connect, less any base closure between the stations (does not include tail surfaces).

*Fuselage boattail angle* ($B_{11}$)—Chord line from customer connect to maximum nacelle/fuselage cross-section area station.

*Fuselage boattail drag* (*aftbody boattail drag*)—Total drag (pressure plus friction) on the aftbody (does not include base drag).

*Gross thrust* ($F_g$) (*uninstalled gross thrust*)—The static force vector generated by the engine operating with Mil Spec total pressure recovery at the compressor face and no customer bleed or horsepower extractions. Here the static force will include the total momentum of all propulsion airstreams at their respective exit station and all static friction and pressure forces on any external surfaces affected by the propulsion streams.

*Highlighter point*—A point at which a line normal to the inlet reference line is tangent to the inlet leading-edge contour.

*Inlet airflow*—All air that enters the inlet at the cowl lip station; includes bleed taken out aft of the lip station but not bleed removed forward of this station.

*Inlet force model*—Wind tunnel model used to determine the relative forces and moments of the aerodynamic reference, operating reference, and operating inlet configurations.

*Inlet lip area* $(A_i)$—Physical duct cross-sectional area measured from the most aft highlight point (usually the cowl centerline highlight point) to the ramp/centerbody at an angle normal to the ramp/centerbody. Pitot (open-nose) inlets are considered as an axisymmetric inlet with a 0-deg centerbody for the purpose of this definition.

*Installed gross thrust* $(F_{ginst})$—The static force vector generated by the engine operating with the installed total pressure recovery at the compressor face and with the installed customer bleed air and horsepower extractions. Static force is defined the same as in gross thrust.

*Installed net thrust* $(F_{ninst})$—Installed gross thrust vector minus the ram drag vector $(\dot{m}_{eng}V)$ of all engine air. (This definition of installed net thrust *does not* include throttle-dependent drag increments.)

*Integrated pressure force*—Force obtained by assigning surface area increments to selected model static-pressure taps and summing the pressure/area components.

*Internal drag* $(D_{INT})$—Pressure plus friction force, positive in the wind direction, on the internal surfaces of the aircraft. $(D_{INT} = -F_{INT})$.

*Internal surfaces*—All internal surfaces from the locus of inlet stagnation points (stagnation points on the cowl lip, sideplates, and forward ramp or boundary-layer gutter leading edge) to the respective propulsion air exit stations at which exit momentum is defined.

*Internal thrust* $(F_{INT})$—The pressure plus friction force developed on all internal surfaces.

*Lift* $(L)$—External pressure plus friction force exerted on the aircraft in a direction normal to the wind direction in the plane of symmetry.

*Lip suction increment*—Change in the external drag (usually a reduction) due to a change in inlet airflow from baseline to operating conditions with the inlet operating at zero external bleed flow.

*Net thrust* $(F_n)$ *(uninstalled net thrust)*—Uninstalled gross thrust vector minus the ram drag vector $(\dot{m}_{eng}V_0)$ of all engine air.

*Nozzle boattail*—Closure on the nozzle from customer connect to nozzle exit stations.

*Nozzle boattail angle* $(\beta)$—Chord line from nozzle exit to customer connect.

*Nozzle drag* *(nozzle boattail drag)*—Total drag (pressure plus friction) on the external surfaces of the nozzle boattail.

*Operating configuration*—Configuration representing an actual aircraft operating condition for which performance is being predicted.

*Operating reference configuration* *(baseline configuration)*—Configuration from which the drag polar is derived and from which all propulsion installation force increments are taken. Usually a function of flight Mach number only.

*Operating reference configuration forces*—The external forces and the additive forces at operating reference conditions resolved into lift, drag, and side-force components.

*Operating reference configuration moments*—The moments associated with external forces, the ram drag, the additive forces, and the gross thrust at operating reference conditions resolved into pitching, yawing, and rolling moments.

*Propulsion airstream*—All exit streams used in the calculation of gross thrust. Since the exact definition of this will vary from configuration to configuration (e.g., if engine air is used for jet flaps or spanwise blowing) this must be well defined for each specific configuration.

*Propulsive force* ($F_p$) (*installed propulsive force*)—Installed net thrust vector minus all throttle-dependent force increment vectors.

*Ram drag* ($D_{ram}$)—Product of the flight velocity ($V_0$) and the engine air mass flow rate ($\dot{m}_{eng}$).

*Reference capture area* ($A_{ref}$)—The design capture area projected in the direction of the inlet reference line. Used in inlet drag coefficient.

*Reference configuration*—Aerodynamic reference configuration.

*Scrubbing drag*—Incremental force (pressure plus friction) in the wind direction on all external surfaces between the zero nozzle flow condition and the operating nozzle flow condition.

*Side force*—Pressure plus friction force exerted on the aircraft in a direction normal to the wind direction and normal to the plane of symmetry.

*Spillage drag* ($D_{spill}$)—Incremental change in additive drag plus lip suction due to a change in inlet airflow from operating reference to operating conditions with the inlet operating at zero external bleed flow.

*Subsonic duct angle* ($\theta_{sd}$)—Equivalent cone expansion half-angle of the subsonic diffuser from the throat station to the compressor face station neglecting the engine bullet nose.

*Throat area, inlet* ($A_t$)—Minimum physical duct cross-sectional area.

*Throttle-dependent exhaust increments*—Force and moment differences between operating reference and operating exhaust system configurations which are allocated to the propulsive force.

*Throttle-dependent external drag increments* ($D_{TD}$)—Throttle-dependent exhaust drag increment plus the throttle-dependent inlet drag increment.

*Throttle-dependent inlet increments*—Force and moment differences between the operating reference and operating inlet configurations which are allocated to the propulsive force.

*Throttle-dependent moments*—The power setting-dependent moment increments between operating and operating reference conditions.

*Total momentum*—Sum of the momentum flux (velocity × mass flow rate, $mV$) and pressure forces [area × pressure increment above ambient $-A(P - P_0)$].

*Trimmed drag polar*—The drag polar of a series of configurations (due to varying control surfaces) for which all moments about the center of gravity are zero at each individual point.

## Bibliography for Chapter V

[1] Munniksma, B. and Jaarsma, F., "Jet Interference on a Podded Engine Installation of a Twin Engined Wide Body Aircraft at Cruise Conditions," AGARD CP-150, Sept. 1974.

[2] Brazier, M. E. and Ball, W. H., "Accounting of Aerodynamic Forces on Airframe/Propulsion Systems," AGARD CP-150, Sept. 1974.

[3] Thornley, S. A. and Carter, E. C., "The Measurement of the Transonic Spillage Drag of a Supersonic Intake," AGARD CP-150, Sept. 1974.

[4] Coombes, T. D., "A Model Technique for Exhaust System Performance Testing," AGARD CP-150, Sept. 1974.

[5] Antonatos, P. P., "Inlet/Airplane Interference and Integration," AGARD LS-53, May 1972.

[6] Aulehla, F. and Lotter, K., "Nozzle/Airframe Interference and Integration," AGARD LS-53, May 1972.

[7] Fuhs, A. E., "Engine Integration and Thrust/Drag Definition," AGARD LS-53, May 1972.

[8] Harper, L. R., "The Subsonic Performance of Practical Military Variable Area Convergent Nozzles," AGARD CP-301, May 1981.

[9] Hunt, B. L. and Gowadia, N. S., "Determination of Throttle-Dependent Drag for Fighter Aircraft," AIAA Paper No. 81-1692, Aug. 1981.

[10] Rooney, E. C., "Development of Techniques to Measure Inflight Drag of a U.S. Navy Fighter Airplane, and Correlation of Flight Measured Drag with Wind Tunnel Data," AGARD CP-124, April 1973.

[11] Arnaiz, H., "Techniques for Determining Propulsion System Forces for Accurate High Speed Vehicle Drag Measurements in Flight," AIAA Paper No. 75-964, Aug. 1975.

[12] Oates, G., "The Aerothermodynamics of Aircraft Gas Turbine Engines," AFPAL-TR-78-52, July 1978.

[13] Anderson, R. D., Martens, R. E., and Lee, C. C., "A Thrust/Drag Accounting Procedure Applicable to Engine Cycle Selection Studies," McDonnell Douglas Report MDC A1197, June 1971.

[14] Marshall, F. L., "Element Performance Integration Techniques for Predicting Airplane System Performance," Boeing Document D162-10449-1, June 1972.

[15] Burcham, F. W. et al., "Recent Propulsion System Flight Tests at NASA Dryden Flight Research Center," AIAA Paper No. 81-2438, Nov. 1981.

[16] Webb, L. D. and Janssen, R. L., "Preliminary Flight and Wind Tunnel Comparisons of the Inlet/Airframe Interactions of the F-15 Airplane," AIAA Paper No. 79-0102, July 1978.

[17] Niedling, L. G., "F-15 Wind Tunnel/Flight Correlation," NASA Wind Tunnel/Flight Correlation, Nov. 1981.

[18] Craig, R. E. and Reich, R. J., "Flight Test Aerodynamic Drag Characteristics Development and Assessment of Inflight Propulsion Analysis Methods for AGM-109 Cruise Missile," AIAA Paper No. 81-2423, Nov. 1981.

[19] Doonan, J. G. and Davis, W. H., "Advanced Exhaust Nozzle Concepts Using Spanwise Blowing for Aerodynamic Lift Enhancement," AIAA Paper No. 82-1132, June 1982.

[20] Aerospace Engineering, Society of Automotive Engineers, Inc., March 1983.

[21] Englar, R. J., "Development of an Advanced No-Moving-Parts High-Lift Airfoil," ICAS-82-6.5.4, Aug. 1982.

[22] Aulehla, F. and Besigk, G., "Reynolds Number Effects on Fore and Aftbody Pressure Drag," AGARD CP-150, Sept. 1974.

[23] Glidewell, R. J. and Fanning, A. E., "Twin Jet Exhaust System Test Techniques," AGARD CP-150, Sept. 1974.

[24] Compton, W. B., "An Experimental Study of Jet Exhaust Simulation," AGARD CP-150, Sept. 1974.

[25] Robinson, C. E. et al., "Exhaust Plume Temperature Effects on Nozzle Afterbody Performance over the Transonic Mach Number Range," AGARD CP-150, Sept. 1974.

[26] Ewald, B., "Engine Jet Simulation Problems in Wind Tunnel Tests," AGARD

CP-150, Sept. 1974.

[27] AGARD/Flight Mechanics Panel, "Ground/Flight Test Techniques and Correlation," Symposium program and abstracts, Oct. 1982.

[28] Binion, T. W. et al., "Progress in Wind Tunnel Test Techniques and in the Corrections and Analysis of the Results," AGARD Symposium, "Ground/Flight Test Techniques and Correlation," Oct. 1982.

[29] Peterson, J. B., "Wind Tunnel/Flight Correlation Program on the XB-70-1," NASA Wind Tunnel/Flight Correlation, Nov. 1981.

[30] Smith, R. H., "Problems in Correlation Caused by Propulsion Systems," NASA Wind Tunnel/Flight Correlation, Nov. 1981.

[31] Carter, E. C., "Experimental Determination of Inlet Characteristics and Inlet and Airframe Interference," AGARD LS-53, May 1982.

[32] Jaarsma, F., "Experimental Determination of Nozzle Characteristics and Nozzle Airframe Interference," AGARD LS-53, May 1982.

[33] Ross, J. A., McGregor, I., and Priest, A. J., "Some RAE Research on Shielded and Unshielded Fuselage Mounted Air Intakes at Subsonic and Supersonic Speeds," AGARD CP-301, May 1981.

[34] Harris, A. E. and Carter, E. C., "Wind Tunnel Test and Analysis Techniques Using Powered Simulators for Civil Nacelle Installation Drag Assessment," AGARD CP-301, May 1981.

[35] Aulehla, F. and Besigk, G., "Fore and Aftbody Flow Field Interaction with Consideration of Reynolds Number Effects," Messerschmitt Boelkow Blohm UFE 1279, Oct. 1976.

[36] Kennedy, T. L., "An Evaluation of Wind Tunnel Test Techniques for Aircraft Nozzle Afterbody Testing at Transonic Mach Number," AEDC TR-80-8, Nov. 1980.

[37] Harper, L. R., "The Subsonic Performance of Practical Military Variable Area Convergent Nozzles," AGARD CP-301, May 1981.

[38] Pozniak, O. M., "A Review of the Effect of Reynolds Number on Afterbody Drag," AGARD CP-301, May 1981.

[39] Voogt, N. et al., "Aerodynamic Aspects of a High Bypass Ratio Engine Installation on a Fuselage Afterbody," AGARD CP-301, May 1981.

[40] Quermann, J. K., "Subsonic Base and Boattail Drag: An Analytical Approach," AGARD CP-150, Sept. 1974.

[41] Moulden, T. H. et al., "On Some Problems Encountered in a Theoretical Study of the External Flow Over a Nozzle Configuration in Transonic Flight," AGARD CP-150, Sept. 1974.

[42] Ahmed, S. R., "Prediction of the Optimum Location of a Nacelle Shaped Body on the Wing of a Wing-Body Configuration by Inviscid Flow Analysis," AGARD CP-150, Sept. 1974.

[43] Krenz, G., "Interference Between Wing and Intake/Jet," AGARD CP-150, Sept. 1974.

[44] Grashof, J. and Schmidt, W., "Computational Methods for Two-Dimensional and Three-Dimensional Inlets in Subsonic and Supersonic Flow," ICAS-82-4.3.1, Aug. 1982.

[45] Herrick, P. W., "Predicting Propulsion Related Drag of Jet Aftbodies," SAE 751088, Nov. 1975.

[46] Antonatos, P. P. et al., "Assessment of the Influence of Inlet and Aftbody/Nozzle Performance on Total Aircraft Drag," AGARD CP-124, April 1973.

[47] Knight, D. D., "Improved Numerical Simulation of High Speed Inlets Using the Navier-Stokes Equations," AIAA Paper No. 80-0383, Jan. 1980.

[48] Haberland, C. et al., "Calculation of the Flow Field Around Engine-Wing Configuration," ICAS-80-4.1, Oct. 1980.

[49]Putnam, L. and Mace, J., "A Survey of Aft Body Flow Prediction Methods," AIAA Paper No. 81-1694, Aug. 1981.

[50]Mace, J. and Cosner, R., "Analysis of Viscous Transonic Flow Over Aircraft Forebodies and Afterbodies," AIAA Paper No. 83-1366, June 1983.

[51]Pyle, J. S. and Saltzman, E. J., "Review of Drag Measurements from Flight Tests of Manned Aircraft with Comparisons to Wind Tunnel Predictions," AGARD CP-124, April 1973.

[52]Walker, S. C., "Isolating Nozzle Afterbody Interaction Parameters and Size Effects —A New Approach," AGARD CP-150, Sept. 1974.

[53]Schoelen, F. J., "B-1 Inlet and Nozzle Flight Performance Determination Program," AIAA Paper No. 81-1852, Aug. 1981.

[54]Richey, G. K. et al., "Wind Tunnel/Flight Test Correlation Program on the B-1 Nacelle Afterbody/Nozzle at Transonic Conditions," AIAA Paper No. 78-989, July 1978.

[55]Richey, G. K. et al., "Wind Tunnel/Flight Test Correlation on the B-1 Nacelle Afterbody/Nozzle," AIAA Paper No. 76-673, July 1976.

[56]Leyland, D. C., "Lessons from Tornado Afterbody Development," British Aerospace.

[57]Whitfield, J. D. et al., "Overview of Flight and Ground Testing with Emphasis on the Wind Tunnel," AIAA Paper No. 81-2474, Nov. 1981.

[58]Nugent, J. and Webb, L. D., "Selected Results of the F-15 Propulsion Interaction Program," NASA Ames Tetwog Meeting, May 1982.

[59]Webb, L. D. and Nugent, J., "Selected Results of the F-15 Propulsion Interactions Program," AIAA Paper No. 82-1041, June 1982.

[60]Ayers, T. G., "Review of the 1980 Wind Tunnel/Flight Correlation Panel," NASA Wind Tunnel/Flight Correlation, Nov. 1981.

[61]Ayers, T. G., "Report of the Wind Tunnel/Flight Correlation Panel," NASA Dryden Flight Research Center, 1981.

[62]Lucas, E. J. et al., "Comparison of Nozzle and Afterbody Surface Pressures from Wind Tunnel and Flight Test of the YF-17 Aircraft," AIAA Paper No. 78-992, June 1978.

[63]Ferri, A., "Introductory Lecture—Engine Airplane Interference—Definition of the Problem and Related Basic Fluid Dynamic Phenomena," AGARD LS-53, May 1972.

[64]Round Table Discussion, AGARD CP-150, Sept. 1974.

[65]Smith, G. D. et al., "Analytical and Experimental Investigation of Ejector-Powered Engine Simulators for Wind Tunnel Models," AEDC-TR-76-128, Jan. 1977.

[66]Lucas, E. J., "Evaluation of Wind Tunnel Nozzle Afterbody Test Techniques Utilizing a Modern Twin Engine Geometry at Mach Numbers from 0.6 to 1.2," AEDC-TR-79-63, Oct. 1980.

[67]Bower, W. W., "An Analytical Procedure for the Calculation of Attached and Separated Diffuser Flows," AIAA Paper No. 74-1173, 1973.

# Chapter VI

# Precision and Propagation of Error

Robert B. Abernethy*

*Pratt & Whitney Engineering Division, West Palm Beach, Florida*

## I. Introduction

$T$HE objective of this chapter is to summarize a standard method of treating measurement error or uncertainty for gas turbine engine performance parameters, such as thrust, airflow, and thrust specific fuel consumption. The need for a standard is obvious to those who have reviewed the several methods currently used. The subject is complex and involves both engineering and statistics. In this chapter, only one method is summarized. In a sense we recommend a single standard method, which is required to make comparisons between engine manufacturers and facilities. However, the reader must recognize that no single method will give a rigorous, scientifically correct answer for all situations. Further, for even a single set of data, the task of finding and proving one method correct is usually impossible. In any event, the method selected is believed to have the widest application. We have made a serious effort to use simple prose with a minimum of jargon.

This summary is really an outline illustrating the techniques for estimating measurement uncertainty. Section II describes a mathematical model for uncertainty and the propagation of error. Section III gives the errors associated with the measurement of force, fuel flow, pressure, and temperature, and Sec.

EDITOR'S NOTE: This material has been abstracted from Abernethy's writing. In particular, two sources were used extensively. The AEDC Handbook 73–5, "Uncertainty in Gas Turbine Measurements" by R. B. Abernethy and J. W. Thompson, 1973,[1] and the material developed by the Society of Automotive Engineers, E-33 Study Committee, chaired by Abernethy.[2] The latter deals with in-flight propulsion measurement. The results from that committee are Aeronautical Recommended Practice 1678 and Aerospace Instruction Report 1703. The latter documents may be purchased from the Society of Automotive Engineers, 400 Commonwealth Drive, Warrendale, Pa. 15096. Thus the credit for the material in this chapter is due to Abernethy. The errors were introduced by my editing. (EEC)

*Man Reliability, Pratt & Whitney Aircraft Group.

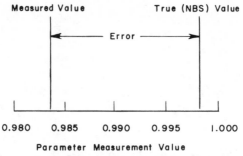

Fig. VI.1   Measurement error.

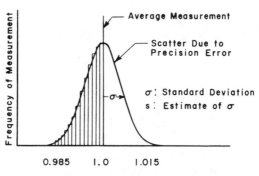

Fig. VI.2   Precision error.

IV treats net thrust and specific fuel consumption. Section V contains concluding remarks.

**Measurement Error**

All measurements of any property contain measurement errors in their value. These errors are the differences between the measured value of the parameter and the true value defined to be that measured by the National Bureau of Standards (NBS), as shown in Fig. VI.1. Uncertainty is the maximum error which might reasonably be expected and is a measure of accuracy, i.e., the closeness of the measurement to the NBS or defined true value. Measurement error has two components: a fixed error, sometimes called bias, and a random error.

*Precision (Random Error)*

Random error is seen in repeated measurements. Each measurement does not and is not expected to agree exactly with others made previously (Fig. VI.2). There are always numerous small effects that cause the difference. The variation between repeated measurements is called precision error. The standard deviation ($\sigma$) is used as a measure of the precision error. A large standard deviation means large scatter in the measurements. The statistic that is calculated to estimate the standard deviation is sometimes called the precision

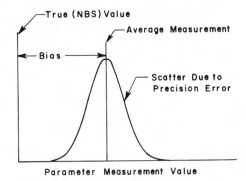

Fig. VI.3   Bias error.

| | Known Sign and Magnitude | Unknown Magnitude | |
|---|---|---|---|
| Large | (1) Calibrated Out | (3) Assumed to be Eliminated | |
| Small | (2) Negligible, Contributes to Bias Limit | (4) Unknown Sign<br>Contributes to Bias Limit | (5) Known Sign |

Fig. VI.4   Five types of bias errors.

index

$$s = \sqrt{\sum_{i=1}^{N} (X_i - \overline{X})^2 / (N - 1)}$$

where $N$ is the number of measurements made and $\overline{X}$ is the arithmetic average or mean value of individual measurements $X_i$.

*Bias (Fixed Error)*

The second component, bias, is the systematic error (Fig. VI.3).[†] In repeated measurements, at the *same* conditions, each measurement has the same bias. The bias cannot be determined unless the measurements are compared with the true value of the quantity measured. Note that in some cases the bias may be dependent upon environmental factors. The control of the environment is essential if the bias error is to be constant.

Bias is categorized into five classes: 1) large known biases, 2) small known biases, 3) large unknown biases, and 4) small unknown biases which may have unknown sign ($\pm$) or 5) known sign as shown in Fig. VI.4.

---

[†] There has been a controversy over the method of accounting for unknown bias errors. One group holds linear addition is correct, the other favors a root-mean-square approach. Both are presently accepted (see Eisenhart, Ku, and Colle in the list of quality-control chart references for details).

*Large Known Biases*

The large known biases are eliminated by comparing the instrument with a standard instrument and obtaining a correction. This process is called calibration.

*Small Known Biases*

Small known biases may or may not be corrected, depending on the difficulty of the correction and the magnitude and importance of the bias.

*Large Unknown Biases*

Unknown biases are not correctable; that is, they may exist, but the magnitude of the bias is not known, and perhaps even the sign is not known.

Every effort must be made to eliminate all large unknown biases. The introduction of such errors converts the controlled measurement process into an uncontrolled worthless effort. Large unknown biases usually come from human errors in data processing, incorrect handling and installation of instrumentation, and unexpected environmental disturbances such as changes in humidity, a barometric pressure that was not expected or allowed for, or maybe uncontrolled processes induced such as shock waves and bad flow profiles. In a well-controlled measurement process, the assumption is that there are no large unknown biases. To ensure that a controlled measurement process exists, all measurements should be monitored with statistical quality control charts. A list of references describing the use of statistical quality-control charts is included at the end of this section. Drifts, trends, and movements leading to out-of-control situations should be identified and investigated. Histories of data from calibrations are required for effective control. It is assumed throughout the following discussion that these precautions have been and are observed so that the measurement process is under control; if not, the methods contained herein are invalid.

*Small Biases, Unknown Sign, and Unknown Magnitude*

In most cases, the bias error is equally likely to be plus or minus about the measurement. That is, it is not known if the limit is positive or negative, and the estimate reflects this. The bias limit is estimated as an upper limit on the maximum fixed error.

It is both difficult and frustrating to estimate the limit of an unknown bias. To determine the exact bias in a measurement, it would be necessary to compare the true value and the measurements. This is almost always impossible. An effort must be made to obtain special tests or data that will provide bias information. The following are examples of such data:

1) Interlab, interfacility, intercompany tests on measurement devices, test rigs, and full scale engines.
2) Flight-test data vs altitude-test-chamber data vs ground-test data.
3) Special comparisons of standards with those instruments in the actual test environment.

Table VI.1   Nonsymmetrical bias limits

| Bias limits | Explanation |
|---|---|
| 0, +10 deg | Bias will range from 0 to + 10 deg |
| −5, +15 lb | Bias will range from − 5 to +15 lb |
| +3, + 7 psia | Bias will range from + 3 to +7 psia |
| −8, − 3 deg | Bias will range from − 8 to − 3 deg |

4) Ancillary or concomitant functions that provide the same performance parameter; i.e., in an altitude engine test, airflow may be measured with an orifice and a bellmouth, estimated from compressor speed-flow rig data, estimated from turbine flow parameter, and measured with jet nozzle calibrations.

5) When it is known that a bias results from a particular cause, special calibrations may be performed, allowing the cause to vary through its complete range, to determine the range of bias.

If there is no source of data for bias, the expert judgment of the most knowledgeable instrumentation engineer must be sought on the measurement. However, without data, the upper limit on the largest possible bias error must reflect the lack of knowledge, i.e., the value is an informed guess. (Usually the user will guess a larger value than the instrumentation engineer.)

*Small Biases, Known Sign, and Unknown Magnitude*

Sometimes the physics of the measurement system provides knowledge of the sign but not the magnitude of the bias. For example, thermocouples radiate and conduct energy to indicate lower temperatures. The resulting bias limits are nonsymmetrical, i.e., not of the form $\pm B$. They are of the form $^{+b}_{-c}$ where both limits may be positive or negative or the limits may be of mixed sign as indicated. Table VI.1 lists several nonsymmetrical bias limits for illustration.

In summary, measurement systems are subject to two types of errors—bias and precision error (Fig. VI.5). One sample standard deviation is used as the precision index. The bias limit is estimated as an upper limit on the maximum fixed error.

## Measurement Uncertainty

Usually a single number—some combination of bias and precision—is presented to express a reasonable limit for error. This single number is called the measurement uncertainty. The measurement uncertainty must have a simple interpretation—the largest error reasonably expected for some fraction of the (repeated) measurements—and be useful without complex explanation. It is impossible to define a single rigorous statistic because the (unknown) bias is an upper limit based on judgment. A judgment of necessity has unknown statistical characteristics. Any function of these two numbers must be a hybrid

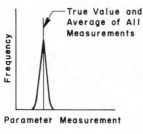

a. Unbiased, Precise, Accurate

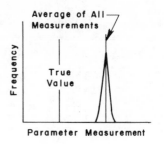

b. Biased, Precise, Inaccurate

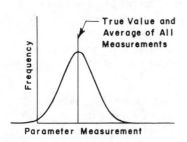

c. Unbiased, Imprecise, Inaccurate

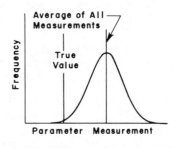

d. Biased, Imprecise, Inaccurate

Fig. VI.5   Measurement error (bias, precision, and accuracy).

combination of an unknown quantity (bias) and a statistic (precision). However, the need for a single number to measure error is so great that the adoption of an arbitrary standard is warranted. The standard most widely used is the bias limit plus a multiple of the precision index. This method is recognized and recommended by the NBS[3,4] and has been widely used in industry.

Uncertainty (Fig. VI.6) may be centered about the measurement and is defined herein as

$$U = \pm ( B + t_{95}S )$$                                    (VI.1)

where $B$ is the bias limit, $S$ is the precision index, and $t_{95}$ is the 95th percentile point for the two-tailed Student's "$t$" distribution (such tables are widely available; see, for example, Table E-1, Appendix E of Abernethy and Thompson's AEDC TR-73-5[1]). The $t$ value is a function of the number of degrees of freedom (df) used in calculating $S$. For small samples, $t$ will be large, and for larger samples, $t$ will be smaller, approaching 1.96 as a lower limit. The use of the $t$ arbitrarily inflates the limit $U$ to reduce the risk of underestimating $S$ when a small sample is used to calculate $S$. Since 30 degrees of freedom yield a $t$ of 2.04 and infinite degrees of freedom yield a $t$ of 1.96, an arbitrary selection of $t = 2$ for values of df from 30 to infinity was made, i.e., $U = \pm(B + 2S)$, when df $\geq 30$.[‡]

In a sample, the number of degrees of freedom is the size of the sample. When a statistical property is calculated from the sample, the degrees of freedom associated with the statistic are reduced by one for every estimated parameter used in calculating the statistic. For example, from a sample of size $N$, $\overline{X}$ is calculated

$$\overline{X} = \sum_{i=1}^{N} X_i \bigg/ N \qquad \text{(VI.2)}$$

which has $N$ degrees of freedom and

$$S = \sqrt{\sum_{i=1}^{N} (X_i - \overline{X})^2 \bigg/ (N-1)} \qquad \text{(VI.3)}$$

which has $N - 1$ degrees of freedom because $\overline{X}$ (based on the same sample of data) is used to calculate $S$. In calculating other statistical characteristics, more than one degree of freedom may be lost. For example, in calculating the standard error of a curve fit, the number of degrees of freedom lost is equal to the number of coefficients required to estimate the shape of the curve.

It is recommended that the uncertainty parameter $U$ be used for simplicity of presentation; however, the three components (bias, precision, and degrees of

---

[‡] The matter of a sample size is a vexing one. It is clear that we would like a large number of samples. Roughly 30 or so are needed to ensure the proper sort of information about the nature of the probability distribution function for the error at each point of interest. However, this is clearly excessive from a cost standpoint. Hence alternate procedures must be used. One such procedure is to record more than enough data to generate an interpolation formula, or curve fit. Another is to consider the nature of the errors incurred in small sample sizes. One such analysis is H. H. Josephs and V. S. Brook, Post Office Engineering Department Research Report No. 13801: "On the Probable Error of Samples of Small Sizes" (Post Office Research Station, Dollis Hill, London). The results from this report show that one records three data points, then one calculates the average from the two extreme values and calculates the estimate of the standard deviation from half the difference of the extreme values. It turns out there is little significant increase in the precision of the estimate until the sample size exceeds 8–10 points. This is particularly true if probability distribution has a small kurtosis (i.e., $\beta_2$ of 1.8–2, which implies a flat-topped distribution). In this case, the extreme case is very reliable. If the distribution is sharply peaked ($\beta_2 \sim 5$ or more), the efficiency of this median relative to the mean improves. For skewed distribution curves there is little to choose between the arithmetic mean or the extreme mean. Generally speaking, the use of an $N$ of 3 and the extreme difference is recommended unless one has reason to believe the probability distribution is normal. In that case the arithmetic average and

$$A^2 = \sum_{i=1}^{\beta} \frac{X_i^2}{N-1} - (\overline{X})^2$$

are the most reliable estimate.

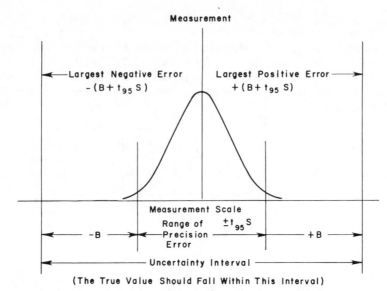

Fig. VI.6   Measurement uncertainty, symmetrical bias.

freedom) should be available in an appendix or in supporting documentation, as shown below. These three components may be required 1) to substantiate and explain the uncertainty value, 2) to provide a sound technical base for improved measurements, and 3) to propagate the uncertainty from measured parameters to performance parameters, and from performance parameters to other more complex performance parameters (i.e., fuel flow to Thrust Specific Fuel Consumption (TSFC), TSFC to aircraft range, etc.). Although uncertainty is not a statistical confidence interval, it is an arbitrary substitute for that interval and is probably best interpreted as the largest error expected. Under any reasonable assumption for the distribution of bias, the coverage of $U$ is greater than 95%, but this cannot be proved since the distribution of bias is both unknown and unknowable.

If there is a nonsymmetrical bias limit (Fig. VI.7), the uncertainty $U$ is no longer symmetrical about the measurement. The upper limit of the interval is defined by the upper limit of the bias interval ($B^+$). The lower limit is defined by the lower limit of the bias interval ($B^-$).

The uncertainty interval $U$ is

$$U^- = B^- - t_{95}S \quad \text{to} \quad U^+ = B^+ + t_{95}S$$

Table VI.2 shows the uncertainty $U$ for the nonsymmetrical bias limits of Table VI.1. The $S$ and $t_{95}$ are assumed to be 1 unit and 2 units for each case.

The proper method for combining elemental measurement uncertainty values is to determine the root-sum-square values of the elemental bias limits and elemental precision indices separately.[§] Such a combination is an equi-probability combination if the errors are represented by a Gaussian distribution. One then applies the uncertainty formula to the combined bias limits and precision limits [Eq. (VI.1)]. In some cases the same value will be obtained if the uncertainties are root-sum-squared directly. However, this is not a general rule, and large errors in the combined uncertainty (10–25%) can result. Further, the root-sum-squared uncertainty value will usually be smaller (more favorable) than the proper uncertainty estimate, and the estimate is likely to underestimate significantly the true measurement error, as shown in the example that follows. This calculation shows that, combining the following uncertainties, the root-sum-square of the uncertainties was 18.38 units. The correct value was 23.21 units.

### List of Statistical Quality-Control Chart References

**Basic References**

"ASTM Manual on Quality Control of Materials," ASTM STP 15-C. (Available from American Society for Testing Materials, 1916 Race Street, Philadelphia, Pa. 19103.)
ASQC Standard B1, B2-1958. "Guide for Quality Control and Control Chart Method for Analyzing Data," ANSI Standard Z1.1, Z1.2, 1958.
ASQC Standard B3-1958. "Control Chart Method of Controlling Quality During Production," ANSI Standard Z1.3, 1958. (Available from American Society for Quality Control, 161 West Wisconsin Avenue, Milwaukee, Wisc. 53203, or from American National Standards Institute, 1430 Broadway, New York, N.Y. 10018.)

---

[§]When precision indices of elemental error sources are combined (root-sum-square), the degrees of freedom of the result must be determined. The Welch-Satterthwaite formula is used for this purpose. The Welch-Satterthwaite formula is an approximation as there is no exact solution. The accuracy of the approximation was compared to other approximations and reported in the "ICRPG Handbook for Estimating the Uncertainty in...Rocket Engine Systems," CPIA 180 (AD 855130). It is a function of the degrees of freedom and magnitude of the elemental precision indices.
If $S_l = \sqrt{S_{1l}^2 + S_{2l}^2 + S_{3l}^2}$ with degrees of freedom $\nu_{1l}, \nu_{2l}, \nu_{3l}$, then

$$\nu_l = \frac{\left(S_{1l}^2 + S_{2l}^2 + S_{3l}^2\right)^2}{\left(S_{1l}^4 / \nu_{1l}\right) + \left(S_{2l}^4 / \nu_{2l}\right) + \left(S_{3l}^4 / \nu_{3l}\right)}$$

The general form being

$$\nu_R = \left[\sum_i S_i^2\right]^2 \bigg/ \left[\sum_i \left(S_i^4 / \nu_i\right)\right]$$

---

(§ footnote continued on the following page.)

Cowden, D. J., *Statistical Methods in Quality Control*, Prentice-Hall, Inc., Englewood Cliffs, N.J., 1957.

Duncan, A. J., *Quality Control and Industrial Statistics*, 3rd Ed., Richard D. Irwin, Inc., Homewood, Ill., 1965.

Eisenhart, C., Ku, H. H., and Colle, R., "Expression of the Uncertainties of Final Measurement Results," reprints, National Bureau of Standards Report, Jan. 1983.

Juran, J. M., ed., *Quality Control Handbook*, 2nd Ed., McGraw-Hill Book Company, New York, N.Y., 1962.

When precision indices for measurement are propagated to the final test result with influence coefficients, the precision index of the result will be obtained from an equation like

$$S_R = \sqrt{\frac{\partial R}{\partial X} S_x^2 + \frac{\partial R}{\partial Y} S_y^2 + \frac{\partial R}{\partial Z} S_z^2}$$

then

$$\nu_R \cong \left[ \left(\frac{\partial R}{\partial X} S_x\right)^2 + \left(\frac{\partial R}{\partial Y} S_y\right)^2 + \left(\frac{\partial R}{\partial Z} S_z\right)^2 \right]^2 \Bigg/ \frac{\left(\frac{\partial R}{\partial X} S_x\right)^4}{\nu_x} + \frac{\left(\frac{\partial R}{\partial Y} S_y\right)^4}{\nu_y} + \frac{\left(\frac{\partial R}{\partial Z} S_z\right)^4}{\nu_z}$$

In this manner, the degrees of freedom are propagated to the final test result and used to determine the $t_{95}$ value for the uncertainty interval. Note that it is conservative to round $\nu_R$ down if the $\nu_R$ is not a whole number.

| Bias limit ($B$) | Precision index ($S$) | Uncertainty |
|---|---|---|
| 1 | 6 | $\pm 13$ |
| 11 | 1 | $\pm 13$ |

where uncertainty = $\pm(B + 2S)$, for a sample of roughly 30 pieces of data.
    Now the bias limit for the combined parameter is the root-sum-square of 1 and 11,

$$B = \sqrt{1^2 + 11^2} = \sqrt{122} = 11.05$$

The precision index for the combined parameter is the root-sum-square of 6 and 1,

$$S = \sqrt{6^2 + 1^2} = \sqrt{37} = 6.08$$

The combined uncertainty is thus

$$U = \pm(B + 2S) = \pm[11.05 + 2(6.08)] = \pm 23.21$$

The root-sum-square of the original uncertainties is

$$\sqrt{(13)^2 + (13)^2} = \sqrt{169 + 169} = \sqrt{338} = 18.38$$

Now

$$[(23.21 - 18.38)/18.38] \times 100 = 26.3\%$$

and over 25% error has been introduced just because the wrong propagation of error formula was used.

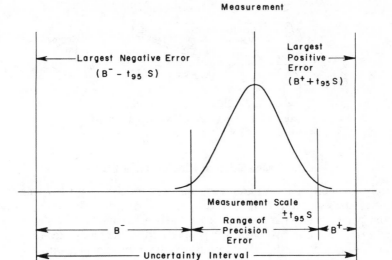

Fig. VI.7   Measurement uncertainty, nonsymmetrical bias.

Table VI.2   Examples of uncertainty intervals defined by nonsymmetrical bias limits

| $B^-$ | $B^+$ | $t_{95}S$ | $U^-$ (Lower limit for $U$) | $U^+$ (Upper limit for $U$) |
|-------|-------|-----------|------------------------------|------------------------------|
| 0 deg | + 10 deg | 2 deg | − 2 deg | + 12 deg |
| − 5 lb | + 15 lb | 2 lb | − 7 lb | + 17 lb |
| + 3 psia | + 7 psia | 2 psia | + 1 psia | + 9 psia |
| − 8 deg | − 3 deg | 2 deg | − 10 deg | − 1 deg |

## Examples of Control Charts in Metrology

Ku, H. H., "Statistical Concepts in Metrology," Chap. 2, *Handbook of Industrial Metrology*, American Society of Tool and Manufacturing Engineers, Prentice-Hall, Inc., New York, N.Y., 1967. (Reprinted in NBS SP 300-Vol. 1, "Precision Measurement and Calibration—Statistical Concepts and Procedures." (Available from the Superintendent of Documents, U.S. Government Printing Office, Washington, D.C. 20402.)

Pontius, P. E., "Measurement Philosophy of the Pilot Program for Mass Calibration," NBS Technical Note 288. (Available from the Superintendent of Documents, U.S. Government Printing Office, Washington, D.C. 20402.)

Pontius, P. E. and Cameron, J. M., "Realistic Uncertainties and the Mass Measurement Process," NBS Monograph 103. (Available from the Superintendent of Documents, U.S. Government Printing Office, Washington, D.C. 20402.)

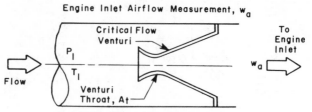

Fig. VI.8   Flow through a choked venturi.

## II.   Propagation of Measurement Errors

Rarely are performance parameters measured directly; usually more basic quantities such as temperature, force, pressure, and fuel flow are measured, and the performance parameter is calculated as a function of the measurements. Error in the measurements is propagated to the parameter through the function. The effect of the propagation may be approximated with the end of Taylor's series methods.

### Example I—Engine Inlet Airflow

Engine inlet airflow is determined by the use of a choked venturi and measurements of upstream temperature and stagnation pressure (Fig. VI.8). The flow ($W_a$) is calculated from

$$W_a = F_A C^* A_{\text{eff}} P_I / \sqrt{T_I} \qquad (\text{VI.4})$$

where

$F_A$ = the factor to account for thermal expansion of the venturi if the working fluid temperature differs from the ambient temperature

$A_{\text{eff}}$ = the effective venturi throat area, i.e., allowing for boundary-layer effects on the geometric area

$P_I$ = the total (stagnation) pressure upstream

$T_I$ = the total temperature upstream

$C^*$ = the factor to account for the properties of the air (critical flow constant)

The precision index for the flow ($S_{W_a}$) is calculated using the Taylor's series expansion

$$S_{W_a} = \left[ \left( \frac{\partial W_a}{\partial F_A} S_{F_A} \right)^2 + \left( \frac{\partial W_a}{\partial C^*} S_{C^*} \right)^2 + \left( \frac{\partial W_a}{\partial A_{\text{eff}}} S_{A_{\text{eff}}} \right)^2 + \left( \frac{\partial W_a}{\partial P_I} S_{P_I} \right)^2 + \left( \frac{\partial W_a}{\partial T_I} \right)^2 \right]^{\frac{1}{2}}$$

$$(\text{VI.5})$$

**Table VI.3    Flow data, $W_2 = F_A C^* A_{\text{eff}} P_1 / \sqrt{T_1}$**

| Parameter | Units | Nominal | Precision index (one standard deviation) | Bias limit |
|---|---|---|---|---|
| $F_A$ | — | 1.00 | 0.0 | 0.001 |
| $C^*$ | $\dfrac{\text{lbm-R}^{\frac{1}{2}}}{\text{lb/s}}$ | 0.532 | 0.0 | 0.000532 |
| $A_{\text{eff}}$ | in.$^2$ | 296.0 | 0.148 | 0.592 |
| $P_1$ | psia | 36.8 | 0.05 | 0.05 |
| $T_1$ | °R | 545.0 | 0.3 | 0.3 |
| $\therefore W_a$ | lbm/s | 248.23 | 0.3658 | 0.6987 |

where $\partial W_a / \partial F_A$ denotes the partial derivative of $W_a$ with respect to $F_A$. Taking the necessary derivatives gives

$$S_{W_a} = \left[ \left( \frac{C^* A_{\text{eff}} P_1}{\sqrt{T_1}} S_{F_A} \right)^2 + \left( \frac{F_A A_{\text{eff}} P_1}{\sqrt{T_1}} S_{C^*} \right)^2 + \left( \frac{F_A C^* P_1}{\sqrt{T_1}} S_{A_{\text{eff}}} \right)^2 \right.$$

$$\left. + \left( \frac{F_A C^* A_{\text{eff}}}{\sqrt{T_1}} S_{P_1} \right)^2 + \left( \frac{F_A C^* A_{\text{eff}} P_1}{-2\sqrt{T_1^3}} S_{T_1} \right)^2 \right]^{\frac{1}{2}} \tag{VI.6}$$

By inserting the nominal values and precision errors from Table VI.3 into Eq. (VI.6), the precision index of 0.3658 lb/s for engine airflow is obtained.

The bias limit in the flow calculation is propagated from the bias limits of the measured variables. Using the Taylor's series formula gives

$$B_f = \left[ \left( \frac{\partial f}{\partial x_1} B_{x_1} \right)^2 + \left( \frac{\partial f}{\partial x_2} B_{x_2} \right)^2 + \left( \frac{\partial f}{\partial x_3} B_{x_3} \right)^2 + \cdots + \left( \frac{\partial f}{\partial x_n} B_{x_n} \right)^2 \right]^{\frac{1}{2}} \tag{VI.7}$$

For example, where $W_a = F_A C^* A_{\text{eff}} P_1 / \sqrt{T_1}$:

$$B_{W_a} = \left[ \left( \frac{\partial W_a}{\partial F_A} B_{F_A} \right)^2 + \left( \frac{\partial W_a}{\partial C^*} B_{C^*} \right)^2 + \left( \frac{\partial W_a}{\partial A_{\text{eff}}} B_{A_{\text{eff}}} \right)^2 + \left( \frac{\partial W_a}{\partial P_1} B_{P_1} \right)^2 + \left( \frac{\partial W_a}{\partial T_1} B_{T_1} \right)^2 \right]^{\frac{1}{2}}$$

$$\tag{VI.8}$$

Taking the necessary derivatives gives

$$B_{W_a} = \left[ \left( \frac{C^* A_{\text{eff}} P_l}{\sqrt{T_l}} B_{F_A} \right)^2 + \left( \frac{F_A A_{\text{eff}} P_l}{\sqrt{T_l}} B_{C^*} \right)^2 + \left( \frac{F_A C^* P_l}{\sqrt{T_l}} B_{A_{\text{eff}}} \right)^2 \right.$$

$$\left. + \left( \frac{F_A C^* A_{\text{eff}}}{\sqrt{T_l}} B_{P_l} \right)^2 + \left( \frac{F_A C^* A_{\text{eff}}}{-2\sqrt{T_l^3}} B_{T_l} \right)^2 \right]^{\frac{1}{2}} \qquad \text{(VI.9)}$$

By inserting the nominal values and bias limits of the measured parameters from Table VI.3 into Eq. (VI.9), a bias limit of 0.6987 lb/s is obtained for a nominal engine airflow of 248.23 lb/s.

To propagate nonsymmetrical bias limits, the bias limit portion of the analysis must be completed for both the upper and the lower limits. Then, the two results are combined as illustrated in Table VI.2.

### Example II—Thrust Specific Fuel Consumption (TSFC)

The goal of any analysis of measurement system errors is to determine the resulting errors in the reduced parameters, for example TSFC, which is calculated as the ratio of fuel-flow rate ($W_f$) to net thrust ($F_N$); TSFC = $W_f / F_N$. Net thrust and TSFC uncertainty calculations are described below. This technique for relating the errors of measurement to the errors in the reduced parameters is again based on a Taylor's series expansion from the calculus. The Taylor's expression for errors in thrust specific fuel consumption is

$$\Delta \text{TSFC} = \frac{\partial \text{TSFC}}{\partial W_l} \Delta W_f + \frac{\partial \text{TSFC}}{\partial F_N} \Delta F_N = \frac{1}{F_N} \Delta W_f - \frac{W_f}{F_N^2} \Delta F_N \qquad \text{(VI.10)}$$

Where $\partial \text{TSFC}/\partial W_f$ and $\partial \text{TSFC}/\partial F_N$ are the partial derivatives of thrust specific fuel consumption with respect to fuel flow and net thrust. Note that all other variables are held constant while differentiating. The precision index is approximated by

$$S_{\text{TSFC}} = \sqrt{\left( \frac{\partial \text{TSFC}}{\partial W_f} S_{W_f} \right)^2 + \left( \frac{\partial \text{TSFC}}{\partial F_N} S_F \right)^2} = \sqrt{\left( \frac{1}{F_N} S_{W_f} \right)^2 + \left( \frac{-W_f}{F_N^2} S_{FN} \right)^2}$$

$$\text{(VI.11)}$$

For example, the following hypothetical data were used to estimate Thrust

Specific Fuel Consumption uncertainty:

| Parameter | Nominal | Bias limit | Precision index | Degrees of freedom | Uncertainty limit |
|---|---|---|---|---|---|
| Thrust, $F_N$, lbf | 10,000 | 18.1 | 37.8 | 57 | 93.7 |
| Fuel flow, $W_f$, lbm/h | 10,000 | 50 | 50 | 60 | 150 |

The nominal thrust specific fuel consumption is calculated from $W_f/F_N$:

$$\frac{W_f}{F_N} = \frac{10,000}{10,000} \frac{\text{lbm/h}}{\text{lbf}} = 1.0 \text{ lbm/lbf-h}$$

The precision index of thrust specific fuel consumption is

$$S_{\text{TSFC}} = \sqrt{\left(\frac{1}{F_N} S_{W_f}\right)^2 + \left(\frac{-W_f}{F_N^2} S_{F_N}\right)^2} = \sqrt{\left(\frac{50}{10,000}\right)^2 + \left(\frac{-10,000}{10,000^2} \times 37.8\right)^2}$$

$$= \pm 0.0063 \text{ lbm/lbf-h}$$

The propagation formula is similar for bias

$$B_{\text{TSFC}} = \sqrt{\left(\frac{\partial \text{TSFC}}{\partial W_f} B_{W_f}\right)^2 + \left(\frac{\partial \text{TSFC}}{\partial F_N} B_{F_N}\right)^2} \tag{VI.12}$$

$$B_{\text{TSFC}} = \sqrt{\left(\frac{1}{F_N} B_{W_f}\right)^2 + \left(\frac{-W_f}{F_N^2} B_{F_N}\right)^2} \tag{VI.13}$$

$$B_{\text{TSFC}} = \sqrt{\left(\frac{50}{10,000}\right)^2 + \left(\frac{-10,000}{10,000^2} \times 18.1\right)^2} = \pm 0.0053 \text{ lbm/lbf-h}$$

The degrees of freedom for the TSFC precision index can be found using the so-called Welch-Satterthwaite technique. In this situation, the partial derivative weighting factors, which are used in the calculation of the precision index, must also be used in the Welch-Satterthwaite formula. It is not necessary to calculate the degrees of freedom for TSFC since the degrees of freedom for thrust and fuel flow are 57 and 60, respectively. The expected minimum result would be 57. The $t$ multiple is essentially 2.0 for degrees of freedom greater than 30. When the degrees of freedom for each component are greater than 30, the Welch-Satterthwaite procedure can be omitted and $t = 2.0$ can be used.

$$df_{TSFC} = \frac{\left[\left(\frac{\partial TSFC}{\partial W_f}S_{W_f}\right)^2 + \left(\frac{\partial TSFC}{\partial F_N}S_{F_N}\right)^2\right]^2}{\frac{\left(\frac{\partial TSFC}{\partial W_f}S_{W_f}\right)^4}{df_{W_f}} + \frac{\left(\frac{\partial TSFC}{\partial F_N}S_{F_N}\right)^4}{df_{F_N}}} \tag{VI.14}$$

$$= \frac{\left[\left(\frac{1}{F_N}S_{W_f}\right)^2 + \left(\frac{-W_f}{F_N^2}S_{F_N}\right)^2\right]^2}{\frac{\left(\frac{1}{F_N}S_{W_f}\right)^4}{df_{W_f}} + \frac{\left(\frac{-W_f}{F_N^2}S_{F_N}\right)^4}{df_{F_N}}} \tag{VI.15}$$

$$= \frac{\left[\left(\frac{1}{10,000}\times 50\right)^2 + \left(\frac{-10,000}{10,000^2}\times 37.8\right)^2\right]^2}{\left[\left(\frac{1}{10,000}\times 50\right)^4\Big/60\right] + \left[\left(\frac{-10,000}{10,000^2}\times 37.8\right)^4\Big/57\right]} = 110$$

The $t$ value is 2, and the uncertainty is

$$U = \pm(B + t_{95}S) = \pm[0.0053 + (2.0)(0.0063)] = \pm 0.0179 \text{ lbm/lbf-h}$$

The results of the error analysis are presented in Table VI.4.

The uncertainty limit as a percent of the nominal value may be calculated by dividing the uncertainty limit in engineering units by the corresponding nominal value and then multiplying by 100.

The propagation of error formulas used in this section are derived and discussed in Appendix B of Abernethy and Thompson.[1]

**Measurement Process**

In making uncertainty analyses, definition of the measurement process is of utmost importance. Uncertainty statements are based on a well-defined mea-

**Table VI.4   Uncertainty components**

| Parameter | Nominal value | Bias limit | Precision error | Degrees of freedom | Uncertainty |
|---|---|---|---|---|---|
| Thrust, $F_N$, lbf | 10,000 | 18.1 | 37.8 | 57 | 93.7 |
| Fuel flow, $W_f$, lbm/h | 10,000 | 50 | 50 | 60 | 150 |
| Thrust specific fuel consumption, lbm/lbf-h | 1.0 | 0.0053 | 0.0063 | 110 | 0.018 |

surement process. A typical process is the measurement of thrust specific fuel consumption (TSFC) for a given gas turbine engine at a given test facility. The uncertainty of this measurement process will contain precision errors due to variations between installations, test stands, and measurement instruments. This uncertainty will be greater than the uncertainty for comparative tests to measure TSFC on a single test stand for a single engine, a different measurement process. The single-stand, single-engine, back-to-back test would assume that most installation and calibration errors would be biases rather than precision errors. If properly understood, bias errors may be ignored in comparative testing in that the same equipment is used for all testing, and bias errors do not affect the comparison of one test with another (the test objective being to determine if a design change is beneficial). The single-stand, single-engine model and other comparative tests treated in Sec. 8.3 of Abernethy and Thompson[1] are worth detailed reading.

Because the definition of the measurement process is a prerequisite to defining the mathematical model, all the elemental bias and precision error sources that affect the measurements must be listed. Then, it must be determined how the bias and precision errors are related to the engine performance parameter. Based on this defined measurement process, the errors may be biases or precision errors.

The bias and precision errors related to the defined measurement process for thrust specific fuel consumption are listed in Sec. II. Uncertainty analyses should be repeated periodically. Continuous validation is essential.

**Reporting Error**

The definitions of the components, bias limit, precision index, and the limit $(U)$ suggest a format for reporting measurement error. The format will describe the components of error, which are necessary to estimate further propagation of the errors, and a single value $(U)$, which is the largest error expected from the combined errors. Additional information, degrees of freedom for the estimate of $S$, is required to use the precision index. These numbers provide all the information necessary to describe and use the measurement error. The reporting format is:

1) $S$, the estimate of the precision index, calculated from data.

2) df, the degrees of freedom associated with the estimate of the precision index $(S)$.

3) $B$, the upper limit of the bias error of the measurement process or $B^-$ and $B^+$ if the bias limit is nonsymmetrical.

4) $U = \pm(B + t_{95}S)$, the uncertainty limit, beyond which measurement errors would not reasonably fall. The $t$ value is the 95th percentile of the two-tailed Student's "$t$" distribution.

5) $U$, the interval between $U^- = B^- - t_{95}S$ and $U^+ = B^+ + t_{95}S$. These limits should be reported when the bias limit is nonsymmetrical.

The model components $S$, df, $B$, and $U$ are required to report the error of any measurement process. As previously recommended, for simplification, the first three components may be relegated to the detailed sections of uncertainty reports and presentations. The first three components, $S$, df, and $B$, are necessary to propagate the errors further, to propagate the uncertainty to more complex parameters, and to substantiate the uncertainty limit.

### Traceability

In recent years the demanding requirements of military and commercial aircraft have led to the establishment of extensive hierarchies of standards laboratories within the military and the aerospace industry. The NBS is at the apex of these hierarchies, providing the ultimate reference for each standards laboratory. It has become commonplace for U.S. Government contracting agencies to require contractors to establish and prove traceability of their measurements to the NBS. This requirement has created even more extensive hierarchies of standards within the individual standards laboratories. At each level of these hierarchies, formal calibration procedures are used. These procedures not only define calibration methods and intervals but also specify just what information must be recorded during a calibration, i.e., meter model, serial number, calibration data, etc., in addition to actual data from the measurement.

The measurement process takes place over a long period of time. During this period, many calibrations occur at each level. Therefore, the precision errors of each comparison are precision errors affecting the measurement process. The overall effect on the measurement of force is a random (precision error) one. Therefore, the resultant overall precision index is the root-sum-square of the individual precision indices. For each comparison, the resultant calibration value is usually the average of several readings. The associated precision index would be a standard error of the mean (or standard error of estimate) for that number of readings. The precision index is

$$S = \sqrt{s_1^2 + s_2^2 + s_3^2 + s_4^2} \qquad (VI.16)$$

for four steps in the calibration process.

The degrees of freedom for each precision index may be combined using the Welch-Satterthwaite formula to provide an estimate of the degrees of freedom for the combined precision index.

$$df = \frac{\left(s_1^2 + s_2^2 + s_3^2 + s_4^2\right)^2}{\left(s_1^4/df_1\right) + \left(s_2^4/df_2\right) + \left(s_3^4/df_3\right) + \left(s_4^4/df_4\right)} \qquad (VI.17)$$

This technique was simulated for various sample sizes and provides the best-known estimate of the equivalent degrees of freedom. If nonintegral values of df result from the Welch-Satterthwaite estimate, appropriate Student's "$t$" values can be found by interpolating (cf. Appendix E of Abernethy and Thompson[1]).

The unknown bias error limit for the end instrument is usually a function of many elemental bias limits, perhaps 10 or 20. It is unreasonable to assume that all of these biases are cumulative. There must be a canceling effect because some are positive and some negative. For this reason, the arbitrary rule that the bias limit $B$ will be the root-sum-square of the elemental bias limit estimates was adopted:

$$B = \sqrt{b_1^2 + b_2^2 + b_3^2 + \cdots + b_L^2} \qquad (VI.18)$$

where $L$ is the number of sources of bias.

In combining elemental nonsymmetrical bias limits, the upper limits should be root-sum-squared to determine the combined upper limit. The lower limits should be root-sum-squared to determine the combined lower limit. The resulting $B^-$ and $B^+$ will be nonsymmetrical bias limits.

The uncertainty in the measurement instrument due to calibration is calculated using the uncertainty formula:

$$U = \pm(B + t_{95}S) \qquad (VI.19)$$

where $S$ is the precision index calculated from Eq. (VI.16).

### III.  Uncertainty Model

Terms such as bias, precision error, uncertainty, standard deviation, NBS, traceability, calibration, and degrees of freedom, and the statistical concepts and mathematical procedures to be employed, were introduced in Sec. I. This section will describe the mathematical model with words, illustrations, and an example.

It is intended that the examples given will closely fit typical applications. However, the model is general, and if specific calibration hierarchies are more or less extensive than the examples, simply add or omit levels and apply the model as shown; if specific measurement systems are different from the examples, substitute the bias limit and precision index terms for the system components and apply the model.

To review briefly, there are two types of measurement error: precision and bias. Precision error is the variation of repeated measurements of the same quantity. The sample standard deviation ($S$) is used as an index of the precision. Bias is the difference between the true value and the average of many repeated measurements. A limit ($B$) for the bias is estimated based on judgment, experience, and testing. The formula for combining these into uncertainty ($U$) is repeated from Eq. (VI.1) for clarity.

$$U = \pm(B + t_{95}S) \qquad (VI.20)$$

or

$$U^- = B^- - t_{95}S \quad \text{and} \quad U^+ = B^+ + t_{95}S$$

when nonsymmetrical biases are present.

Note that throughout this handbook lower-case notation always indicates elemental errors, i.e., $s$ and $b$ for elemental precision and bias, and upper-case notation indicates the root-sum-square (RSS) combination of several errors, e.g., $S = \pm \sqrt{s_1^2 + S_2^2 + s_3^2}$ and $B = \pm \sqrt{b_1^2 + B_2^2 + b_3^2}$, where $S_2 = \pm \sqrt{\Sigma_i s_i^2}$ and $B_2 = \pm \sqrt{\Sigma_i b_i^2}$.

The remainder of this section is devoted to illustration of a typical measurement uncertainty analysis and the propagation of errors to performance parameters.

**Measurement Error Sources**

For purposes of illustration, the elemental error sources for the force measurement system will be treated in this section. These error sources fall into three categories: 1) calibration hierarchy errors, 2) data-acquisition errors, and 3) data reduction errors.

Elemental error sources for other measurements will be enumerated in the section dealing with each measurement.

*Calibration Hierarchy Errors*

To demonstrate traceability of measurements to the NBS, whose standards are by definition the "truth," it is necessary to establish calibration hierarchies. Each level in the hierarchy, including NBS, constitutes an error source which contributes to the error in the final measurement. Calibration of all measurement instruments at the NBS is possible; however, such calibrations would be inconvenient, time-consuming, and very expensive. The purpose here is to illustrate a typical hierarchy and to enumerate the error sources within. Figure VI.9 is a typical force transducer calibration hierarchy. Associated with each comparison in the calibration hierarchy is a pair of elemental errors. These errors are the unknown bias and the precision index in each process. Note that these elemental errors are independent, e.g., $b_{21}$ is not a function of $b_{11}$. The error sources are listed in Table VI.5.

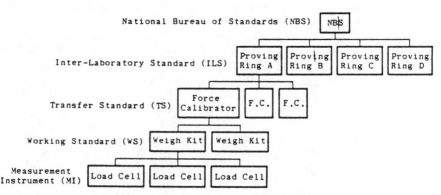

**Fig. VI.9   Force measurement calibration hierarchy.**

**Table VI.5    Calibration hierarchy error sources**

| Calibration | Bias limit | Precision index | Degrees of freedom |
|---|---|---|---|
| NBS–ILS | $b_{11}$ | $s_{11}$ | $df_{11}$ |
| ILS–TS | $b_{21}$ | $s_{21}$ | $df_{21}$ |
| TS–WS | $b_{31}$ | $s_{31}$ | $df_{31}$ |
| WS–MI | $b_{41}$ | $s_{41}$ | $df_{41}$ |

## Test Stand Force Measurement—An Example

The purpose of this section is to illustrate the overall measurement uncertainty model, Fig. VI.10, and logical decision diagram, Table VI.6, as applied to a typical force measuring system. Figure VI.11 illustrates the system and will be the basis for all examples in this section. The propagation of force measurement and other parameter measurement uncertainties to obtain net thrust uncertainty is discussed below.

## Force Measurement Error Sources

Error sources for any measurement fall into three categories: 1) calibration, 2) data acquisition, and 3) data reduction. These categories will be discussed in this order, describing methods for determining values for the elemental errors.

### Force Transducer Calibration Hierarchy

Figure VI.12 illustrates a typical force transducer calibration hierarchy.

Associated with each comparison in the calibration hierarchy is a pair of elemental errors. These errors are the unknown bias and the precision index in each process. Note that these elemental errors are not cumulative, e.g., $b_{21}$ is not a function of $b_{11}$. The error sources are listed in Table VI.7.

The bias limit for the calibration process is the RSS of the elemental precision indices in the preceding steps of the process.

$$B_1 = \pm \sqrt{b_{11}^2 + b_{21}^2 + b_{31}^2 + b_{41}^2} \qquad (VI.21)$$

Calibrations are accomplished by applying known forces to the instrument being calibrated and recording the output. The output may be in inches deflection, millivolts, pounds force, or other measurement units. Several known forces (usually eleven or more) are applied over the range of the instrument being calibrated as indicated in Table VI.8. The forces are applied first going upscale and then going down to determine the hysteresis errors.

The data from Table VI.8 are plotted in Fig. VI.13. The calculated values are determined from a polynomial curve fit of the data. All four curves demonstrate the nonlinearity of the calibration (exaggerated for this example). The difference between the upscale and downscale lines demonstrates the hysteresis of the system. Note that the contribution to overall uncertainty by transducer

nonlinearity and hysteresis can be minimized by considering system performance over a relatively small range near nominal.

*Precision Index*

A least-squares polynomial curve ($Y$) is fitted to the calibration data,

$$Y = A_0 + A_1 F + A_2 F^2 + \cdots + A_k F^k \tag{VI.22}$$

where

$Y$ = calculated force with $F$ pounds force applied
$k$ = degree of curve fit, i.e., largest exponent of $F$
$A_0 \ldots A_k$ = curve coefficients

Figure VI.14 illustrates a plot of calibration data with the dashed line
representing a least-squares polynomial curve fit. The $Y$ value is the force
calculated from the curve fit and corresponding to the applied force $F$ in Eq.
(VI.22). The $Y_j$ values are calculated from curve fits of previous calibrations
establishing a set of $Y_j$, $j = 1, \ldots, N$, where each $Y_j$ represents a calibration and
curve fit. The precision index at any particular point $F_i$ on the curve is
calculated by

$$s_i = \pm \sqrt{\sum_{j=1}^{N} (Y_j - \overline{Y})^2 / (N-1)} \tag{VI.23}$$

where

$\overline{Y}$ = the average of all $Y_j$ values for this transducer at $F_i$
$s_i$ = the precision index for $y$ with $F_i$ pounds force applied
$N$ = the number of calibrations in the estimate

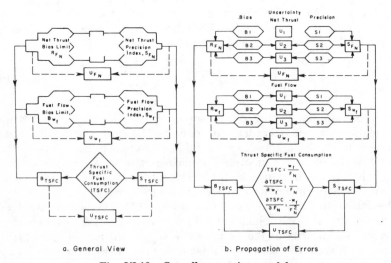

a. General View                    b. Propagation of Errors

**Fig. VI.10   Overall uncertainty model.**

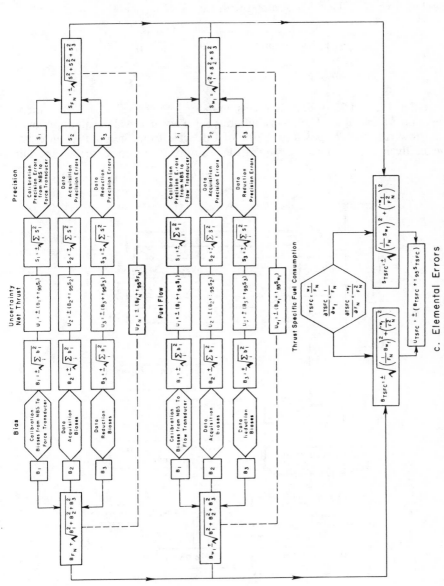

**Fig. VI.10 (continued) Overall uncertainty model.**

Equation (VI.23) yields the precision index for any value of $F_i$ for any number of calibrations. For example, Fig. VI.13 exhibits the data from two calibrations of one device, say, an interlab standard. Table VI.8 lists calculated values of $\overline{Y}$ from the data from both calibrations. At $F_i = 4$, $\overline{Y} = 2.675$ and the $Y_j$ values are 2.65 and 2.70. These data, along with Eq. (VI.23), yield

$$s_i = \pm \sqrt{\left[\left(2.65 - 2.675\right)^2 + \left(2.70 - 2.675\right)^2\right]\big/\left(2 - 1\right)}$$

$$s_i = \pm 0.034$$

Note that Eq. (VI.23) yields the precision index for each value of $F_i$, but there are many values of $F_i$. The precision index for a given $F_i$ value will apply over a narrow range of $F$, possibly $\pm 10\%$ of full scale from the point of interest. Generally it should not be assumed that the precision index for 80% of full

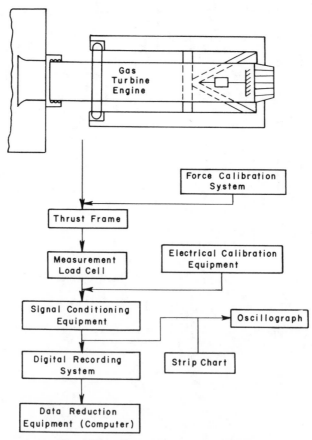

Fig. VI.11   Force measurement system.

**Table VI.6  Logic decision diagram**

| To estimate | Use | Formula |
|---|---|---|
| *Bias limit* | (For treatment of signed biases, see Sec. 1.) | |
| 1) Elemental ($b$) | Judgment supported by special test data | Estimate a reasonable limit for each bias error |
| 2) Measurement bias | Estimated elementals | $B_j = \sqrt{\sum_i b_i^2}$ |
| 3) Performance parameter bias | Measurement bias and the propagation of error (Taylor series) | $B_F = \sqrt{\sum_j \left(\frac{\partial F}{\partial j} B_j\right)^2}$    $F$ denotes performance parameter function |
| *Precision index* | | |
| 1) Elemental $s_i$ | Data from multiple measurements | $S_i = \sqrt{\dfrac{\sum_L \left(X_L - \bar{X}\right)^2}{N_L - 1}}$, $\mathrm{df} = N_L - 1$ |
| 2) Measurement precision index | Calculated elementals and data | $S_j = \sqrt{\sum_i S_i^2}$, $\mathrm{df} = \dfrac{\left[\sum_i S_i^2\right]^2}{\sum_i \left[\dfrac{S_i^4}{\mathrm{df}_i}\right]}$ |

**(continued)**

**Table VI.6 (continued)  Logic decision diagram**

| To estimate | Use | Formula |
|---|---|---|
| *Bias limit* | (For treatment of signed biases, see Sec. 1.) | |
| 3) Performance parameter precision index | Measurement precision indices and the propagation of error (Taylor series) | $S_F = \sqrt{\sum_j \left(\dfrac{\partial F}{\partial j}\right) S_j^2}$ |
| | | $df_F = \dfrac{\left[\sum_i \left(\dfrac{\partial F}{\partial j}\right) S_i^2\right]^2}{\sum_j \left[\dfrac{\left(\dfrac{\partial F}{\partial j} S_j\right)^4}{df_j}\right]}$ |
| $t_{95}$ value | Degrees of freedom less than 30 (df < 30) | Interpolate in two-tailed Student's "$t$" table for $t$ |
| | Degrees of freedom greater than or equal to 30 (df ≥ 30) | Use $t = 2.0$ |
| *Uncertainty* | | |
| 1) Elemental | Elemental bias limit and precision index | $U_i = \pm[B_i + t_{95}S_i]$ |
| 2) Measurement | Measurement bias limits and precision indices | $U_j = \pm[B_j + t_{95}S_j]$ |
| 3) Performance parameter | Performance parameter bias limit and precision index | $U_F = \pm[B_F + t_{95}S_F]$ |

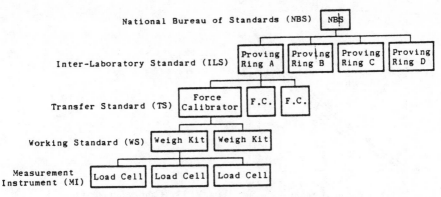

Fig. VI.12    Force transducer calibration hierarchy.

Table VI.7    Calibration hierarchy error sources

| Calibration | Bias limit | Precision index | Degrees of freedom |
|---|---|---|---|
| NBS–ILS | $b_{11}$ | $s_{11}$ | $df_{11}$ |
| ILS–TS | $b_{21}$ | $s_{21}$ | $df_{21}$ |
| TS–WS | $b_{31}$ | $s_{31}$ | $df_{31}$ |
| WS–MI | $b_{41}$ | $s_{41}$ | $df_{41}$ |

Table VI.8    Calibration data

| Applied force levels, $F$ | Calibration one | | | Calibration two | | |
|---|---|---|---|---|---|---|
| | Measured force | | Calculated force (from curve fit of calibration 1 data) | Measured force | | Calculated force (from curve fit of calibration 2 data) |
| | Up | Down | | Up | Down | |
| 0 | 0.00 | 0.20 | 0.105 | 0.00 | 0.20 | 0.11 |
| 2 | 1.00 | 1.60 | 1.30 | 1.25 | 1.50 | 1.38 |
| 4 | 2.20 | 3.10 | 2.65 | 2.50 | 2.90 | 2.70 |
| 6 | 3.75 | 4.82 | 4.29 | 4.00 | 4.60 | 4.30 |
| 8 | 5.77 | 6.77 | 6.29 | 5.90 | 6.50 | 6.20 |
| 10 | 9.30 | | 9.30 | 9.30 | | 9.29 |

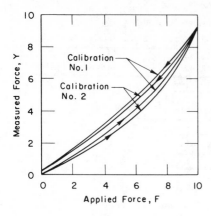

Fig. VI.13   Calibration curves.

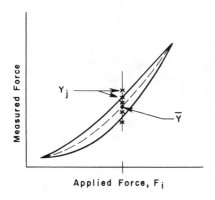

Fig. VI.14   Scatter in measured forces.

scale is the same as that for 10, 20, 30%, and so on. Equation (VI.23) applies in all these cases, but the differences $(Y_j - \bar{Y})^2$ must be summed for the different values of $F_i$ and the largest reported as the precision index for the calibration process. Equations (VI.22) and (VI.23) and all the rules presented in the preceding section will apply to each level in the calibration hierarchy. The precision index for the complete hierarchy is calculated as indicated by Eq. (VI.24):

$$s_l = \pm \sqrt{\Sigma_i s_i^2} \qquad\qquad (VI.24)$$

where $s_l$ = calibration hierarchy precision index and $s_i$ = precision index for the individual levels in the hierarchy.

*Degrees of Freedom*

The degrees of freedom $(df_i)$ associated with the precision index $(s_i)$ is one less than the number $(N)$ of $Y$ observations used to determine the precision index

$$df_i = N - 1 \qquad\qquad (VI.25)$$

Degrees of freedom $(df_i)$ may be calculated at each level in the calibration hierarchy. When the precision index $(S_l)$ is calculated for the complete hierarchy, also calculate $df_l$ for the hierarchy utilizing the Welch-Satterthwaite technique if any of the degrees of freedom are less than 30.

*Bias*

The bias for each level in the calibration hierarchy Fig. VI.15 should be reported as plus or minus the largest unknown fixed error expected. At the first level in the hierarchy (NBS vs interlab standard), the NBS will state an upper limit of bias for the deadweights, e.g., "the errors of the applied loads did not exceed 0.002 percent," quoted for the calibration of an interlab standard with a 60,000-lb capacity. This, however, is only the bias in the applied forces. There will be additional biases resulting from the calibration process. The estimate of the bias limits for the calibration process must be based on careful analysis of the calibration data and any other available information tempered by engineering judgment. For example, the data from an extremely large calibration history may lead to a bias estimate no greater than that reported at the preceding level in the hierarchy. On the other hand, if only one calibration is available to use as a guide to the bias estimate, the estimate from the preceding level in the hierarchy might be increased by an order of magnitude.

When bias limits $(b_i)$ have been established for each level of the calibration hierarchy, a bias limit $B_l$ for the total hierarchy may be calculated, i.e.,

$$B_l = \pm \sqrt{\sum_i b_i^2} \qquad \text{(VI.26)}$$

*Uncertainty*

Uncertainty in the calibration process is now obtained by a simple combination of the precision index and bias limit. As indicated in Fig. VI.16,

$$U_l = \pm ( B_l + t_{95} S_l ) \qquad \text{(VI.27)}$$

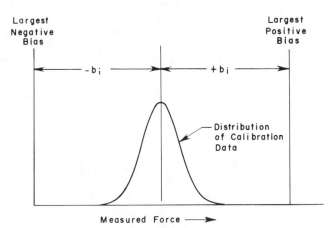

Largest Negative Bias

Largest Positive Bias

$-b_i$    $+b_i$

Distribution of Calibration Data

Measured Force ⟶

Fig. VI.15  Calibration hierarchy elemental bias.

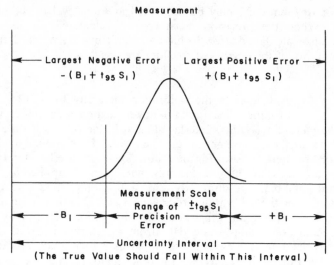

**Fig. VI.16   Calibration process uncertainty parameter $U_1 = \pm(B_1 + t_{95}S_1)$.**

where

$U_1$ = calibration hierarchy uncertainty
$B_1$ = calibration hierarchy bias limit
$S_1$ = calibration hierarchy precision index
$t_{95}$ = 95th percentile of the Student's "$t$" distribution with df$_1$ degrees of freedom

*Data Acquisition and Data Reduction Errors*

The best method of evaluating the net effect of data-acquisition and data-reduction errors is to perform periodic applied load tests. These tests will evaluate all data acquisition and data reduction errors including errors due to force transducer temperature variations, fuel-line temperature and pressure variations, and fuel-flow and environmental effects on the thrust bed.

All data acquisition error sources are listed in Table VI.9. Symbols for the elemental bias and precision errors and for the degrees of freedom are also shown.

The bias limit for the data acquisition process is

$$B_2 = \pm\sqrt{b_{12}^2 + b_{22}^2 + b_{32}^2 + b_{42}^2 + b_{52}^2 + b_{62}^2 + b_{72}^2} \qquad (VI.28)$$

The precision index for the data acquisition process is

$$S_2 = \pm\sqrt{s_{12}^2 + s_{22}^2 + s_{32}^2 + s_{42}^2 + s_{52}^2 + s_{62}^2 + s_{72}^2} \qquad (VI.29)$$

Table VI.9 Data acquisition error sources

| Error source | Bias limit | Precision index | Degrees of freedom |
|---|---|---|---|
| Excitation voltage | $b_{12}$ | $s_{12}$ | $df_{12}$ |
| Electrical simulation | $b_{22}$ | $s_{22}$ | $df_{22}$ |
| Signal conditioning | $b_{32}$ | $s_{32}$ | $df_{32}$ |
| Recording device | $b_{42}$ | $s_{42}$ | $df_{42}$ |
| Force transducer | $b_{52}$ | $s_{52}$ | $df_{52}$ |
| Thrust bed mechanics | $b_{62}$ | $s_{62}$ | $df_{62}$ |
| Environmental effects | $b_{72}$ | $s_{72}$ | $df_{72}$ |

Table VI.10 Data reduction error sources

| Error source | Bias limit | Precision index | Degrees of freedom |
|---|---|---|---|
| Calibration curve fit | $b_{13}$ | $s_{13}$ | $df_{13}$ |
| Computer resolution | $b_{23}$ | $s_{23}$ | $df_{23}$ |

The computer operates on the raw data to produce output in engineering units. The errors in this process stem from the calibration curve fits and the computer resolution.¶

Symbols for the data reduction error sources are listed in Table VI.10. These errors are often negligible in each process.

The bias limit for the data reduction process is

$$B_3 = \pm \sqrt{b_{13}^2 + b_{23}^2} \qquad (VI.30)$$

The precision index for the data reduction process is

$$S_3 = \pm \sqrt{s_{13}^2 + s_{23}^2} \qquad (VI.31)$$

*Applied Load Tests*

Applied load tests are performed as follows:

1) Install an engine in the test stand.
2) Connect all service lines (fuel, instrumentation, water, etc.) and other restraints to the engine.

---

¶Note that the computer introduces two kinds of errors on its own, truncation errors and roundoff errors. The latter are due to an excess number of digits in a number, which may result from multiplying numbers several times. The former are due to finite step size in differential processes, or sampling notes. Usually these errors are small compared to instrumentation errors in a well-controlled situation.

3) Perform an end-to-end calibration of the force measurement system in the usual manner (with all service systems operating at nominal levels if practical).

4) Evaluate the calibration data and perform a curve fit according to normal procedures.

5) Perform the usual prerun setup procedures as if preparing for an engine run. Just prior to an engine run is an excellent time to perform an applied load test.

6) Apply a known force by means of the force calibration system equal to the nominal expected when the engine is delivering rated thrust.

7) Record digital data at a sampling rate and for a period which is normal for steady-state engine conditions; also, record the applied force ($X$) indicated by the standard.

8) Make several (ten or more) recordings as defined in instruction 7 above.

9) Reduce the data by means of the engine data reduction program.

10) Regardless of the number of measuring devices used (multiple bridge load cells, multiple load cells, etc.), calculate multiple sample average $Y_{kj}$ for each steady-state recording. The average of the ten or more recordings ($\overline{Y}_j$) for the $j$th bridge or transducer is

$$\overline{Y}_j = \sum_{k=1}^{M} Y_{kj} \bigg/ M \qquad (VI.32)$$

where $M$ = number of steady-state recordings. The grand average $\overline{Y}$ is then calculated for all load cells

$$\overline{Y} = \sum_{i=1}^{N} \overline{Y}_j \bigg/ N \qquad (VI.33)$$

where $N$ = number of bridges.

11) This grand average ($\overline{Y}$) represents a measurement of the known input. The precision index ($s_i$) for this average is estimated from the precision error of the bridges and/or load cells ($s_b$):

$$s_b = \pm \sqrt{\sum_{j=1}^{N} (\overline{Y}_j - \overline{Y})^2 \bigg/ (N-1)} \qquad (VI.34)$$

The estimate of $s_i$ is $s_b$ divided by the square root of the number of bridges ($N$):

$$s_i = \pm \frac{s_b}{\sqrt{N}} = \pm \sqrt{\sum_{j=1}^{N} (\overline{Y}_j - \overline{Y})^2 \bigg/ [N(N-1)]} \qquad (VI.35)$$

12) If several applied load tests are performed, the precision index of the data acquisition and data reduction process is calculated by pooling the estimates $(s_i)$ from each test:

$$s_{ip} = \pm\sqrt{\sum s_i^2 / K} \qquad (\text{VI.36})$$

where $K$ tests have been performed.

13) The bias limit for data acquisition and data reduction may be estimated through careful analysis of ancillary data such as applied load test histories tempered by the judgment of the most knowledgeable force measurement engineer.

Applied load tests performed in the preceding manner will evaluate the net effect of the following error sources:

1) Excitation voltage.
2) Electrical calibration of the data-recording system.
3) Analog-to-digital conversion.
4) Stand mechanics.
5) Calibration curve fit.
6) Computer resolution.

Performance of additional applied load tests at the conclusion of engine runs will provide evaluation of the net effects of:

1) Environmental effects on force transducer.
2) Effects of fuel-line temperature and pressure variations.
3) Environmental effects on the thrust bed.

Postrun applied load tests are performed immediately after engine shutdown as follows:

1) Ensure that fuel lines are at nominal run temperatures and pressures.
2) If practical, flow fuel at nominal flow rates.
3) Ascertain that the force measurement system is at run temperature.
4) Apply a known force as described in the prerun applied load test.
5) Record test data as described in prerun test procedures.
6) Record an electrical calibration of the data-recording system to ensure no change since the prerun calibration.
7) Reduce data as described in prerun test procedures.
8) Calculate precision index using Eqs. (VI.34)–(VI.36).
9) Estimate bias limit.

*Elemental Error Evaluation*

If it is undesirable or impractical to perform applied load tests frequently, the alternative is to evaluate individually all elemental errors and combine them statistically. The complete list of data acquisition and data reduction elemental errors is:

1) Stand mechanics.
2) Fuel-line temperature variations.

3) Fuel-line pressure variations.
4) Force transducer temperature variations.
5) Excitation voltage.
6) Recording-system electrical calibration.
7) Analog-to-digital converter nonlinearity and drift.
8) Recording-system resolution.
9) Electrical noise.
10) Tare evaluation (thrust stand losses).
11) Computer resolution.

These errors are discussed in enough detail in the following sections to evaluate them and combine them statistically.

*Stand Mechanics*

Several mechanical features of the thrust stand (Fig. VI.17) must be considered and their effects evaluated. These features are:

1) Design of thrust bed support system.
2) Flexure design.
3) How much deflection can be expected in the thrust bed and measurement linkage when a force equal to nominal rated thrust is applied at the point(s) of engine support?
4) How are fuel lines supported?
5) How are fuel lines and other restraints oriented with relation to the axial force vector?
6) What is the effect of engine and instrumentation weight on force measurement?

Some of these are simply considerations to be made, with improvements to minimize the effects being the obvious course of action. Errors caused by stand mechanics will always exist to some degree.

If in-place calibrations are not performed, i.e., only laboratory calibrations of the force measurement transducers are performed, then applied load tests

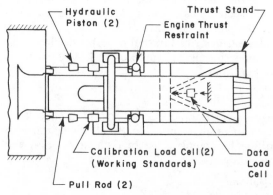

Fig. VI.17   Gas turbine thrust measurement system calibration configuration.

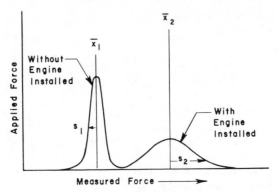

**Fig. VI.18   Precision errors.**

are required to evaluate errors caused by stand mechanics. For example, if applied load tests are performed with and without the weight of the engine and instrumentation on the thrust bed and if the load cell output is measured by means of a laboratory potentiometer to eliminate recording-system errors, then the difference between measured force and applied force with no engine installed is the error due to thrust bed design. The difference between measured force and applied force with an engine and instrumentation installed is the error caused by test stand design and the additional weight. If several tests of this nature are performed, distributions of data can be developed and handled statistically. Figure VI.18 illustrates data distribution $X_1$ without the engine installed, and data distribution $X_2$ with the engine installed. The precision index for tests without the engine installed is

$$s_1 = \sqrt{\sum ( X_i - \overline{X}_1)^2 /[( N - 1)]} \qquad (VI.37)$$

The following outline shows the same detail (in outline form) for fuel flow, pressure and temperature, and airflow.

FUEL-FLOW MEASUREMENT
General
Fuel-Flow Measurement Error Sources
   Calibration Errors
      Volumetric Calibration
         Calibration of the Interlab Standard
         Uncertainty in the Working Standard
      Gravimetric Calibration
      Calibration by Comparison
   Data Acquisition Errors
      Multiple Instruments
   Data Reduction Errors
      Density Determination Errors
      Computer Resolution

## IV.   Net Thrust and Net Thrust Specific Fuel Consumption

**General**

This section details an error analysis for net thrust and net thrust specific fuel consumption at an altitude test facility. In order to calculate net thrust, engine gross (jet) thrust must first be determined. Two independent techniques for the determination of gross thrust can be used: 1) external forces, or scale-force method, and 2) internal forces, or momentum balance method. Both techniques are utilized for most gas turbine engine performance programs when the determination of net thrust and thrust specific fuel consumption is a primary requirement. Performance determined by each method can be compared for agreement in order to improve the confidence in thrust data. Error analysis is presented for only the external forces or scale-force method. The relationship between gross thrust and net thrust will be shown below.

The measurements associated with the determination of net thrust and net thrust specific fuel consumption include pressure, temperature, force, fuel-flow, and airflow measurements. Error analyses of measurement systems have been presented in Abernethy and Thompson[1] as follows: temperature and pressure measurements (Sec. 5), force measurement (Sec. 3), fuel-flow measurement (Sec. 4), and airflow measurement (Sec. 6).

**Gross Thrust Measurement Techniques**

*Scale-Force Method*

The engine assembly and engine support mount are installed on a thrust stand which is flexure-mounted on a model support cart. The engine inlet duct system contains a zero leakage labyrinth-type air seal. Resultant axial forces are measured by a strain-gage load cell. This installation permits the defining of a control volume (Fig. VI.19), which allows the calculation of gross (jet) thrust ($F_G$) from easily measurable parameters.

The free-body diagram associated with the scale-force method of thrust determination is shown in Fig. VI.19.

The derivation of gross thrust ($F_G$) from Fig. VI.19 is

$$\sum F_X = 0 = \frac{W_{AI} V_I}{g} + A_I p_I + F_S - p_0 \int_{A_J}^{A_I} dA - \frac{W_J V_J}{g} - A_J p_J$$

Rearranging and combining terms gives

$$\frac{W_J V_J}{g} + A_J (p_J - p_0) = \frac{W_{AI} V_I}{g} + A_I (p_I - p_0) + F_S$$

$W_{A1}$- engine inlet airflow rate, lbm/sec; $V_1$- engine inlet airflow velocity, ft/sec; g- dimensional constant, 32.174 lbm-ft/lbf-sec$^2$; $A_1$- engine inlet duct cross-sectional area (OD), in$^2$; $P_1$- engine inlet duct static pressure, lbf/in$^2$; $F_S$- force measuring transducer output (scale force), lbf; $W_J$- engine exhaust nozzle exit gas flow rate, lbm/sec; $V_J$- engine exhaust nozzle exit gas flow velocity, ft/sec; $A_J$- engine exhaust nozzle exit area, in$^2$; $P_J$- engine exhaust nozzle exit static pressure, lbf/in$^2$; $P_W$- nozzle wall static pressure, lbf/in$^2$; $P_0$- freestream (ambient) static pressure, lbf/in$^2$; $W_{TH}$- gas flow rate at nozzle throat, lbm/sec; $V_{TH}$- gas flow velocity at nozzle throat, ft/sec; $A_{TH}$- nozzle throat area, in$^2$; $P_{TH}$- nozzle throat static pressure, lbf/in$^2$

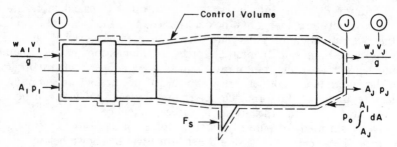

Fig. VI.19  Freebody diagram for external forces (scale force) method of determining engine gross (jet) thrust.

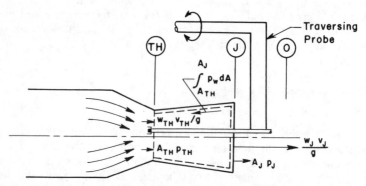

Fig. VI.20  Freebody diagram for internal forces (momentum balance) method of determining engine gross (jet) thrust.

Engine gross (jet) thrust in pounds force is, by definition,

$$F_G = (W_J V_J/g) + A_J(p_J - p_0)$$

therefore,

$$F_G = (W_{A1} V_1/g) + A_1(p_1 - p_0) + F_S \qquad (\text{VI.38})$$

*Momentum Balance Method*

The momentum balance method of thrust determination utilizes the nozzle throat total pressure and temperature profiles obtained with a traversing probe and a mathematical flowfield integration to determine the nozzle stream thrust.

Theoretical calculations utilized in the momentum balance method assume the fluid to be inviscid, thermally perfect, and non-heat-conducting. The method of calculation consists of direct numerical integration of the equations for continuity, momentum, and energy to define the gas state at the nozzle throat. The gas properties at the nozzle throat are then used to determine the exit momentum and pressure area forces required to obtain the nozzle stream thrust.

The free-body diagram associated with the momentum balance method of thrust determination is shown in Fig. VI.20.

The derivation of gross thrust from Fig. VI.20 is

$$\sum F_X = 0 = \int_0^{A_{TH}} \frac{W_{TH} V_{TH}/g}{A} \, dA + \int_0^{A_{TH}} p_{TH} \, dA + \int_{A_{TH}}^{A_J} p_W \, dA$$

$$- \int_0^{A_J} \frac{W_J V_J/g}{A} \, dA - \int_0^{A_J} p_J \, dA$$

Engine gross (jet) thrust in pounds force is, by definition,

$$F_G = \int_0^{A_J} \frac{W_J V_J/g}{A} \, dA + \int_0^{A_J} p_J \, dA - A_J p_0$$

therefore,

$$F_G = \int_0^{A_{TH}} \frac{W_{TH} V_{TH}/g}{A} \, dA + \int_0^{A_{TH}} p_{TH} \, dA + \int_{A_{TH}}^{A_J} p_W \, dA - A_J p_0 \qquad \text{(VI.39)}$$

**Propagation of Errors to Net Thrust**

When $F_G$ has been obtained by either the scale-force or the momentum balance method, or both, identical equations are used to obtain net thrust. The external force (scale-force) method for measuring net thrust is used as the example, and the derivation of this method as shown. The Taylor's series method of propagating error to net thrust is used.

The relationship of net thrust ($F_N$) to gross thrust ($F_G$) is

$$F_N = F_G - F_R \qquad \text{(VI.40)}$$

where

$$F_R = W_{AI} V_0/g \qquad \text{(VI.41)}$$

and $V_0$ is the aircraft freestream velocity in ft/s, $W_{AI}$ is the engine inlet airflow rate in lbm/s, and $g$ is a dimensional constant. Combining Eqs. (VI.38)–(VI.41) then results in the following equation for engine net thrust:

$$F_N = \frac{W_{AI}}{g} (V_I - V_0) + A_I (p_I - p_0) + F_S \qquad \text{(VI.42)}$$

where

$$V_I = \sqrt{\frac{2KgRT_I}{K-1}\left[1-\left(\frac{p_I}{P_I}\right)^{(K-1)/K}\right]}$$   (VI.43)

$$V_0 = \sqrt{\frac{2KgRT_I}{K-1}\left[1-\left(\frac{p_0}{P_I}\right)^{(K-1)/K}\right]}$$   (VI.44)

$P_I$ = engine inlet duct total pressure, $\text{lbf/in.}^2$

$K$ = ratio of specific heats at $T_I$

$R$ = gas constant for air at $T_I$, $\text{ft-lbf/lbm-}°R$

All other parameters are as designated in Fig. VI.19.

Net thrust in terms of pressure, temperature, area, airflow, a gas constant, a ratio of specific heats, a dimensional constant, and force measurement now becomes

$$F_N = \frac{W_{AI}}{g}\left(\frac{2KgRT_I}{K-1}\right)^{\frac{1}{2}}\left[\sqrt{1-\left(\frac{p_I}{P_I}\right)^{(K-1)/K}} - \sqrt{1-\left(\frac{p_0}{P_I}\right)^{(K-1)/K}}\right]$$

$$+ A_I(p_I - p_0) + F_S$$   (VI.45)

The propagation formulas for the bias limit and precision index are derived from Eq. (VI.45) for $F_N$.

The bias limit propagation formula is the weighted RSS of the bias limits for $W_{AI}$, $g$, $R$, $K$, $T_I$, $P_I$, $p_I$, $p_0$, $A_I$, and $F_S$:

$$B_{F_N}^2 = \pm\left\{\left(\frac{\partial F_N}{\partial W_{AI}}B_{W_{AI}}\right)^2 + \left(\frac{\partial F_N}{\partial g}B_g\right)^2 + \left(\frac{\partial F_N}{\partial K}B_K\right)^2 + \left(\frac{\partial F_N}{\partial T_I}B_{T_I}\right)^2\right.$$

$$+\left(\frac{\partial F_N}{\partial R}B_R\right)^2 + \left(\frac{\partial F_N}{\partial P_I}B_{P_I}\right)^2 + \left(\frac{\partial F_N}{\partial p_0}B_{p_0}\right)^2 + \left(\frac{\partial F_N}{\partial P_I}B_{P_I}\right)^2 + \left(\frac{\partial F_N}{\partial A_I}B_{A_I}\right)^2 + \left.\left(\frac{\partial F_N}{\partial F_S}B_{F_S}\right)^2\right\}$$

$$(VI.46)$$

In the same way, the precision index propagation formula is the weighted RSS of the precision indices of $W_{Al}$, $g$, $K$, $R$, $T_1$, $P_1$, $p_1$, $p_0$, $A_1$, and $F_S$:

$$S_{F_N}^2 = \pm \left\{ \left( \frac{\partial F_N}{\partial W_{Al}} S_{W_{Al}} \right)^2 + \left( \frac{\partial F_N}{\partial g} S_g \right)^2 + \left( \frac{\partial F_N}{\partial K} S_K \right)^2 + \left( \frac{\partial F_N}{\partial T_1} S_{T_1} \right)^2 \right.$$

$$\left. + \left( \frac{\partial F_N}{\partial R} S_R \right)^2 + \left( \frac{\partial F_N}{\partial P_1} S_{P_1} \right)^2 + \left( \frac{\partial F_N}{\partial p_0} S_{p_0} \right)^2 + \left( \frac{\partial F_N}{\partial P_1} S_{P_1} \right)^2 + \left( \frac{\partial F_N}{\partial A_1} S_{A_1} \right)^2 + \left( \frac{\partial F_N}{\partial F_S} S_{F_S} \right)^2 \right\}$$

$$(VI.47)$$

Errors associated with values of $g$, $K$, and $R$ are generally assumed negligible. The uncertainty for net thrust $(F_N)$ is calculated using the uncertainty formula

$$U = \pm (B + t_{95} S) \tag{VI.48}$$

The following list of partial derivatives is given as an aid to the analyst:

$$\frac{\partial F_N}{\partial W_{Al}} = \frac{1}{g} \left[ \frac{2KgRT_1}{K-1} \right]^{\frac{1}{2}} \left[ B_1^{\frac{1}{2}} - B_2^{\frac{1}{2}} \right]$$

$$\frac{\partial F_N}{\partial T_1} = \frac{W_{Al}}{g} \left[ \frac{2KgR}{(K-1)T_1} \right]^{\frac{1}{2}} \left[ B_1^{\frac{1}{2}} - B_2^{\frac{1}{2}} \right]$$

$$\frac{\partial F_N}{\partial A_1} = p_1 - p_0$$

$$\frac{\partial F_N}{\partial F_S} = 1.0$$

$$\frac{\partial F_N}{\partial p_0} = W_{Al} \left( \frac{1}{p_0} \right)^{1/K} \left( \frac{1}{P_1} \right)^{(K-1)/K} \left( \frac{(K-1)RT_1}{2gKB_2} \right)^{\frac{1}{2}} - A_1$$

$$\frac{\partial F_N}{\partial p_1} = - W_{Al} \left( \frac{1}{p_1} \right)^{1/K} \left( \frac{1}{P_1} \right)^{(K-1)/K} \left( \frac{(K-1)RT_1}{2gKB_1} \right)^{\frac{1}{2}} + A_1$$

$$\frac{\partial F_N}{\partial P_1} = W_{Al} \left( \frac{(K-1)RT_1}{2gK} \right)^{\frac{1}{2}} (P_1)^{(1-2K)/K} \left[ \left( (p_1)^{(K-1)/K} B_1^{-\frac{1}{2}} \right) - \left( (p_0)^{(K-1)/K} B_2^{-\frac{1}{2}} \right) \right]$$

where

$$B_1 = 1 - \left(\frac{p_1}{P_1}\right)^{(K-1)/K} \quad \text{and} \quad B_2 = 1 - \left(\frac{p_0}{P_1}\right)^{(K-1)/K}$$

By using Eqs. (VI.46) and (VI.47), the partial derivative equations above, and the example values listed in Table VI.11, the propagation of error to net thrust can be determined.

In this example, values for the partial derivative terms above were approximated through the basic net thrust equation. First, the thrust level is determined from the measured values using performance Eq. (VI.45). Then each measured value in Eq. (VI.45) is changed independently by the amount of its precision index, and the resulting change in $F_N$ is obtained. This process, repeated for each measured value, provides good approximate numerical values for each term in Eq. (VI.47). For example, to obtain the approximate value for the term $[(\partial F_N/\partial W_{A1})S_{W_{A1}}]$ of Eq. (VI.47), determine $F_N$ [Eq. (VI.45)] with the measured values. Then, determine the change in $F_N(\Delta F_N)$ by changing airflow $(W_{A1})$ by the amount of its precision index. The change in $F_N$ resulting from the change in $W_{A1}$ is approximately equal to the term $[(\partial F_N/\partial W_{A1})S_{W_{A1}}]$. The identical process is repeated for bias limits to obtain numerical values for the terms in Eq. (VI.46).

By using Eqs. (VI.43)–(VI.45) and the measurement values and their uncertainty components listed in Table VI.12, the propagation of error to $V_0$, $V_1$, $F_R$, $F_N$, and TSFC are shown in Table VI.12.

**Propagation of Error to Net Thrust Specific Fuel Consumption**

$$\text{TSFC} = \frac{W_F}{F_N} , \frac{\text{lbm/h}}{\text{lbf}} \tag{VI.49}$$

where $W_F$ is total engine fuel flow (lbm/h), and $F_N$ is net thrust (lbf).

The error analysis for fuel flow has been outlined, and the error analysis for net thrust is shown above.

The bias limit propagation formula is the RSS of the bias limits for $W_F$ and $F_N$ weighted by the partial derivatives

$$B^2_{\text{TSFC}} = \pm \left[ \left( \frac{\partial \text{TSFC}}{\partial W_F} B_{W_F} \right)^2 + \left( \frac{\partial \text{TSFC}}{\partial F_N} B_{F_N} \right)^2 \right] \tag{VI.50}$$

In the same way, the precision index is calculated as the RSS of the precision indices for $W_F$ and $F_N$,

$$S^2_{\text{TSFC}} = \pm \left[ \left( \frac{\partial \text{TSFC}}{\partial W_F} S_{W_F} \right)^2 + \left( \frac{\partial \text{TSFC}}{\partial F_N} S_{F_N} \right)^2 \right] \tag{VI.51}$$

**Table VI.11 Typical measurement and uncertainty values used in net thrust for supersonic afterburning turbofan engine**

| | Flight conditions: 30,000-ft altitude, Mach number 0.9, military power | | | | |
|---|---|---|---|---|---|
| Component | Nominal value | Bias limit | Precision index | Uncertainty, $U$ | Degrees of freedom[a] |
| $F_S$, scale force, lbf | 4388 | 7.90 | 3.95 | 15.80 | 105 |
| $g$, dimensional constant, $\dfrac{\text{lbm-ft}}{\text{lbf-s}^2}$ | 32.174 | b | b | b | b |
| $A_l$, inlet duct area (OD), in.$^2$ | 984 | 0.050 | 0.050 | 0.160 | 12 |
| $p_l$, inlet duct static pressure, psia | 6.50 | 0.0065 | 0.0065 | 0.0202 | 15 |
| $p_0$, freestream static pressure, psia | 4.31 | 0.0099 | 0.0099 | 0.0310 | 15 |
| $R$, gas constant for air at $T_l$, $\dfrac{\text{ft-lbf}}{\text{lbm-}°\text{R}}$ | 53.329 | b | b | b | b |
| $T_l$, inlet duct total temperature, °R | 477 | ±2.48 | ±0.19 | ±2.86 | 100 |
| $K$, ratio of specific heats at $T_l$ | 1.4034 | b | b | b | b |
| $P_l$, inlet duct total pressure, psia | 7.43 | ±0.0074 | ±0.0111 | ±0.0312 | 15 |
| $W_f$, total fuel flow, lbm-h | 4662 | ±6.06 | ±5.13 | ±16.32 | 35 |

[a] Degrees of freedom determined by the Welch-Satterthwaite formula [Eq. (I.11)].
[b] Bias, precision, and uncertainty considered as negligible in thrust uncertainty.

**Table VI.12    Derived measurement uncertainty values**

| Parameter | Nominal | Bias limit | Precision index | Uncertainty | Degrees of freedom |
|---|---|---|---|---|---|
| $V_0$, freestream velocity, ft/s | 908.4 | ±3.049 | ±2.119 | ±7.414 | 26 |
| $V_1$, inlet duct velocity, ft/s | 463.4 | ±2.684 | ±3.057 | ±8.981 | 26 |
| $W_{A1}$, inlet duct airflow, lbm-s | 115.5 | ±0.531 | ±0.139 | ±0.825 | 16 |
| $F_R$, ram drag, lbf | 3261.0 | ±18.563 | ±8.559 | ±35.681 | 38 |
| $F_N$, net thrust, lbf | 4945.0 | ±12.498 | ±7.471 | ±27.440 | 59 |
| TSFC, net thrust specific fuel consumption, lbm-lbf-h | 0.943 | ±0.0027 | ±0.0018 | ±0.0063 | 94 |

The partial derivatives for the terms in Eqs. (VI.50) and (VI.51) are

$$\frac{\partial \text{TSFC}}{\partial W_F} = \frac{1}{F_N} \quad \text{and} \quad \frac{\partial \text{TSFC}}{\partial F_N} = -\frac{W_F}{F_N^2}$$

Equations (VI.13) and (VI.14) are now evaluated using the above partials and the measurement values and uncertainty components from Tables VI.11 and VI.12 for $W_F$ and $F_N$, respectively.

$$B_{\text{TSFC}} = \pm \sqrt{[(0.0002022)(6.06)]^2 + [(0.0001906)(12.498)]^2}$$

$$= \pm 0.0027 \text{ lbm/h/lbf}$$

$$S_{\text{TSFC}} = \pm \sqrt{[(0.0002022)(5.13)]^2 + [(0.0001906)(7.471)]^2}$$

$$= \pm 0.0018 \text{ lbm/h/lbf}$$

The uncertainty for TSFC is calculated using the uncertainty formula

$$U = \pm (B + t_{95}S)$$

Here $t_{95}$ is = 2.0 since the degrees of freedom are greater than 30.

$$U = \pm [0.0027 + 2(0.0018)]$$

$$= \pm 0.0063 \text{ lbm/h/lbf}$$

The previous examples have concentrated on ground-test procedures. However, one can imagine they apply equally well to flight testing. Tables VI.13 and VI.14, from the March 1983 draft of ARP-1678 (used with Dr. Abernethy's permission), show part of the logic associated with flight testing. Note that one expects the flight-testing procedures to be more demanding if the precision is to be comparable with ground testing.

**Table VI.13** **In-flight thrust measurement uncertainty activities**

| Program divisions | Related activities | Relevant section |
|---|---|---|
| Program definition: Planning | Establish program in-flight thrust uncertainty requirements. | 4.1 |
| | Identify potential thrust measurement math-model options. | 4.1.1 |
| | Conduct thrust math-model options performance survey and select options for use in the program. | 4.1.2 and 4.1.3 |
| Ground test | *Pretest* | 4.2.1 |
| | Estimate engine and facility measurement errors $(S_{gi}, B_{gi})$ $(S_{gt}, B_{gt})$. | |
| | Estimate selected math-model(s) errors $(S_{mm}, B_{mm})$. | |
| | *Posttest* | 4.2.2 |
| | Confirm pretest estimates, adjust as necessary. | |
| | Calibrate math model(s) and derive errors $(S_{mm}, B_{mm})$. | |
| Flight test | *Pretest* | 4.3.1 |
| | Estimate flight measurement errors $(S_{fi}, B_{fi})$. | |
| | Estimate selected math-model(s) errors (may be same as ground-test estimates) $(S_{mm}, B_{mm})$. | |
| | Estimate the in-flight thrust errors $(S, B)$. | |
| | *Posttest* | 4.3.2 |
| | Confirm pretest estimates, adjust as necessary. | |
| | Calculate in-flight thrust measurement uncertainty $\pm U = \pm(B + 2S)$. | |
| Program results: Analysis and reporting | Examine consistency of ground- and flight-test data. | 4.4 |
| | Compare thrust options performance results with program predictions. | 4.4.1 |
| | Resolve test data and performance discrepancy problems. | 4.4.1 |
| | Report test results and supportive uncertainties. | 4.4.3 |

**Table VI.14  Example net thrust option table[a]**

| Option | 1 | 2 | 3 | 4 |
|---|---|---|---|---|
| Airflow, WA | Fan map characteristics | | | ↑ |
| Velocity, VO | Airstream measurements | | | ↑ |
| Gross thrust, FG | Gross thrust parameter $FG = \left[FGP \cdot \dfrac{P2}{PS2} - 1\right] *(A8)(PS2)$ | Mass-weighted velocity coefficient dual stream $FG = CV_2 = M_{i,\text{ideal}}$ | Mass-weighted velocity coefficient mixed stream $FG = CV_2 = M_{i,\text{ideal}}$ | Area-weighted velocity coefficient mixed stream $FG = CV = M_{i,\text{ideal}}$ |
| Basic measurements required | Low rotor speed<br>High rotor speed<br>Inlet static pressure<br>Inlet delta pressure<br>Inlet total temperature | Low rotor speed<br>Inlet static pressure<br>Inlet delta pressure<br>Inlet total temperature<br>Turbine exit pressure<br>Duct disk pressure<br>Turbine exit temperature<br>Duct disk temperature<br>Fuel flow<br>Burner pressure | ↑ | ↑ |

N.B. The terminology in this table may be inconsistent with AIR 1703 but is used to be consistent with AEDC-TR-81-2, the example source.

[a] $M_{i,\text{ideal}}$ is the ideal nozzle exit momentum. See AEDC-TR-81-2 for more detail.

## V.   Concluding Remarks

As indicated at the end of the previous section, the procedures developed for the ground-test stand are quite reliable and can be generalized to flight testing as well. It is interesting to look at two actual cases to see how well these procedures work in practice. Smith and Wehofer[5] have discussed the results from the application of these procedures to ground testing. They conclude that this examination of the state-of-the-art capabilities for measurement of thrust of jet engines operating in altitude ground test facilities may be summarized as follows:

1. There are three competitive state-of-the-art thrust measurement methods available for use today. For convenience these three methods are called scale force method, momentum balance method, and component stacking method.
2. Many factors, including the specific engine cycle and geometry, the engine environmental operating conditions, the required measurement uncertainties, and the program costs and schedules must be carefully considered and trade-offs made before an optimum thrust measurement method can be selected.
3. The scale force system is relatively independent of the specific engine geometry, but it requires a detailed audit of the thrust stand operating environment, and meticulous calibration procedures.
4. The momentum balance method requires a complex probe system designed for high-temperature application and complex, analytical, flow-field computer codes. Particular attention must be paid to the design of the probe system to avoid interference with the gas-path flow field or the engine matchpoint conditions established through the engine power control systems.
5. The component performance stacking method provides engine thrust estimates during the early phases of engine development when only component

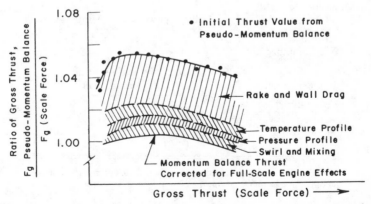

Fig. VI.21   **Comparison of methods offers insight into system behaviors (nonafterburning turbojet).**

**Table VI.15   Scale-force/momentum balance steady-state measurement uncertainty (low-bypass turbofan, planar nozzle)**

| Simulated flight condition | Military power | Military power | Maximum power |
|---|---|---|---|
| Altitude, ft | Sea-level static | 30,000 | 40,000 |
| Mach number | | 0.9 | 2.2 |
| Force parameter | Measurement uncertainty, percent of level | | |
| Gross thrust | ±0.65 | ±0.54 | ±0.45 |
| Net thrust | ±0.65 | ±0.75 | ±0.79 |

**Table VI.16   Scale-force/momentum balance steady-state measurement uncertainty (high-bypass turbofan, nonplanar nozzle)**

| Simulated flight condition | Cruise power | Cruise power |
|---|---|---|
| Altitude | 10,000 | 35,000 |
| Mach number | 0.37 | 0.85 |
| Force parameter | Measurement uncertainty, percent of level | |
| Gross thrust | ±0.30 | ±0.50 |
| Net thrust | ±0.40 | ±0.90 |

**Table VI.17   Component-stacking steady-state measurement uncertainty (turbojet, ejector nozzle)**

| Simulated flight condition | Military power | Military power |
|---|---|---|
| Altitude, ft | 35,000 | 75,000 |
| Mach number | 0.7 | 0.7 |
| Force parameter | Measurement uncertainty, percent of level | |
| Gross thrust | ±3 (±2)* | ±7 (±3)* |
| Net thrust | ±5 (±2)* | ±10 (±3)* |

( )* After resolution of model nozzle coefficient inputs and Reynolds number influences on the compressor and turbine.

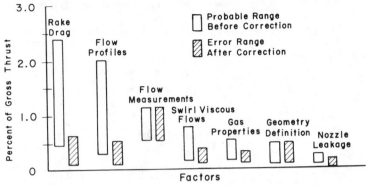

**Fig. VI.22 Magnitude of errors—momentum balance method (nonafterburning turbojet).**

data are available. As real engine cycle data become available from full-scale engine development tests, the measurement uncertainty of these thrust values can be greatly reduced.

6. In general, the measurement uncertainty values attainable with either the scale force method or the momentum balance method are in the range of $\pm 1$ percent of the thrust value. The measurement uncertainty attainable with the component stacking system is substantially larger, and values of several percent are not uncommon during the early phases of engine development.

7. The simultaneous availability of thrust values from two or three of these relatively independent thrust measurement methods provides an extremely effective tool for understanding of the total system thrust behavior and for diagnosing and quantifying second-order and third-order factors which affect the thrust performance of a given system. Some of these factors include instrumentation drag, flow separation, nonuniform temperature and pressure profiles, and exhaust gas swirl. (Fig. VI.21)

Tables VI.15–VI.17 show the uncertainties for a low-bypass turbofan, a high-bypass turbofan, and a turbojet.

These three types of engines represent a spread in classes of engines, and the results show that the errors are consistent in size. Note that achievement of low errors requires corrections for several physical phenomena as shown in Fig. VI.22.

Similarly Brown and Bradley[6] have applied Abernethy and Thompson's procedures to flight test and results are given in Table VI.18.

Further, they conclude the primary source of error in drag measurement is in engine performance. This is clear from the sensitivity analysis, which shows a 1% error. The large angle-of-attack error, when applied to the lift coefficient, makes a smaller contribution to the drag error than thrust.

Table VI.18    Comparison of the computed and actual AGM-86B
flight-test variance

| Nominal Mach number | 0.65 | | 0.70 | |
|---|---|---|---|---|
| Nominal altitude | 500 ft | | 500 ft | |
| Maneuver | Maximum acceleration | | Cruise | |
| Parameter | Computed, % | Actual, % | Computed, % | Actual, % |
| Mach number | 0.281 | 0.138 | 0.055 | 0.086 |
| Angle of attack | 17.461 | 16.358 | 15.131 | 17.384 |
| Bypass airflow | 3.547 | 0.275 | 0.516 | 0.177 |
| Primary airflow | 0.955 | 0.817 | 0.387 | 0.383 |
| Ideal gross thrust | 1.545 | 0.978 | 0.181 | 0.392 |
| Vectored gross thrust | 1.647 | 0.990 | 0.595 | 0.395 |
| Nozzle pressure ratio | 0.108 | 0.569 | 0.082 | 0.257 |
| Drag coefficient | 2.462 | 6.707 | 1.135 | 3.678 |
| Lift coefficient | 0.652 | 2.533 | 0.294 | 2.257 |

# References for Chapter VI

[1]Abernethy, R. B. and Thompson, J. W., "Uncertainty in Gas Turbine Measurements," AEDC Handbook 73-5, 1973.

[2]Society of Automotive Engineers, E-33 Study Committee, R. B. Abernethy, Chairman.

[3]Eisenhart, C., "Expression of Uncertainties of Final Results, Precision Measurement and Calibration," NBS Handbook 91, Vol. 1, Feb. 1969, pp. 69–72.

[4]Ku, H. H., "Expressions of Imprecision, Systematic Error, and Uncertainty Associated with a Reported Value, Precision Measurement and Calibration," NBS Handbook 91, Vol. 1, Feb. 1969, pp. 73–78.

[5]Smith, R. E. Jr. and Wehofer, S., "Measurement of Engine Thrust in Altitude Ground Facilities," AIAA Paper 82-0572, 12th Aerodynamic Testing Conference, March 22–24, 1982.

[6]Brown, S. W. and Bradley, D., "Uncertainty Analysis in Flight Test Performance Calculation," AIAA Paper 81-2390, First Flight Testing Conference, Las Vegas, Nev., Nov. 11–13, 1981.

# Summary and Conclusions

Eugene E. Covert

$I$N this chapter we plan to provide a brief summary of the results given in Chaps. II–VI. Then we will discuss pertinent subjects in more detail. Following a brief discussion of the state of the art, we will offer some conclusions and recommendations.

In the previous chapters we have discussed the procedures for thrust and drag accounting, the so-called bookkeeping procedures (Chap. II). We have concluded that there is no one correct procedure that can be recommended at this time. The most commonly accepted and most widely used are discussed in enough detail to allow one to adapt them to any particular situation. Two clear recommendations have emerged:

1) The thrust-drag accounting procedure must be carefully designed to ensure that each force is only accounted for once.

2) The thrust-drag accounting procedure must be established as early as possible in a program with the idea in mind that it will be useful throughout the flight-test program.

The logic behind these recommendations is simple. Early definition of the drag accounting procedures leads to early definition of the reference conditions, such as shown in Table VII.1. With proper planning, the wind-tunnel test program will be better able to support the flight-test program. Further, the resulting data will have the merit of offering a more reasonable comparison between model and flight-test data at less cost.

In Chap. III we discussed the problems of determining the performance of jet engines. Three procedures were described: the scale-force measurement, the momentum balance, and the component-stacking method. The first two seem preferable to the third for application to ground testing. For flight testing, the use of calibrated engines is preferable, with the momentum balance a second choice. Under ideal circumstances the thrust can be measured to a little better than $\pm 1\%$ on the ground, and perhaps with errors about twice that large in flight.

**Table VII.1   Summary of aerodynamic reference conditions and operating reference conditions**

| | Airframe | Inlet | Exhaust system |
|---|---|---|---|
| Aerodynamic reference conditions | Aeromodel tested at these conditions. Criteria for selection include: 1) Can be reliably reproduced in the separate inlet and nozzle test. 2) Facilitates accurate measurement of both aeromodel drag and associated propulsion increments from the separate tests. 3) Represents realistic conditions modified as necessary to satisfy above criteria. | Generally realistic geometry modified as necessary to assure a stable, attached inlet flow. No bleed or bypass flow (which are difficult to reproduce at different scales). Mass flow ratio selected to minimize aeromodel lip separation, which is more accurately simulated on the large scale inlet drag model. Thus there will be minimal spillage drag at this mass flow ratio. | Generally realistic geometry, if feasible, but modified where necessary to eliminate flow separation and to accommodate model support system and inlet airflow requirements. Ram pressure ratio (since aeromodel generally uses flow-through propulsion simulation). |
| Operating reference conditions | The conditions to which the airplane system drag polar corresponds, by definition. The inlet and nozzle geometry are a function of freestream Mach number only. All other geometry corresponds approximately to a specified flight condition but would not be changed for minor configuration variations. | Realistic geometry but no bleed or bypass flow. Inlet operating at critical mass flow ratio supersonically and subsonically at the choking MFR or MFR $=1$, whichever is smaller. | Realistic aft-end geometry for a fixed area nozzle. If aircraft has a variable area nozzle it is in the maximum operational nozzle exit area configuration. Pressure ratio ($P_{T_j}/P_a$) for a perfectly expanded nozzle ($P_{s_e} = P_a$). |

In Chap. IV we dealt with the estimation and measurement of drag. The results discussed there show that under most circumstances the induced drag can be determined accurately enough. Aspects of profile drag that are dependent upon the square of the local lift are not treated in as much depth as may be necessary in the future. However this discrepancy is very small compared to the problems associated with determining the minimum drag of the vehicle as a function of the flight conditions, i.e., Mach number and Reynolds number. This will be discussed in detail below.

In Chap. V we dealt with a special part of the overall drag calculation problem, namely the throttle-induced drag. This particular part of the drag is normally accounted for with the thrust and not the drag because it depends upon the airflow through the engine and the temperature and pressure of the exhaust plume. Because of the attention it has received, this aspect of the drag calculation is better defined than the overall problem. Nevertheless, the present state of the art is not completely acceptable.

Chapter VI contains a detailed discussion of a method of performing an error analysis that is widely used and seems destined to become standard practice.

## Measurement of Engine Performance

For purposes of this discussion, the measurement of engine performance will be separated into ground testing and flight testing. The former can be subdivided into static test stands and altitude test cells. Normally altitude tests are conducted by the direct-connect method, but it is possible to use a freejet or even a closed-jet or open-jet wind tunnel, if provisions are made to scavenge the engine exhaust products. The freejet or wind-tunnel facilities have the potential to measure some of the effects of the inlet and exhaust nozzle on the engine performance. That is, the throttle-dependent drag forces may be determined reasonably close to full scale conditions.

As discussed in Chap. III and in the closing section of Chap. VI, the direct measurement of thrust is probably accurate to $\pm 1\%$ after correction. If the momentum balance method is used, the error may range from 0.5 to 1.5%, depending upon the altitude and Mach number of the test. It was also pointed out that fuel-flow measurement had a scatter band of $\pm 0.45$–$0.85\%$. This implies a scatter in the thrust specific fuel consumption of 1–2.25%. Naturally, all the usual corrections must be made to be able to have data at this level of accuracy. Generally the lower range of values corresponds to tests of pure turbojets, and the upper end of the range corresponds to large, high-bypass turbofans. The test configuration for the latter engines must be carefully selected to avoid interference from the ground or from the test cell. Nominally several fan nacelle diameters' distance from the engine to the nearest solid boundary is needed to reduce interference effects.

The difficulties of making in-flight measurements of thrust are more severe. So far as we know, no one has succeeded in making direct force measurements on an engine in flight. So one is left with making rake measurements, using gas-generator methods (or component data method as it is sometimes called), using existing curves and tables that describe the engine (this has been called the "brochure method"), or using a calibrated engine. The last method has been extremely useful, but it is a complicated procedure, and the "calibration" is based upon the assumption the inlet airstream possesses the same degree of uniformity as the airflow on the test stand. This assumption is not always fully accurate. Through the use of inertial navigation data, the excess thrust can be measured to $\pm 1\%$ or so. No standard method seems to have been developed to allow for the effects of flow nonuniformities, even if they are known. Finally the calibration of the engine deteriorates slowly as the engine itself deteriorates with use. The size of this effect is not known a priori. It may be small in some cases and larger in others.

W. Earl Hall ran a survey at the 7th Joint Propulsion Conference, and the respondents there estimated that the thrust could be measured to $\pm 1$–2% in flight. One respondent noted that inertial navigation systems were necessary to achieve accuracies of this order, since an error of $0.005g$ was roughly equivalent to 10 drag counts or about 1.4% for fighter aircraft. There was a universal reaction that one of the problems in accurate in-flight thrust determination was that of determining the exit area (A9) accurately at any point in flight.

As a general conclusion the value of the thrust may be determined to within 1% on the ground and 1–2% in flight although any given engine-airframe combination may not quite meet these limits of error. In any single case one must examine the test very carefully to determine whether such agreement constitutes independent verification of the data on engine performance, or merely demonstrates consistency in handling the engine data.

An issue that needs further discussion is defining the flow nonuniformity from experimental data and that of accurate determination of the inlet and exhaust momentum for use with the integral method to determine thrust. The procedures are complicated by the accuracy needed to make the measurement, since at low speeds the net thrust turns out to be the difference between two large numbers, the exhaust momentum and the inlet momentum. (At high speeds the problem is still severe but less so because the pressure area force is of increasing importance.) The problem is the selection of the number and location of pressure and temperature probes. This problem is not solved in general. In most cases the best number will undoubtedly turn out to be larger than the project office would like. The question is involved in the tradeoff between cost and accuracy.

### Drag and Its Determination

Any discussion of the determination of the drag must include a variety of factors including the source of the data, its relation to other data, the possibility of unwanted interference, scale effects, and the like. Since this was dealt with in detail in Chap. IV, there is no merit in repeating that information. It should be sufficient here merely to mention it.

The basic nature of the problem of drag determination becomes clear if one notes the lift-drag ratio of a modern airfoil approaches 200:1, and the vehicle itself may have a lift-drag ratio from 6:1 to 20:1. (Modern sailplanes do far better than this, but they are not penalized by having powerplants installed.) It is well known that when an airplane is operating at the maximum lift-drag ratio, the induced drag is equal to the parasite drag. Hence, compared to the lift, the parasite drag may range from 8 to 2.5% of the lift at the maximum lift-drag ratio. Of course as the flight speed increases, the parasite drag becomes more nearly equal to the overall drag coefficient, particularly at low altitudes.

Nevertheless, the parasite drag coefficient of an efficient airplane-type configuration is 0.10 or less of the lift coefficient, so the calculation based upon first principles is the more difficult. Note that under a number of circumstances, but particularly when the separation locus is not affected by configuration changes, an incremental drag can be determined quite accurately. The drag prediction of the absolute value of a slender configuration is still an art, not a science. This is evident when one considers the effects of bypass air, boundary-layer bleed, manufacturing tolerance, steps and gaps, leakage between sections, and the effects of rain, snow, or dust, as well as aeroelastic effects and changes in trim and control rigging upon the parasite drag.

The basic premise in most drag estimation procedures has, for the most part, been unchanged for the last few decades. An ideal sequence would be to determine the external pressure distribution. This is no easy task in modern integrated or closely coupled configurations. The next step is to calculate the viscous drag using boundary-layer theory. One of the results of boundary-layer theory is an estimate of the displacement thickness which represents the increment in shape due to viscous effects. Another way of handling this same issue is to assume that fluid is injected at the wall at a rate so the proper increment in normal velocity corresponds to the shape change due to viscous effects. This latter approach also comes from boundary-layer theory. The next step is recalculation of the external pressure distribution, which leads to a new boundary layer, and so forth. Usually one iteration is sufficient for classical configurations where the geometry is straightforward. At the present time this matter is relatively straightforward on those parts of the surface where the velocity profile is in a plane normal to the surface, i.e., where the flow in the boundary layer is locally two-dimensional. Even here a significant problem exists if one is trying more or less to calculate from first principles. This problem is that of determining the location of the region of transition from laminar to turbulent flow. If the boundary layer were to be calculated, it would be found that in many cases the flow would separate somewhere near the trailing edge. If the depth of the separated region is thin compared to the thickness of the airfoil, or whatever, then experimental evidence suggests that the pressure coefficient is roughly constant from that point downstream. Under these circumstances there are relatively reliable methods for estimating the location of the two-dimensional separation point. The total drag is found by adding the viscous drag to the pressure drag found from integration of the separation-modified pressure distribution. Note that, in this simple case, three appeals to experiment have been made, namely: the location of the transition zone, the location of the separation point, and the observation of constant pressure in a thin separation layer.

Unfortunately modern configurations are complex, and the flow is highly three-dimensional. The correlations that served well in two-dimensional flow are no longer adequate either for defining transition or for defining separation loci. As if that were not challenge enough, the interaction of the boundary layer with the outer flow at wing-body junctions, or other places where the mating surfaces are nonplanar, gives rise to secondary flow with corresponding axial vorticity. This flow element interacts with the geometry and changes the pressure recovery on aft-facing surfaces in subsonic flow. At transonic and supersonic speeds, shock/boundary-layer interaction gives rise to local separation zones and increased uncertainty about drag.

At high angles of attack where large regions of separated flow are present, the situation is the more complicated. Now the separation region is coupled with the outer flow because the large wake, whose thickness may be of the order of the chord, changes the effective shape of the body and hence the pressure distribution acting on the body. It is in these interacting flows that the primary appeals to empiricism are necessary.

Recent advances in computational fluid dynamics have allowed one to make substantial progress in two areas. First, when the slope of the lift curve is linear, the induced drag can be computed reasonably well even for quite complicated wings with partial span slats and flaps. Second, the aerodynamic designers are able to design airfoils whose drag creep is small, and the airfoils can be incorporated into wings without appreciable deterioration of their desirable characteristics. One notes that even in these latter cases the designer does not compute the drag directly. Rather it is incremental changes of drag due to shape changes that are calculated. Determination of the absolute value of the drag is no simpler than in our earlier description.

While direct solution of the Navier-Stokes equation, large eddy simulation, and turbulence modeling have considerable promise for the future, the direct application to design aerodynamics for the purpose of calculating aerodynamic drag is not available today.

In Chap. V the so-called throttle-dependent drag was discussed. This drag comes about because the airflow requirements of the engine depend upon speed, altitude, and throttle setting in a very complicated way. As yet the ideal inlet, which can conform to diverse airflows and not introduce drag, has not been announced so it is assumed not to exist. This part of the drag comes about not only by local spillage and changes in local stagnation line around the inlet but also due to aerodynamic interference downstream from the inlet. This interference can result in unexpected regions of separation with their attendant drag either on the wing, fuselage, or nacelle/pylon/wing region. Similarly, the nozzle/afterbody region is one where adverse interactions can and do take place. For the case of simple geometries, the results from both experiment and calculation compare reasonably well and are reasonably accurate. Unfortunately most aircraft configurations are sufficiently complex that the estimate of nozzle/afterbody drag lacks the needed accuracy. One problem that must be addressed here involves an agreed definition of the geometry of the base region.[1] In several cases it is easy to show that the divergence between seemingly comparable results is due to the location of the part line. Once this sort of detail is agreed upon, the matter is still very complicated by upstream influence and other interference. Finally the methods by which one can simulate this throttle-dependent drag for wind-tunnel use are still in a primitive state at best. There is much room for advances in the state of the art in this activity.

It should be noted in passing that the practicing aerodynamicist has developed empirical means of estimating nonlifting drag (the so-called parasite drag). These methods are based upon wind-tunnel test results, flight-test results, and much considered judgment. The data base that is the foundation of this judgment is considered proprietary. That no procedure based upon fundamental aerodynamics is widely accepted can be judged from the proprietary nature of many of these empirical approaches.

## Conclusions

The general perception that one gains from this survey is that more accuracy is generally needed than is available. At the present time the following exceptions to the statement exist:

1) Determination of the induced drag from wind-tunnel tests, from the application of large scale computation, or from empirical handbook procedures is generally adequate throughout the linear range of the lift coefficient–angle-of-attack curve.

2) Ground measurement of thrust and specific fuel consumption of an uninstalled turbojet or afterburning turbojet either at sea level or in altitude facilities is adequate, but improvements are desirable.

3) A self-consistent method of estimating experimental errors and their propagation through an analysis has been given in Chap. VI. If commonly adopted, a number of needless arguments about the meaning of the accuracy of experimental data would be avoided. This measure of adequacy is only relative to the other factors in the performance equation. As skills are developed in treating drag due to inlet spillage, and the other terms contributing to the drag, the existing accuracy in estimating the accuracy of these two categories will no longer be good enough. Continual improvement in the technologies listed above is therefore needed.

## Closing Remarks

What remains to be done? In the area of propulsion, the effects of nonuniform inlet flows, both steady and unsteady, upon the engine performance and stability are poorly understood. The precise way that the engines average these nonuniformities and unsteady flows seems to be unknown. The interaction of deflected thrust and high aspect ratio nozzle geometries is not well understood either. Handling of the complications introduced by the relative motion between rotors and stators has yet to be properly treated. The value of a means of direct in-flight thrust measurements is very high. This problem has received much attention, without an overwhelming degree of success. What is needed here are new ideas, or better execution of old ones.

The external aerodynamic knowledge needed to compute drag is in somewhat better condition, if for no other reason than the absence of relative motion between the parts for most of the flight path. Yet even in steady flow, accurate calculations have yet to be made from "first principles." The success-ful practitioner of the art of estimating drag depend upon empiricism to a large extent. So it is not surprising that we still do not understand well the drag due to maneuvering, the drag due to gusts, including both the increment due to altered profile drag and that due to altered lift distribution, which increases the induced drag, and the drag due to rain, snow, dust, or insects. Indeed, it is only recently that the serious study of the scaling laws governing icing and its effect on the airfoil shape and hence drag has been studied.[2]

To add to the list, the data base on the drag increment due to aileron or inlet buzz, buffeting, and of course, poststall flight is so limited that estimates are not very precise. The same statement may be applied to the estimate of drag deterioration within service use such as dents, gaps, leaks, surface wrinkles, and add-ons like external doublers.

It may be argued the list contains many second-order effects that are so small that it hardly seems worth the engineering effort to come to grips with them. This may be true in some cases; but if the price of fuel continues to rise, some of these higher-order effects may have a first-order effect on life-cycle costs.

Last but not least, the authors and editors feel that the whole subject of aerodynamic interference and its effect on drag is worthy of study. This is important not only for drag of external stores but also for airframe, pylon, and nacelle drag. Such drag can be calculated as an increment in some cases; for the latter it may be used to actually reduce the drag in some cases.

All engineers interested in drag prediction should follow the efforts to control skin friction to provide laminar flow when the drag is to be reduced and to provide turbulent flow when its advantages are needed.

Thus, following recommendations are offered:

1) That the method of error analysis and propagation of error outlined in Chap. VI be widely adopted.

2) That studies be continued to determine a means of direct measurement of in-flight thrust.

3) That additional studies be conducted to determine a relation between the degree of flow nonuniformity and the number and location of probes needed to define the flow in an engine inlet.

4) That studies be continued in an effort to define the locus of boundary-layer transition as a function of the geometry, flight, surface conditions, and other environmental conditions.

5) That continued studies to define two- and three-dimensional separation loci in steady and unsteady flow be enlarged to provide the base for successful prediction of drag under these circumstances.

6) That studies of complicated geometrical shapes be continued with the goal of understanding the aerodynamic interference not only for passive parts of the configuration but also for the zones where active interference occurs, like that which occurs at engine inlet and exhaust zones.

7) That studies be continued to develop new and imaginative procedures to simulate the effects of engine-airframe coupling on drag over the entire flight regime.

These studies should be carried out in a three-pronged campaign using theory, computations, and experiments.

## References for Chapter VII

[1] Putnam, L., NASA Langley, personal communication, 1981.
[2] Hunt, J., Sverdrup Technology, Inc., AEDC Div., personal communication, 1979.

# PROGRESS IN ASTRONAUTICS AND AERONAUTICS
## SERIES VOLUMES

### VOLUME TITLE/EDITORS

**\*1. Solid Propellant Rocket Research** (1960)
Martin Summerfield
*Princeton University*

**\*2. Liquid Rockets and Propellants** (1960)
Loren E. Bollinger
*The Ohio State University*
Martin Goldsmith
*The Rand Corporation*
Alexis W. Lemmon Jr.
*Battelle Memorial Institute*

**\*3. Energy Conversion for Space Power** (1961)
Nathan W. Snyder
*Institute for Defense Analyses*

**\*4. Space Power Systems** (1961)
Nathan W. Snyder
*Institute for Defense Analyses*

**\*5. Electrostatic Propulsion** (1961)
David B. Langmuir
*Space Technology Laboratories, Inc.*
Ernst Stuhlinger
*NASA George C. Marshall Space Flight Center*
J.M. Sellen Jr.
*Space Technology Laboratories, Inc.*

**\*6. Detonation and Two-Phase Flow** (1962)
S.S. Penner
*California Institute of Technology*
F.A. Williams
*Harvard University*

**\*7. Hypersonic Flow Research** (1962)
Frederick R. Riddell
*AVCO Corporation*

**\*8. Guidance and Control** (1962)
Robert E. Roberson
*Consultant*
James S. Farrior
*Lockheed Missiles and Space Company*

**\*9. Electric Propulsion Development** (1963)
Ernst Stuhlinger
*NASA George C. Marshall Space Flight Center*

**\*10. Technology of Lunar Exploration** (1963)
Clifford I. Cummings and Harold R. Lawrence
*Jet Propulsion Laboratory*

**\*11. Power Systems for Space Flight** (1963)
Morris A. Zipkin and Russell N. Edwards
*General Electric Company*

**\*12. Ionization in High-Temperature Gases** (1963)
Kurt E. Shuler, Editor
*National Bureau of Standards*
John B. Fenn, Associate Editor
*Princeton University*

**\*13. Guidance and Control—II** (1964)
Robert C. Langford
*General Precision Inc.*
Charles J. Mundo
*Institute of Naval Studies*

**\*14. Celestial Mechanics and Astrodynamics** (1964)
Victor G. Szebehely
*Yale University Observatory*

**\*15. Heterogeneous Combustion** (1964)
Hans G. Wolfhard
*Institute for Defense Analyses*
Irvin Glassman
*Princeton University*
Leon Green Jr.
*Air Force Systems Command*

**\*16. Space Power Systems Engineering** (1966)
George C. Szego
*Institute for Defense Analyses*
J. Edward Taylor
*TRW Inc.*

**\*17. Methods in Astrodynamics and Celestial Mechanics** (1966)
Raynor L. Duncombe
*U.S. Naval Observatory*
Victor G. Szebehely
*Yale University Observatory*

**\*18. Thermophysics and Temperature Control of Spacecraft and Entry Vehicles** (1966)
Gerhard B. Heller
*NASA George C. Marshall Space Flight Center*

**\*19. Communication Satellite Systems Technology** (1966)
Richard B. Marsten
*Radio Corporation of America*

---

\*Out of print.

(Other Volumes are planned.)